ECSTASY: THE CLINICAL, PHARMACOLOGICAL AND NEUROTOXICOLOGICAL EFFECTS OF THE DRUG MDMA

TOPICS IN THE NEUROSCIENCES

Other books in the series:

Rahamimoff, Rami and Katz, Sir Bernard, eds.: *Calcium, Neuronal Function and Transmitter Release.* ISBN 0-89838-791-4.

Fredrickson, Robert C.A., ed.: *Neuroregulation of Autonomic, Endocrine and Immune Systems.* ISBN 0-89838-800-7.

Giuditta, A., et al., eds.: *Role of RNA and DNA in Brain Function.* ISBN 0-89838-814-7.

Stober, T., et al.,: *Central Nervous System Control of the Heart.* ISBN 0-89838-820-1.

Kelly J., et al., eds.: *Polyneuropathies Associated with Plasma Cell Dyscrasias.* ISBN 0-89838-884-8.

Galjaard, H. et al., eds.: *Early Detection and Management of Cerebral Palsy.* ISBN 0-89838-890-2.

Ferrendelli, J., et al., eds.: *Neurobiology of Amino Acids, Peptides and Trophic Factors.* ISBN 0-89838-360-9.

ECSTASY: THE CLINICAL, PHARMACOLOGICAL AND NEUROTOXICOLOGICAL EFFECTS OF THE DRUG MDMA

Edited by
STEPHEN J. PEROUTKA
Stanford University Medical Center

KLUWER ACADEMIC PUBLISHERS
BOSTON/DORDRECHT/LONDON

Distributors

for North America: Kluwer Academic Publishers, 101 Philip Drive, Assinippi Park, Norwell, MA, 02061, USA

for all other countries: Kluwer Academic Publishers Group, Distribution Centre, Post Office Box 322, 3300 AH Dordrecht, The Netherlands

Library of Congress Cataloging-in-Publication Data

Ecstasy: the clinical, pharmacological, and neurotoxicological effects of the drug MDMA / edited by Stephen J. Peroutka.
p. cm. — (Topics in the neurosciences ; TNSC9)
Includes bibliographies and index.
ISBN 0-7923-0305-9
1. MDMA (Drug) 2. Central nervous system—Effect of drugs on.
I. Peroutka, Stephen J. II. Series.
[DNLM: 1. Ampheamines—analogs & derivatives. 2. Amphetamines—pharmacology.
3. Nervous System—drug effects. W1 TO54VF v. 9 / QV 102 E19]
RM666.M35E37 1989
615'.785—dc 20
DNLM/DLC
for Library of Congress

PRINTED IN THE UNITED STATES OF AMERICA

CONTENTS

LIST OF CONTRIBUTORS

Efrain C. Azmitia
Department of Biology
New York University
New York, NY 10003

James B. Bakalar
Department of Psychiatry
Harvard Medical School
Massachusetts Mental Health Center
74 Fenwood Road
Boston, MA 02115

George Battaglia
Department of Pharmacology
Loyola University Medical Center
Stritch School of Medicine
2160 South First Avenue
Maywood, IL 60153

Jerome Beck
School of Public Health
University of California, Berkeley, CA
and

Institute for Scientific Analysis
2235 Lombard Street
San Francisco, CA 94123

Lloyd Bush
Department of Pharmacology and Toxicology
University of Utah
Salt Lake City, UT 84112

Errol B. De Souza
Chief, Laboratory of Neurobiology
NIDA Addiction Research Center
P.O. Box 5180
Baltimore, MD 21224

Graeme P. Dowling
Office of the Chief Medical Examiner
P.O. Box 2257
Edmonton, Alberta T5J 2P4
Canada

James W. Gibb
Professor
Department of Pharmacology and Toxicology
University of Utah
Salt Lake City, UT 84112

George Greer
3 Azul Drive
Santa Fe, NM 87505

Lester Grinspoon
Associate Professor of Psychiatry
Harvard Medical School
Massachusetts Mental Health Center
74 Fenwood Road
Boston, MA 02115

Glen R. Hanson
Department of Pharmacology and Toxicology
University of Utah
Salt Lake City, UT 84112

Michel Johnson
Department of Pharmacology and Toxicology

University of Utah
Salt Lake City, UT 84112

Anita A. Letter
Department of Pharmacology and Toxicology
University of Utah
Salt Lake City, UT 84112

Herbert Y. Meltzer
Department of Psychiatry
School of Medicine
Case Western Reserve University
Cleveland, OH 44106

Kalpana M. Merchant
Department of Pharmacology and Toxicology
University of Utah
Salt Lake City, UT 84112

J. Frank Nash
Department of Psychiatry
School of Medicine
Case Western Reserve University
Cleveland, OH 44106

David E. Nichols
Professor of Medicinal Chemistry
Department of Medicinal Chemistry and Pharmacognosy School of
Pharmacy and Pharmacal Sciences
Purdue University
West Lafayette, IN 47907

Robert Oberlender
Department of Medicinal Chemistry and Pharmacognosy
School of Pharmacy and Pharmacal Sciences
Purdue University
West Lafayette, IN 49707

Stephen J. Peroutka
Assistant Professor of Neurology
Departments of Neurology and Pharmacology
Stanford University School of Medicine
Stanford, CA 94305

Christopher J. Schmidt
Merrell Dow Research Institute
2110 E. Galbraigh Road
Cincinnati, OH 45215

Alexander Shulgin
1483 Shulgin Road
Lafayette, CA 94549

Donna Stone
Department of Pharmacology and Toxicology
University of Utah
Salt Lake City, UT 84112

Vicki L. Taylor
Merrell Dow Research Institute
2110 E. Galbraith Road
Cincinnati, OH 45215

Requa Tolbert, M.S.N.
3 Azul Drive
Santa Fe, NM 87505

Patricia M. Whitaker–Azmitia, Ph.D.
Department of Psychiatry
SUNY
Stony Brook, NY 11794

Robert Zaczek, Ph.D.
Laboratory of Neurobiology
Neuroscience Branch
Addiction Research Center
National Institute on Drug Abuse
Baltimore, MD 21224

PREFACE

The variety of viewpoints expressed in this book illustrate the many controversies surrounding MDMA [1]. On the one hand, the proponents of MDMA use believe this agent offers a unique psychoactive effect that may have important clinical applications, especially in the field of psychotherapy. On the other hand, the scientific data concerning the neurotoxic effects of the drug are unequivocal. The most striking feature of the human information of MDMA is the paucity of data that has been generated on the drug since it was patented in 1914.

As pointed out by Beck (Chapter 6) and others, a clear need exists for better epidemiological and clinical data on MDMA. In the absence of such data, arguments both for and against the cotinued use of MDMA with humans will be difficult to support. Unfortunately, the currently available data must be used to develop rational policies for potential human users of MDMA.

At the present time, there are no data indicating that recreational doses of MDMA permanently damage the human brain. Nonetheless, based on a review of the contents of this book as well as on informal discussions with approximately 200 recreational users of MDMA, the following personal observations suggest that MDMA is radically different from other recreational drugs.

MDMA IS NOT "ADDICTING."

The most frequent use of MDMA has been reported to occur in the first few months following the initial experience [2]. It is extremely rare to find individuals who have taken large quantities of this drug. Again, this is quite different from most recreational drugs, which tend to be either psychologically or physically addicting. There are simply no reports of individuals who take frequent and large amounts of MDMA for extended periods of time. If MDMA is such an outstanding psychoactive agent, why is the drug not used in large quantities for prolonged periods of time?

MDMA USERS OFTEN DELAY REPEAT DOSES OF THE DRUG.

Recreational users of MDMA state that they usually wait at least two to three weeks between doses of the drug. The reason given for this unusual pattern of recreational drug use is that the "good" effects of the drug appear to diminish, while the "negative" side-effects of the drug appear to increase, if the drug is taken too frequently. For example, taking a double dose of MDMA does not "double" the supposed "good" effects of the drug but simply increases the "negative" effects of the drug. Long-term MDMA users have also been reported to suffer from prolonged "burn-out" periods of one to two days after MDMA use [2].

THE EFFECTS OF MDMA CHANGE WITH TIME.

The majority of people who have taken more than five individual doses of MDMA state that the "good" effects of the drug change with successive doses. As stated by one college student, "freshmen love it; sophomores like it; juniors are ambivalent; and seniors are afraid of it." These observations are of concern since no other drugs are known that, when taken at very infrequent intervals (i.e., every month or so), cause different effects with successive doses. As reported by Beck [2], "long-term users often describe increasingly uncomfortable and prolonged 'burn-out' periods" following MDMA use.

THE LACK OF AN ACUTE NEUROPSYCHIATRIC "SYNDROME" FOLLOWING MDMA USE DOES NOT PRECLUDE NEUROTOXICITY.

An analogy between MDMA and 1-methyl-4-phenyl-1,2,3,6-tetrahydropyridine (MPTP) may be appropriate. MPTP is a selective neurotoxin to the substantia nigra, which was sold as "synthetic heroin" to recreational drug users [3–6]. Approximately 400 people in the San Jose, California, area are known to have been exposed to MPTP. Importantly, only seven of these patients currently have clinical evidence of Parkinson's Syndrome [4, 5]. However, positron emission tomography on four of the clinically normal patients has documented significant depletions of dopamine [7]. These studies demonstrate that significant dopaminergic toxicity due to recreational MPTP use can exist in the absence of clinical deficits. Similarly, the neurotoxic effects of MDMA may not be significant enough to produce an overt clinical syndrome. Conceivably, lesions of 5-HT nerve terminals may not become clinically apparent for many years following a single dose of MDMA.

MDMA USE IS NOT "SAFE."

Deaths and toxic reactions have occurred in both recreational and therapeutic users of MDMA (see Dowling, this volume). Unfortunately, MDMA users are rarely aware of this fact. However, it has been my impression that when this type of information is conveyed to recreational users, MDMA use is often reduced or stopped by many of them. No data exist concerning the exact risk of death to a human user from a single dose of MDMA.

SPECIES DIFFERENCES EXIST IN THE NEUROTOXIC EFFECTS OF MDMA.

The initial studies of MDMA-induced neurotoxicity were performed in rats and guinea pigs. Subsequently, data were generated that indicated that mice are relatively insensitive to MDMA-induced changes in 5-HT terminal density [8–10]. By contrast, the monkey appears to be even more sensitive that the rat to the neurotoxic effects of MDMA [11–14]. These species studies suggest that the human central nervous system is likely to be most similar to the monkey in terms of its sensitivity to the neurotoxic effects of MDMA.

CONCLUSIONS

At the present time, definitive evidence of neurotoxicity has not been detected in human users of MDMA. However, more thorough clinical evaluations are necessary to determine if any human neurotoxicity from this drug exists. Indeed, the data derived from MPTP users suggest that the lack of overt clinical toxicity in recreational users of MDMA does not rule out mild to moderate neurotoxicity to human serotonergic pathways. Moreover, the clinical sequelae of neurotoxicity to human serotonergic neurons is unknown. Whether any long-term clinical effects will occur in the recreational users of MDMA is a critical question that will be answered in the years ahead.

MDMA is radically different from all other recreational drugs. As outlined above, its pharmacological effects in humans are unusual. Why do people tend to wait two to three weeks between doses? Why do many people report that the "good" effects of the drug "decrease" with time and usage? The scientific evidence would appear to suggest that these unusual effects of the drug may relate to its long-term and potentially damaging effects on the human brain.

Clearly, MDMA would never be approved for human use by the Food and Drug Administration because of its toxic effects on animal brains. Given our present knowlede, a reasonable and informed conclusion is that recreational use of MDMA should be avoided. Human use should be restricted to carefully controlled clinical trials that are designed to assess both the acute and long-term effects of MDMA on the human central nervous system.

ACKNOWLEDGEMENTS

I thank Bruce G. McCarthy for his helpful comments. This work was supported in part by the McKnight Foundation.

REFERENCES

1. Barnes, D.M., 1988. New data intensify the agony over ecstasy. Science 239:864–866.
2. Beck, J. and Morgan, P.A., 1986. Designer drug confusion: A focus on MDMA. J. Drug Education 16:287–302.
3. Davis, G.C., Williams, A. C., Markey, S.P., et al., 1979. Chronic Parkinsonism secondary to intravenous injection of meperidine analogues. Psychiat. Res. 1:249–254.
4. Langston, J.W., Ballard, P., Tetrud, J.W., and Irwin, I., 1983. Chronic Parkinsonism in humans due to a product of meperidine-analog synthesis. Science 219:979–980.
5. Langston, J.W. and Ballard, P., 1984. Parkinsonism induced by 1-methyl-4phenyl-1,2,3,6-tetrahydropyridine (MPTP): Implications for treatment and pathogenesis of Parkinson's Disease. Can. J. Neurol. Sci. 11:160–165.
6. Snyder, S.H., 1984. Clues to aetiology from a toxin. Nature 311:514.
7. Calne, D.B, Langston, J.W., Martin, W.R.W., et al., 1985. Positron emission tomography after MPTP: Observations relating to the cause of Parkinson's disease. Nature 317:246–248.
8. Stone, D.M., Hanson, G.R., and Gibb, J.W., 1987. Differences in the central serotonergic effects of methylenedioxymethamphetamine (MDMA) in mice and rats. Neuropharmacology 26:1657–1661.
9. Logan, B.J., Laverty, R., Sanderson, W.D., and Yee, Y.B., 1988. Differences between rats and mice in MDMA (methylenedioxymethylamphetamine) neurotoxicity. Eur. J. Pharmacol. 152:227–234.
10. Peroutka, S.J., 1988. Relative insensitivity of mice to 3,4-methylenedioxymethamphetamine (MDMA) neurotoxocity. Res. Commun. Sub. Abuse 9:193–205.
11. Slikker Jr, W., Ali, S.F., Scallet, A.C., Frith, C.H., Newport, G.D., and Bailey, J.R., 1988. Neurochemical and neurohistological alterations in the rat and monkey produced by orally administered methylenedioxymethamphetamine (MDMA). Toxicol. Appl. Pharmacol. 94:448–457.
12. Ricaurte, G.A., Forno, L.S., Wilson, M.A., DeLanney, L.E., Irwin, I., Molliver, M.E., and Langston, J.W., 1988. (±)3,4-Methylenedioxymethamphetamine selectively damages central serotonergic neurons in nonhuman primates. JAMA 260:51–55.
13. Ricaurte, G.A., DeLanney, L.E., Irwin, I., and Langston, J.W., 1988. Toxic effects of MDMA on central serotonergic neurons in the primate: Importance of route and frequency of drug administration. Brain Res. 446:165–168.
14. Ricuarte, G.A., DeLanney, L.E., Wiener, S.G., Irwin, I., and Langston, J.W., 1988. 5-Hydroxyindoleacetic acid in cerebrospinal fluid reflects serotonergic damage induced by 3,4-methylenedioxymethamphetamine in CNS of non-human primates. Brain Res. 474:359–363.

ECSTASY: THE CLINICAL, PHARMACOLOGICAL AND NEUROTOXICOLOGICAL EFFECTS OF THE DRUG MDMA

1. HISTORY OF MDMA

ALEXANDER T. SHULGIN

1. INTRODUCTION

There can never be a complete history of any intensely controversial topic whose proponents and skeptics state their beliefs with equal confidence. Some historical facts will rest uncontested. Many facts will be clothed in opinions that will color the way the facts are to be interpreted. Other facts will never be publicly known, for reasons of legality or privacy. And some facts may simply be irretrievably lost.

Most important, no history can be complete if it concerns a topic that is alive and developing. The subject of MDMA is very much alive and developing today. The story of its neurochemical effects is still unfolding and is being widely published. The story of its psychotherapeutic value is also unfolding and, although not being published, is nonetheless being widely distributed. This very volume is part of the developing history of MDMA in that it brings together spokesmen for all aspects of this history. In this opening chapter, I will attempt to present a number of historical facts with as little interpretation as possible and with available documentation.

The organization of this review largely follows the historical record. The chemical synthesis of MDMA (1912) was followed by the Army-sponsored toxicological studies (1953). The initial therapeutic exploration of MDMA in humans (1976) was followed by its popularization outside of the medical area (1981) and by the initial legal moves by the DEA to establish control (1984). The current flood of animal study (biochemistry, pharmacology, and especially neuotransmitter research) had its start in 1985.

2. CHEMICAL HISTORY

2.1. Chemical nomenclature

MDMA are the initials of the synthetic base 3,4-methylenedioxymethamphetamine. It has a number of correct chemical synonyms. With amphetamine as a stem, the principal name is N-methyl-3,4-methylenedioxyamphetamine. With the benzene ring itself as the target of naming, MDMA can be called either N, alpha-dimethyl-3,4-methylenedioxyphenethylamine or N,alpha-dimethyl-homopiperonylamine. Named as an aliphatic amine, it is either N,alpha-dimethyl-beta-(3,4-methylenedioxyphenyl)-ethylamine or N-methylbeta-(3,4-methylenedioxyphenyl)-isopropylamine. The hydrocarbon name is 2-methylamine-1-(3,4-methylenedioxyphenyl)-propane. And finally, named as a heterocycle, there is N,alpha-dimethylbenzodioxole-5-ethylamine (ethanamine in present Chemical Abstracts).

There are many code names and popular terms for MDMA. The shortened MDM stands for methylene-dioxy-methamphetamine. An early street name, "Ecstasy," has given rise to the initials XTC. In Europe it is often called, simply, "E." In the area of clinical psychology, the name "Adam" is common, having been created by the psychologist who first introduced MDMA into psychotherapy. The U.S. Army in its studies assigned it the code EA-1475, wherein EA stands for Edgewood Arsenal.

2.2. Chemical origin

Neither the chemist nor the date of the first synthesis of MDMA is known. The first public recording of the preparation and properties of MDMA was a German patent filed for in 1912 and issued in 1914, assigned to the firm E. Merck [1]. Although the popular press has stated that the compound had been developed as an appetite suppressant, the only claims in the patent are that this compound (among others) can serve as important intermediates for the production of therapeutically active compounds. A second German patent appeared a few years later describing a chemical modification of MDMA [2]. Again, there is no pharmacology mentioned.

2.3. Chemical synthesis

All known synthetic schemes leading to MDMA start with materials that contain the preformed methylendioxyphenyl ring. The principal sources are piperonal, isosafrole, safrole, and piperonylacetone [3]. These sources are, to some extent, interconvertible, in that isosafrole can be made from safrole, and piperonylacetone can be made from any of the other three.

2.3.1. Synthesis via piperonal

Piperonal may be used as a source of piperonylacetone. There are reports that describe a Darzans reaction with ethyl bromopropionate followed by hydrolysis [4], an intermediate glycol followed by hydrolysis [5], or reaction with nitroethane to form an intermediate nitrostyrene followed by reduction with elemental iron [6]. The conversion of piperonylacetone to MDMA is discussed in section 2.3.4.

The last mentioned intermediate, the nitrostyrene, can be reduced by a number of methods to produce 3,4-methylenedioxyamphetamine (MDA). This centrally active base has served as a synthetic starting point for the preparation of MDMA by the addition of the one-carbon methyl group to the amine nitrogen. This has been done via the formate [7] followed by reduction with LAH. This process has been used for the preparation of the optical isomers of MDMA [8]. The intermediate urethane (from ethylchloroformate) can be reduced with Red-Al [9]. The direct methylation with methyl iodide is complicated by the formation of di- and tri-methylated contaminants [10].

2.3.2. Synthesis via isosafrole

All procedures that employ isosafrole require its conversion to piperonlylacetone. The usual oxidant is hydrogen peroxide and formic acid [11], although the use of peracetic acid has been encountered [12]. This ketone can be reductively aminated to MDA, which can be methylated to MDMA, as discussed in section 2.3.1. Again, the conversion of piperonylacetone to MDMA is outlined in section 2.3.4.

2.3.3. Synthesis via safrole

The addition of HBr to safrole followed by reaction with methylamine is the original method described for the preparation of MDMA [1]. This same process was published almost a half century later by two Polish chemists [13], several years following the U.S. Army's first contracted studies on its toxicology. A second procedure employing safrole requires its base isomerization to isosafrole followed by this product's oxidation to piperonylacetone, as described in 2.3.2.

2.3.4. Synthesis via piperonylacetone

The reductive methylamination of piperonylacetone can be achieved with sodium cyanoborohydride [7] or amalgamated aluminum [14]. The cyanoborohydride method has been used for the preparation of tritium-labeled MDMA [15]. The conversion has also been reported using N-methylformamide in a Leuckart reaction with hydrolysis of the intermediate amide to provide MDMA [16].

2.4. Chemical analysis

It was through careful analytical work that the first clues arose that a new drug, MDMA, was coming into general availability. The first reports were in the Midwest, with observations in the early 1970s in Illinois [17, 18] and Indiana [19]. The initial screening tests for methamphetamine are positive for MDMA as well, so the two can be initially confused. Even the presumptive microcrystalline tests give very similar results for the two drugs [20]. The confirmatory analyses of recent "at risk" samples, both from street submissions [21] and the urine of suspected users [22], have shown that the vast majority of these are indeed methamphetamine.

Many of the analytic procedures for MDMA are incorporated as experimental data in papers that are largely directed to another aspect. Below are the major papers that are concerned primarily with analytical procedures per se, with a primary emphasis on MDMA and closely related compounds.

2.4.1. Thin-layer chromatography

Two studies have been made employing TLC techniques. A group of N-methylated amphetamine derivatives was synthesized and characterized spectroscopically [16]. Six distinct TLC systems were evaluated. Another study evaluated two systems and a variety of visualization procedures [10]. A study of a number of phenethylamines had included by misprint the drug MDMA in its title, but the drug actually studied was MMDA [23].

2.4.2. Gas chromatography

The early studies of Bailey and his coworkers [16] covered a number of column materials and conditions for the separation of MDMA from close chemical allies. The use of GC has been used for the establishment of purity of MDMA in toxicology studies [14]. A number of papers have reported studies of large collections of drugs, including MDMA. These papers are intended for screen purposes and are not included here.

2.4.3. Immunoassay

MDMA and its two close homologues, the primary amine MDA and the N-ethyl counterpart MDE, have been compared with each other and with amphetamine or methamphetamine in immunological analyses designed to detect amphetamine. In the homogeneous EMIT assay, all three showed positive crossreactivity but with reduced response [24]. Studies with RIA (Abuscreen) and TDX (a fluorescent polarization immunoassay), as well as EMIT, showed similar crossreactivity but with extremely variable results, depending on the specific assay employed [25].

2.4.4. High-pressure liquid chromatography

Two reports have appeared describing a study of retention characteristics of MDMA and its close relatives MDA and MDE by HPLC [26, 27].

3. TOXICITY

The first report that described any research other than chemistry was a large toxicological study done at the University of Michigan under a classified contract with the U.S. Army [28]. The study was performed in the 1953–54 period, declassified in 1969, and finally published in 1973. It embraced eight phenethylamine bases that had been synthesized at Edgewood Arsenal; all were structural variations of mescaline involving ring substitution patterns and chain length. MDMA was the only N-methyl compound studied. Five laboratory animal species were employed, and the study evaluated toxicity and behavioral effects.

The results of this study, along with more recent animal toxicity data, are presented below, arranged by the specific animal used. A recent review emphasized a close toxicological resemblance between MDMA and MDA [29].

3.1. Mouse

The Army studies by Hardman et al. [28] included toxicity measurements on five laboratory animal species, including the mouse. MDMA was found to have an LD-50 of 97 mg/Kg following i.p. administration. Recent studies [9] have shown almost exactly the same values (106 mg/Kg i.p. in six hours, 98 mg/Kg in 24 hours). The aggregate toxicity phenomenon, well established for amphetamine [30], is still present for MDMA, but to a lesser extent (aggregate LD-50 30 mg/Kg at six hours, 20 mg/Kg at 24 hours). A study of activity cage behavior of mice (crowding conditions not reported) showed an LD-50 of about 20 mg/Kg [31].

3.2. Rat

The Hardman study [28] found an acute i.p. LD-50 for MDMA in Sprague Dawley rats of 32 mg/Kg. Chronic studies employing oral dosages of up to 100 mg/Kg [32] were conducted to assess pathology at necropsy. No treatment-related brain lesion could be found, although some clinical measurements were noted. This study reported no deaths, and coupled with chronic oral studies by Slikker et al. [33] of up to 80 mg/Kg and chronic subcutaneous studies by O'Hearn of 20 mg/Kg [34], it seems that MDMA has relatively low toxicity by these routes in the rat. Another study [35] employed oral chronic administration of MDMA to rats at dosages of up to 300 mg/Kg, with complete blood chemistry and microscopic and histological workup. Kidney changes and possible testicular tubular changes were noted, but there was no evidence of brain damage. Acute oral LD-50 was estimated to be 325 mg/Kg and chronic oral LD-50 about half of this.

3.3. Guinea Pig

The Hardman study [28] reports an i.p. acute LD-50 of 98 mg/Kg in the guinea pig.

3.4. Dog

The acute i.v. LD-50 of MDMA in the dog is 14 mg/Kg [28]. Chronic studies at dosages of up to 15 mg/Kg were conducted [32] with clinical chemistry measurements made during and a complete pathology workup done at the end of the highest dosage exposure. There were signs of testicular atrophy and gross prostatic enlargement in some of the animals at the higher levels of drug administration. Weight loss was also observed, but there were no indications of neuropathological changes.

3.5. Monkey

The Hardman study [28] reports an LD-50 of 22 mg/Kg for the i.v. administration of MDMA to the rhesus monkey *Macaca mulatta*. Several other primate species have been employed in neurotoxicity and behavioral studies.

3.6. Man

The lethal level of MDMA in man can only be inferred from anecdotal data in the published literature. A report describes five deaths in Dallas associated with MDMA or MDE use, with one stated to be due to MDMA specifically [36]. Hayner and McKinney describe two toxic episodes [37], one of which has been presented in detail [38]. A toxic interaction with MDMA use in association with a monoamine oxidase inhibitor has been described [39]. The association between human plasma levels of MDMA and clinical state is unclear, in that levels of 7 ug/ml [38] and 0.1 ug/ml [40] have been associated with non-lethal usage, whereas levels of 1.4 ug/ml [41] and 1.1 ug/ml [36] have been seen in fatalities. No experimental procedures are provided for any of these numbers. A review of the medical literature has been assembled for use by the physician in the emergency room [42]. One investigation of human CSF for evidence following MDMA use reported no abnormalities in the levels of neurotransmitter metabolities [43].

4. HUMAN PHARMACOLOGY

The first report of the pharmacological action of MDMA in humans appeared in 1978 [44], but it made no mention of the exploratory therapy role that had been initiated by clinical psychologists some two years earlier. The written description of the action of MDMA compared it with that of MDA when used at low levels.

MDMA is described as evoking an easily controlled altered state of consciousness with emotional and sensual overtones. "Within the effective dosage range, 75–150 mg orally, the effects are first noted very quickly, usually within a half-hour following administration. With most subjects, the plateau of effects is reported to occur in another half-hour to one hour. The intoxication symptoms are largely dissipated in an additional two hours, except for a mild residual stimulation..."

These properties, the openness of emotional expression and the unusually short duration, established the unique character of MDMA, which made it so promising to therapists and tempting, eventually, to the curious public as well.

4.1. Therapy studies

MDMA was first introduced into clinical practice on the West Coast in the latter part of 1976 and was being used by therapists on the East Coast within a few months. It is impossible to determine accurately the size of the medical following it commanded. By far, the largest body of published clinical work with MDMA has been authored by George Greer, M.D. (see Chapter 2). His first publication [45] describes sessions with 29 patients in a therapeutic setting. He described its value in the treatment of alcohol and drug-abuse problems, in the facilitation of communication and intimacy between people involved in emotional relationships, and as an adjunct to insight-oriented psychotherapy. Further detail and some retrospective evaluation of these 29 patients has been given [46], as well as a recommended protocol for MDMA sessions [47]. A second report [48] presents a discussion of a training experiment designed to familiarize a group of 13 potential clinicians with a first-hand experience of MDMA so that they might evaluate its differences from earlier therapeutic tools, such as LSD.

Two extensive clinical studies are reported from Germany, one involving 11 MDMA-assisted group therapy meetings with a total of 52 patients [49]. There were some positive changes described, and in some cases there was no improvement observed. One extensive report has appeared in considerable detail of a clinical application [50], and several surveys have appeared of MDMA users. These include retrospective psychological interviews [51] and an evaluation of sexual aspects of the drug's effects [52]. A small volume was published containing a collection of several dozen first-hand reports of personal experiences [53].

Another large study [54] employed 21 normal volunteers, who underwent continuous physiological measurement, blood chemistry analyses, and (for some) neurological and electrocardiogram tests. These latter tests were continued for 24 hours. All subjects experienced an elevation of blood pressure and pulse rate, peaking at about one hour, and all experienced loss of appetite, to some degree. Other changes (jaw clench, reflex enhancement, and physical incoordination) were seen in some subjects. At the psychological level, all subjects reported a heightened sensual awareness and three subjects experienced sexual arousal. One retrospective survey has failed to note signs of neurotoxocity in humans [55].

4.2. Public controversy

The first promotional publications concerning MDMA were geared to the drug-oriented population. An anonymous article appeared in the counterculture magazine *Wet* in 1981 [56], and shortly thereafter appeared two privately published tracts that attempted to answer questions concerning the use of MDMA [57]. In 1985 the entire issue became a public controversy, with the appearance of news articles in *Newsweek* [58] and, just after MDMA's proposed emergency scheduling, in *Time* [59]. In the following weeks, all major magazines and newspapers carried articles and editorials that ranged from strong support to open-mindedness to outright condemnation. Reference to writings in *Life* [60], *C&E News* [61], *Harvard Medical School Mental Health*

Letter [62], *Harpers Bazaar* [63], *Alcohol and Addiction* [64], *New Age* [65], *Psychology Today* [66], *Rolling Stone* [67], and the comic strip "Doonesbury" [68] gives a good cross-section of this onslaught of information and opinion.

The controversy intensified. On the medical use side, a small but dedicated group of professional psychologists and psychiatrists maintained that MDMA was too valuable in therapy to simply have it disappear into legal oblivion. Groups such as the Earth Metabolic Design Laboratories formed to champion the cause [69]; a national conference was held in Oakland, California [70]; and several days of testimony were presented at the DEA hearings that addressed the scheduling problem. On the abuse and illegalization side, the Government issued anonymous position papers that emphasized the public health considerations [71], and extended emergency funding to researchers to quickly provide information that dealt with the neurotoxicity subject.

4.3. Future research

Two factors regarding future research into drugs that might have use in psychotherapy are immediate outgrowths of the MDMA controversy.

First, clinical studies with MDMA have come to a complete halt. The recommendations from the HHS (Health and Human Services) to the DEA, which accompanied the emergency scheduling of MDMA, encouraged the facilitation of research with the drug. However, the FDA has yet to approve an IND (Investigational New Drug) license that permits a study in humans. None have been conducted since the effective scheduling by the DEA. The abrupt invocation of the Emergency Scheduling Act in controlling MDMA, even while hearings were underway to examine the very question of scheduling itself, had a chilling effect on research in this area in the pharmaceutical houses [72] and the academic world [73]. The Analogue Substane Act of 1986 specifically outlaws any human research with drugs that have actions or structures substantially similar to MDMA (or any other Schedule I or II drugs) unless permission is received from the FDA (in the form of an IND).

A second aspect that may bear on future research deals with the pharmacological classification of MDMA. Although legally it is classified together with hallucinogenic drugs, it is not similar in action to the usual psychedelic drug. A new classification seems needed [74], and the term "entactogen" has been proposed [75] to emphasize the unique nature that MDMA has shown in therapy. A search for examples of drugs that are psychologically enabling but not hallucinogenic has been started [76].

5. LEGAL HISTORY

This review of history (as with the neurotransmitter story presented in 6. below) will be quite brief, as specific chapters in this volume will cover these subjects in intimate detail.

The first administrative acknowledgment of MDMA was a request from the World Health Organization (WHO) to the Food and Drug Administration

(FDA) for information and comments concerning the abuse potential, actual abuse, and medical usefulness of some 28 stimulants and/or hallucinogens [77]. Just one week later [78], the DEA filed a pro forma request for comments, objections, or requests for hearings, in connection with its intent to place MDMA into Schedule I of the Controlled Substances Act.

A petition requesting hearings on this listing was sent to the DEA [79] and an initial procedural hearing was scheduled [80]. At this time, the law judge assigned to this matter recommended that, as there is no place in the scheduling structure for a drug with no accepted medical use but with less than a high abuse potential, MDMA should either not be scheduled or it should be placed in less severe schedule [81]. The hearings were set to take place in Los Angeles on June 10, in Kansas City on July 10–11, and in Washington D.C. on October 8,9,10, and 11, 1985. On May, 31, 1985, just ten days before the first hearing was to be held, the DEA unilaterally invoked the Emergency Scheduling Act regarding MDMA and effected its placement on a temporary basis into Schedule I, effective July 1, 1985 [82].

The judicial recommendation that followed the hearings was that MDMA had some accepted medial use and should be placed in Schedule III [83]. The DEA took exception to the facts that were presented [84] and maintained that the placement of MDMA in Schedule I was appropriate. The temporary emergency status was extended as required on the first anniversary of the original invocation [85] and then made permanent four months later, effective November 13, 1986 [86]. It has become apparent [87] that the emergency scheduling invoked during this period by the DEA (mid-1985 to late 1986) was not valid, as Congress had invested this authority in the Attorney General, who had never subdelegated it to the DEA.

This final action by the DEA, which was contrary to the opinion and recommendation of the law judge, was appealed by Dr. Grinspoon, and one specific claim concerning the currently accepted use of MDMA in the United States was found valid. It was found [88] that FDA approval was not the sole criterion for determining the acceptability of a drug for medical use, and the DEA was ordered to remove MDMA from Schedule I, pending reconsideration of its medical status. The DEA removed MDMA from Schedule I, effective December 22, 1987 [89], but upon reconsideration replaced it into Schedule I, effective three months later [90].

MDMA now rests soundly as a Schedule I drug. In light of the removal from Schedule I ordered by the Court for documented reasons, the Department of Justice stated [91] that valid challenges may be made to any legal action that had been taken prior to the eventual permanent scheduling (which became final on March 23, 1988).

6. PHARMACOLOGY

Under the general heading of pharmacology are gathered all references to pharmacological studies including behavior and discrimination studies, and at least a brief outline of the development of the serotonin story. Again, as with the legal history section, there are several contributions in this volume that will deal specifically and at length with these matters. Only the historic sequence of findings will be outlined here.

There are a few reports on animal behavior and drug discrimination studies that were in the literature prior to the proposed legal scheduling in 1985, but with this government action there was urgent solicitation of supporting pharmacological data from a number of academic researchers. Several reports were promptly provided and sent in unpublished form directly to the DEA for its use at the hearings. These reports were introduced directly into evidence by the DEA attorneys, as they contained conclusions (MDMA has neurotoxicity [92], MDMA is self-administered in baboons similarly to cocaine and phencyclidine [93], MDMA action in monkeys suggests a high abuse potential [31]) that were felt to support the government's position. Some of these findings have subsequently appeared in the published literature.

6.1. Behavior studies and pharmacological responses

The first studies of animal behavior changes induced by MDMA were included in the original toxicity studies done at the University of Michigan in 1953 [28]. Observations were made in both the dog and monkey of motor activity, including convulsions, rigidity, and tremor, and indications of central activity, such as bizarre body attitudes and fright displays that were interpreted as hallucinations. These studies did, however, employ i.v. dosages that incorporated levels in excess of three times the observed LD-50s for both animals.

Observers of operant behavior of primates have reported interference with grooming and social interactions [94] and the loss of conditioned behavior [95] due to MDMA, and mice studies with the optical isomers have involved specific behavioral regimens [96, 97]. Studies have been made of rotational activity [98], changes in reward thresholds [99, 100], and the voluntary self-administration of MDMA in both rhesus monkeys [101] and baboons [102].

Specific pharmacological responses that have been observed with MDMA administration have included analgesia [103, 104] and hyperactive stimulation [105, 106].

6.2. Discrimination studies

Animals (usually the rat) that are trained to discriminate between a test drug and a saline control have been used to help classify drugs into specific pharmacological categories. The Glennon group at Virginia has worked extensively with MDMA and has generally found that MDMA is "seen" by the test animal more as a stimulant than as a hallucinogen. The earliest studies were conducted before the proposed scheduling of MDMA [107, 108], and their findings were generally consistent with the reported human effects. Other discrimination tests compared MDMA to drugs involved in specific neurotransmitter function [109], and other groups used animals trained to MDMA itself to attempt to classify its optical isomers [110, 111]; the *S* isomer (the isomer effective in man [8]) was the more potent.

Both pigeons [112] and monkeys [113] have also been used as test animals in discrimination studies. In a study with both rats and monkeys trained to discriminate amphetamine from saline, MDMA mimicked amphetamine [114].

6.3. Biochemical studies

The first biochemical studies on MDMA (1986) have shown an interference with the enzyme TPH, tryptophan hydroxylase [115, 116], which is involved with the biosynthesis of the neurotransmitter serotonin. The corresponding enzyme in the biosynthesis of dopamine, TH, (tyrosine hydroxylase), which is disturbed by pretreatment with methamphetamine, is unchanged with MDMA administration [116]. However, the inhibition of the enzyme that deaminates gamma-aminobutyric acid (GABA) protects the rat from toxic doses of methamphetamine and also protects the rat from neurotoxicity due to MDMA [117]. The TPH loss is spared by the prior removal of the adrenal glands [118].

The metabolism of MDMA has been briefly studied. N-demethylation to form 3,4-methylenedioxyamphetamine (MDA) is a minor pathway in humans [41] and in microsome studies [119]. The abstract of a study [120] reports the identification of seven metabolites (including MDA) in the rat.

6.4. In Vitro studies

The first in vitro study of MDMA appeared in 1982, and this was also the first report that serotonin might play some role in the action of MDMA [121]. Both serotonin and dopamine release from striatal slices have been reported [122], and specific binding to receptors has been studied in tissue homogenates and with radioligand assays [15, 123, 124]. The releasing potencies of MDMA for the dopamine-labelled caudate nucleus [125] and striatum [126] have been determined. Using the optical isomers of MDMA, the receptor sites in various regions of the brain have been recorded [127]. MDMA was compared with several close structural analogues in rat brain-slice studies, with dopamine release reflecting a greater dependence on chemical structure than does serotonin release [128]. A warning has been issued concerning binding of radioactive MDMA to non-biological components in these assays [129].

The observed contraction in ilium strips induced by MDMA [130] apparently does not involve serotonin receptors.

6.5. Neurotoxicity

The initial study that was used to support the government placing MDMA into Schedule I of the Controlled Substances Act was conducted by researchers at the University of Chicago. This was an investigation into the serotonin nerve terminal damage caused by MDA (methylenedioxyamphetamine) [131].

Just prior to the effective date of the DEA's emergency scheduling of MDMA (July 1, 1985) and during the period of intense publicity that MDMA was receiving in the popular press, there was a television forum, *the Phil Donahue Show*, which brought together several prominent figures in the controversy. Mr. Gene Haislip (a representative of the DEA), Dr. Charles Schuster (the director of the University of Chicago Drug Abuse Research Center), and Dr. Rick Ingrasci (a psychiatrist with broad clinical experience with MDMA) were on the program. After the show, Dr. Schuster mentioned his unpublished study on MDA, which showed nerve damage [132].

A preprint of that paper was obtained by Mr., Haislip, who used it in justifying the proposed emergency scheduling. The draft, states that other ring-substituted amphetamines (MMDA, TMA, and DOM are specified) are widely abused and that their toxicity need be evaluated. When this paper finally appeared in September, 1985, the drug MMDA had been replaced with the drug name MDMA, and the DEA justified the public health hazard, saying, "research with a similar drug (MDA) showed that a single dose may cause permanent brain damage" [133].

The torrent of serotonin-related research involving MDMA which followed these events will be only outlined briefly below, as this topic is addressed specifically in several chapters in this volume.

6.5.1. *Mouse studies*

The few studies that have investigated the potential neurotoxicity of MDMA in the mouse have agreed that there is very little serotonin involvement — certainly less than that seen in the rat [134]. Chronic treatment of high doses for several days showed, following a seven day rest, no serotonin loss and no nerve damage [135]. A single large dose showed transient effects, but there was no further change following additional exposure [136].

6.5.2. *Rat studies*

By far the most research on the neurotoxicity of MDMA has been conducted on the rat. The studies have employed a wide range of doses administered chronically or acutely by any of several routes, and in general all agree that there is a depletion of serotonin and a long-term axon damage, both of which are dose-dependent [33, 134–144]. The homologue of MDMA with an ethyl group on the nitrogen (MDE) has also been found to be neurotoxic [145, 146].

One study has reported that the administration of MDMA intracerebrally produces no neurotoxicity [147], suggesting that some peripherally formed metabolite might be responsible for the apparent neurotoxicity of MDMA. A study with a number of potential metabolites, however, shows all to be of reduced toxicity [148]. Also, one study involving the chronic administration of MDMA where each application was preceded by the serotonin uptake blocker citalopram showed a complete blockade of all neurodegenerative effects of MDMA [138].

Two studies have found evidence for the involvement of dopamine with MDMA. An effort to explain the rewarding aspect of MDMA, using brain electrodes and specific neurotransmitter inhibitors, has indicated that the reinforcing values may be mediated by dopamine D-2 receptors rather than serotonin 5-HT-2 receptors [149]. And with 6-hydroxydopamine-induced lesions, there was less motor activity following MDMA administration [150].

6.5.3. Guinea pig studies

Two studies have compared the guinea pig directly with the rat [135, 140]. In both, there was a drug-induced decrease in serotonin and in the density of uptake sites.

6.5.4. Cat studies

A single study in the cat [151] covered the dosage range of 0.25 to 5.0 mg/Kg. There was a decrease in serotonin level observed, and this was suppressed by pretreatment with p-chloroamphetamine, suggesting that the action of the two drugs is similar.

6.5.5. Primate studies

The serotonin depletion and neural damage, so well established in the rat for MDMA, appears to occur in primates as well [152, 153]. Subcutaneous administration of MDMA to three species of monkey at dosages of between 2.5 and 5 mg/Kg, twice daily for four consecutive days produces a dose-related depletion of serotonin and its principal metabolite 5-hydroxyindole acetic acid. Also, there was evidence of structural damage to serotonergic nerve fibers. The most recent study by this group has shown that the oral route is less effective than the subcutaneous and that single dosages are less effective than multiple dosages in the depletion of serotonin. However, even a single oral dose of 5 mg/Kg MDMA is effective in producing a long-lived depletion of serotonin [154].

REFERENCES

1. Verfahren zur Darstellung von Alkyloxyaryl-, Dialkyloxyaryl- und Alkylenedioxyaryl-aminopropanen bzw. deren am Stickstoff monoalkylierten Derivaten, 1914. German Patent #274,350, filed December 24, 1912, issued May 16, 1914, and assigned to E. Merck in Darmstadt.
2. Formyl derivatives of secondary bases, 1920. German patent #334,555, assigned to E. Merck. Chem. Abst. 17:1804a.
3. Care must be taken with the term piperonylacetone. This term has been used commercially in reference to two distinct chemical individuals, vis., 1-(3,4-methylenedioxyphenyl)-2-propanone and 1-(3,4-methylenedioxyphenyl)-3-butanone. Only the former compound, also known unambiguously as either 3,4-methylenedioxyphenylacetone or 3,4-methylenedioxybenzyl methyl ketone, gives rise ultimately to MDMA. For chemical and pharmacological details, see: Shulgin, A.T. and Jacob III, P., 1982. Potential misrepresentation of 3,4-methylenedioxyphenyl-amphetamine (MDA). A toxicological warning. J. Anal. Toxicol. 6:71–75. For a toxicological study of the products following the use of the latter (wrong) ketone, see reference 9.

4. Elks, J. and Hey, D.H., 1943. β-3,4-methylenedioxyphenylisopropylamine. J. Chem. Soc. pp. 15–16.
5. Ichiro, H., 1952, α-Methyl-3,4-methylenedioxyphenethylamine, Japanese patent 1770 ('52). Chem. Abst. 48:2097b.
6. Biniecki, S., Muszynski, E., Jagiellowicz, H., and Chojnacka, Z., 1962. Preparation of 1,N-dimethylveratrylethylamine and 1,N-dimethylpiperonylethylamine. Acta Polon. Pharm. 19:31–35, and Chem. Abst. 58:3334f.
7. Braun, U., Shulgin, A.T., and Braun, G., 1980. Centrally active N-substituted analogs of 3,4-methylenedioxyphenylisopropylamine (3,4-methylenedioxyamphetamine). J. Pharm. Sci. 69:192–195.
8. Anderson III, G.M., Braun, G., Braun, U., Nichols, D.E., and Shulgin, A.T., 1978. Absolute configuration and psychotomimetic activity. NIDA Research Monograph #22, pp. 8–15.
9. Davis, W.M. and Borne, R.F., 1984. Pharmacological investigation of compounds related to 3,4-methylenedioxyamphetamine. Subst. Alcohol Action/Misuse 5:105–110.
10. O'Brien, B.A., Bonicamp, J.M., and Jones, D.W., 1982. Differentiation of amphetamine and its major hallucinogenic derivatives using thin layer chromatography. J. Anal. Toxicol. 6:143–147.
11. Fujisawa, T. and Deguchi, Y., 1954. Utilization of safrole as medical raw material. VI. New synthesis of 3,4-methylenedioxybenzyl methyl ketone. 1. J. Pharm. Soc. Jpn. 74:975–977, and Chem. Absts. 44:10958i.
12. Hansson, R.C., 1987. Clandestine laboratories production of MDMA 3,4-methylenedioxymethamphetamine. Analog 9 (3):1–10.
13. Binierki, S. and Krajewski, E., 1960. Preparation of dl-1-(3,4-methylenedioxyphenyl)-2-(methylamino)propane. Acta Polon. Pharm. 17:421–425.
14. Frith, C.H., 1986. Report: 28-day oral toxicity of methylenedioxymethamphetamine hydrochloride (MDMA) in rats. Protocol No. EMD-SC-002. Toxicology Pathology Associates, Little Rock, AR.
15. Gehlert, D.R., Schmidt, C.J., Wu, L., and Lovenberg, W., 1985. Evidence for specific methylenedioxymethamphetamine (Ecstasy) binding sites in the rat brain. Eur. J. Pharmacol. 119:135–136.
16. Bailey, K., By, A.W., Legault, D., and Verner, D., 1975. Identification of the N-methylated analogs of the hallucinogenic amphetamines and some isomers. J. Assoc. Off. Anal. Chem. 58:62–69.
17. Sreenivasan, V.R., 1972. Problems in identification of methylenedioxy and methoxy amphetamines. J. Crim. Law 63:304–312.
18. Gaston, T.R. and Rasmussen, G.T., 1972. Identification of 3,4-methylenedioxymethamphetamine. Microgram 5:60–63.
19. Eichmeier, L.S. and Caplis, M.E., 1975. The forensic chemist; an analytical detective. Anal. Chem. 47:841A–844A.
20. Ruybal, R., 1986. Microcrystalline test for MDMA. Microgram 19:79–80.
21. Renfroe, C.L., 1986. MDMA on the street: Analysis anonymous. J. Psychoactive Drugs 18:363–369.
22. Sedgwick, B., Lo, P., and Yee, M., 1986. Screening and confirmation of 3,4-methylenedioxymethamphetamine (MDMA) in urine. Evaluation of 1000 specimens. Abstracts of the CAT/SOFT Meetings, Oct. 29–Nov. 1, 1986, Reno/Lake Tahoe, NV.
23. Shaw, M.A. and Peel, H.W., 1975. Thin-layer chromatography of 3,4-methylenedioxyamphetamine and other phenethylamine derivatives. J. Chromatog. 104:201–204.
24. Hearn, W.L., Hime, G., and Andollo, W., 1986. Recognizing Ecstasy: Adam and Eve, the MDA derivatives — analytical profiles. Abstracts of the CAT/SOFT Meetings, Oct. 29–Nov. 1, 1986, Reno/Lake Tahoe, NV.
25. Ruangyuttikarn, W. and Moody, D.E., 1988. Comparison of three commercial amphetamine immunoassays for detection of methamphetamine, methylenedioxyamphetamine, methylenedioxymethamphetamine, and methylenedioxyethylamphetamine. J. Anal. Toxicol. 12:229–233.
26. Noggle Jr, F.T., DeRuiter, J., and Long, M.J., 1986. Spectrophotometric and liquid chromatographic identification of 3,4-methylenedioxyphenylisopropylamine and its N-methyl and N-ethyl homologs. J. Assoc. Off. Anal. Chem. 69:681–686.

27. Noggle Jr, F.T., DeRuiter, J., McMillan, C.L., and Clark, C.R., 1987. Liquid chromatographic analysis of some N-alkyl-3,4-methylenedioxyamphetamines. J. Liq. Chromatog. 10:2497–2504.
28. Hardman, H.F., Haavik, C.O., and Seevers, M.H., 1973. Relationship of the structure of mescaline and seven analogs to toxicity and behavior in five species of laboratory animals. Tox. Appl. Pharmacol. 25:299–309.
29. Davis, W.M., Hatoum, H.T., and Waters, I.W., 1987. Toxicity of MDA (3,4-methylenedioxyamphetamine) considered for relevance to hazards of MDMA (Ecstasy) abuse. Alcohol Drug Res. 7:123–134.
30. George, D.J. and Wolf, H.H., 1966. Dose-lethality curves for d-amphetamine in isolated and aggregated mice. Life Sci. 5:1583–1590.
31. Harris, L.S., 1985. Preliminary report on the dependence liability and abuse potential of methylenedioxymethamphetamine (MDMA). Document entered into evidence Re: MDMA Scheduling Docket No. 84–48, U.S. Department of Justice, Drug Enforcement Administration, October 16, 1985.
32. Frith, C.H., Chang, L.W., Lattin, D.L., Walls, R.C., Hamm, J., and Doblin, R., 1987. Toxicity of methylenedioxymethamphetamine (MDMA) in the dog and the rat. Fund. Appl. Toxicol. 9:110–119.
33. Slikker Jr, W., Ali, S.F., Scallet, A.C., and Frith, C.H., 1986. Methylenedioxmethamphetamine (MDMA) produces long lasting alterations in the serotoninergic system of rat brain. Soc. Neurosci. Abst. 12:363.
34. O'Hearn, E., Battaglia, G., DeSouza, E.B., Kubar, M.J., and Mollivar, M.E., 1986. Systemic MDA and MDMA, psychotropic substituted amphetamines, produce serotonin neurotoxicity. Soc. Neurosci. Abst. 12:1233.
35. Goad, P.T., 1985. Acute and subacute toxicity study of methylenedioxymethamphetamine in rats. Project Report. Intox Laboratories, Redfield, AR.
36. Dowling, G.P., McDonough III, E.T., and Bost, R.O., 1987. "Eve" and "Ecstasy": A report of five deaths associated with the use of MDEA and MDMA. JAMA 257:1615–1617.
37. Hayner, G.N. and McKinney, H., 1986. The dark side of Ecstasy. J. Psychoactive Drugs 18:341–347.
38. Brown, C. and Osterloh, J., 1987. Multiple severe complications from recreational ingestion of MDMA (Ecstasy). JAMA 258:780–781.
39. Smilkstein, M.J., Smolinske, S.C., and Rumack, B.H., 1987. A case of MAO inhibitor/MDMA interaction: Agony after Ecstasy. Clin. Toxicol. 25:149–159.
40. Verebey, K., Alrazi, J., and Jaffe, J.H., 1988. The complications of "Ecstasy" (MDMA). JAMA 259:1649–1650.
41. Reynolds, P.C., 1986. Personal communication.
42. Buchanan, J., 1985. Ecstasy in the emergency department. Clin. Toxicol. Update 7:1–4.
43. Peroutka, S.J., Pascoe, N., and Faull, K.F., 1987. Monoamine metabolites in the cerebrospinal fluid of recreational users of 3,4-methylenedioxymethamphetamine (MDMA; "Ecstasy"). Res. Commun. Subst. Abuse 8:125–138.
44. Shulgin, A.T. and Nichols, D.E., 1978. Characterization of three new psychotomimetics. In *The Psychopharmacology of Hallucinogens* (Stillman and Willette, eds). New York: Pergamon Press, pp. 74–83.
45. Greer, G., 1983. MDMA: A new psychotropic compound and its effects in humans. Privately published. 333 Rosario Hill, Sante Fe, NM 87501. Copyright 1983.
46. Greer, G. and Tolbert, R., 1986. Subjective reports of the effects of MDMA in a clinical setting. J. Psychoactive Drugs 18:319–327.
47. Greer, G., 1985. Recommended protocol for MDMA sessions. Privately Published. 333 Rosario Hill, Sante Fe, NM 87501. Copyright 1985.
48. Greer, G., 1985. Using MDMA in psychotherapy. Advances 2:57–59.
49. Tatar, A. and Naranjo, C., 1985. MDMA in der Gruppenpsychotherapie. Symposium: "Uber den derzeitigen Stand der Forschung auf dem Gebiet der psychoaktiven Substanzen," Nov. 29–Dec. 12, 1985, in Hirschhorn/Neckar, Germany.
50. Wolfson, P.E., 1986. Meetings at the edge with Adam: A man for all seasons. J. Psychoactive Drugs 18:329–333.
51. Siegel, R.K., 1986. MDMA, nonmedical use and intoxication. J. Psychoactive Drugs 18:349–354.

52. Buffum, J. and Moser, C., 1986. MDMA and human sexual function. J. Psychoactive Drugs 18:355–359.
53. Adamson, S., 1985. Through the gateway of the heart: Accounts of experiences with MDMA and other empathogenic substances. San Francisco, CA: Four Trees Publications.
54. Downing, J., 1986. The psychological and physiological effects of MDMA on normal volunteers. J. Psychoactive Drugs 18:335–340.
55. Peroutka, S.J., 1987. Incidence of recreational use of 3,4-methylenedioxymethamphetamine (MDMA, "Ecstasy") on an undergraduate campus. N. Engl. J. Med. 317:1542–1543.
56. Ecstasy: Everything looks wonderful when you're young and on drugs, 1981. Wet Magazine, September–October, p. 76.
57. ca. 1984. (a) General Information; MDMA. (b) Ecstasy: 21st Century Entheogen. The former is directed towards therapists; the latter is strictly promotional in nature.
58. Adler, J., 1985. Getting high on "Ecstasy." Newsweek, April 15, p. 96.
59. Toufexis, A., 1985. A crackdown on Ecstasy. Time, June 10, p. 64.
60. Dowling, C.G., 1985. The trouble with Ecstasy. Life Magazine, August, pp. 88–94.
61. Baum, R.M., 1985. New variety of street drugs poses growing problem. Chem. Eng. News, September 9, pp. 7–16.
62. Grinspoon, L. and Bakalar, J.B., 1985. What is MDMA? Harvard Medical School Mental Health Letter 2(2):8.
63. Gertz, K.R., 1985. "HugDrug" alert: The agony of Ecstasy. Harper's Bazaar, November, p.48.
64. Gold, M.S., 1985. Ecstasy, Etc. Alcoholism and Addiction, September–October, p. 11.
65. Abramson, D.M., 1985. Ecstasy: The new drug underground. New Age, October, pp. 35–40.
66. Shafer, J., 1985. MDMA. Psychedelic drug faces regulation. Psychology Today, May, pp. 68–69.
67. O'Rourke, P.J., 1985. Tune in. Turn on. Go to the office late on Monday. Rolling Stone, December 19, p. 109.
68. Trudeau, G.B., 1985. Doonesbury, August 12–August 24, 1985. Universal Press Syndicate.
69. Doblin, R., 1985. MDMA — a multidisciplinary investigation: Reports from the medical, scientific, and regulatory communities. The Earth Metabolic Design Laboratories, Inc., Berkeley, CA.
70. MDMA, A Multidisciplinary Conference. May 17 and 18, 1986, Institute for Addiction Studies, Merritt Peralta Hospital, Oakland, CA. Psychoactive Drugs 18(4).
71. 1985. (a) MDMA. NIDA Capsules. Issued by the Press Office of the National Institute on Drug Abuse, Rockland, MO. July, 1985. (b) Designer drugs: new concern for the drug abuse community. NIDA Notes, December, 1985, pp. 2–3.
72. Schulman, R., 1985. The losing war against "designer drugs." Business Week, June 24, pp. 101–104.
73. Smith, D.E., Wesson, D.R., and Buffum, J., 1985. MDMA. "Ecstasy" as a adjunct to psychotherapy and a street drug of abuse. California Society for the Treatment of Alcoholism and Other Drug Dependencies News 12:1–3.
74. Nichols, D.E., 1985. MDMA represents a new type of pharmacologic agent and cannot be considered to be either a hallucinogenic agent or an amphetamine-type stimulant. Essay distributed to the DEA and the WHO, in connection with the hearings concerning the scheduling of MDMA.
75. Nichols, D.E., 1986. Differences between the mechanism of action of MDMA, MBDB, and the classic hallucinogens. Identification of a new therapeutic class, entactogens. J. Psychoactive Drugs 18:305–313.
76. Nichols, D.E., Hoffman, A.J., Oberlender, R.A., Jacob III, P., and Shulgin, A.T., 1986. Derivatives of 1-(1,3-benzodioxol-5-yl)-2-butanamine: Representatives of a novel therapeutic class. J. Med. Chem. 29:2009–2015.
77. Randolph, W.F., 1984. International drug scheduling; convention on psychotropic substances; stimulant and/or hallucinogenic drugs. Fed. Regis. 49(140):29273–29274.
78. Mullen, F.M., 1984. Schedules of controlled substances. Proposed placement of 3,4-methylenedioxymethamphetamine into Schedule I. Fed. Regis. 49(146):30210–30211.
79. Cotton, R., 1984. Letter from Dewey, Ballantine, Bushby, Palmer & Wood, 1775 Pennsylvania Avenue, N.W., Washington, D.C. to F.M. Mullen Jr., DEA, September 12, 1984.

The four retaining parties were Professor Thomas B. Roberts, George Greer, M.D., Lester Grinspoon, M.D., and Professor James Bakalar.
80. Mullen, F.M., 1984. Schedules of controlled substances. Proposed placement of 3,4-methylenedioxymethamphetamine into Schedule I. Hearing. Fed. Regis. 49(252):50732–50733.
81. Young, F.L., 1985. Opinion and recommended decision on preliminary issue. Docket No. 84–48, June 1.
82. Lawn, J.C., 1985. Schedules of controlled substances. Temporary placement of 3,4-methylenedioxymethamphetamine (MDMA) into Schedule I. Fed. Regis. 50(105):23118–23120.
83. Young, F.L., 1986. Opinion and recommended ruling, findings of fact, conclusions of law and decision of administrative law judge. Docket 84–48, May 22.
84. Stone, S.E. and Johnson, C.A., 1986. Government's exceptions to the opinion and recommended ruling, findings of fact, conclusions of law and decision of the administrative law judge. Docket 84–48, June 13.
85. Lawn, J.C., 1986. Schedules of controlled substances. Extension of temporary control of 3,4-methylenedioxymethamphetamine (MDMA) in Schedule I. Fed. Regis. 51(116): 21911–21912.
86. Lawn, J.C., 1986. Schedules of controlled substances. Scheduling of 3,4-methylenedioxymethamphetamine (MDMA) into Schedule I of the Controlled Substances Act. Fed. Regis. 51(198):36552–36560.
87. Kane, J., 1986. Memorandum and opinion, Case No. 86-CR-153 in the United States District Court for the District of Colorado. Pees and McNeill, Defendents, October 1.
88. Coffin, Torruella, and Pettin, 1987. United States Court of Appeals for the First Circuit. Lester Grinspoon, Petitioner, vs. Drug Enforcement Administration, Respondent, September 18.
89. Lawn, J.C., 1988. Schedules of controlled substances. Deletion of 3,4-methylenedioxymethamphetamine (MDMA) from Schedule I of the Controlled Substances Act. Fed. Regis. 53:2225.
90. Lawn, J.C., 1988. Schedule of controlled substances. Scheduling of 3,4-methylenedioxymethamphetmine (MDMA) into Schedule I of the Controlled Substances Act. Remand. Fed. Regis. 53:5156.
91. Harbin, H., 1988. MDMA. Narcotics, Forfeiture, and Money Laundering Update. U.S. Department of Justice, Criminal Division, Winter, pp. 14–19.
92. Seiden, L.S., 1985. Report of preliminary results on MDMA. Document entered into evidence Re: MDMA Scheduling Docket No. 84-48, U.S. Department of Justice, Drug Enforcement Administration, October 16.
93. Griffiths, R.R., Lamb, R., and Brady, J.V., 1985. A preliminary report on the reinforcing effects of racemic 3,4-methylenedioxymethamphetamine in the baboon. Document entered into evidence Re: MDMA Scheduling Docket No. 84-48, U.S. Department of Justice, Drug Enforcement Administration, October 16.
94. Schlemmer Jr, R.F., Montell, S.E., and Davis, J.M., 1986. MDMA induces behavioral changes in members of primate social colonies. Fed. Proc. 45:1059(#5263).
95. Thompson, D.M., Winsauer, P.J., and Mastropaolo, J., 1987. Effects of phencyclidine, ketamine and MDMA on complex operant behavior in monkeys. Pharmacol. Biochem. Behav. 26:401–405.
96. Glennon, R.A., Little, P.J., Rosecrans, J.A., and Yousif, M., 1987. The effects of MDMA ("Ecstasy") and its optical isomers on schedule-controlled responding in mice. Pharmacol. Biochem. Behav. 26:425–426.
97. Rosecrans, J.A. and Glennon, R.A., 1987. The effect of MDA and MDMA ("Ecstasy") isomers in combination with pirenpirone on operant responding in mice. Pharmacol. Biochem. Behav. 28:39–42.
98. Kulmala, H.K., Boja, J.W., and Schechter, M.D., 1987. Behavioral suppression following 3,4-methylenedioxyamphetamine. Life Sci. 41:1425–1429.
99. Bird, M. and Kornetsky, C., 1986. Naloxone antagonism of the effects of MDMA "Ecstasy" on rewarding brain stimulation. Pharmacologist 28:149 (#319).
100. Hubner, C.B., Bird, M., Rassnick, S., and Kornetsky, C., 1988. The threshold lowering effects of MDMA (Ecstasy) on brain-stimulating reward. Psychopharmacology 95:49–51.
101. Beardsley, P.M., Balster, R.L., and Harris, L.S., 1986. Self administration of methylenedioxymethamphetamine (MDMA) in rhesus monkeys. Drug and Alc. Depend. 18:149–157.

102. Lamb, R.J. and Griffiths, R.R., 1987. Self-injection of dl-3,4-methylenedioxymethamphetamine in the baboon. Psychopharmacology 91:268–272.
103. Braun, U., Shulgin, A.T., and Braun, G., 1980. Prufung auf zentral Aktivitat und Analgesia von N-substituierten Analogen des Amphetamin-Derivates 3,4-Methylenedioxyphenylisopropylamin. Arzneim.-Forsch. 30:825–830.
104. Beaton, J.M., Benington, F., Christian, S.T., Monti, J.A., and Morin, R.D., 1987. Analgesic effects of MDMA and related compounds. Pharmacologist 29:281 (abstract).
105. Callahan, P.M. and Appel, J.B., 1987. Differences in the stimulus properties of 3,4-methylenedioxyamphetamine (MDA) and N-methyl-3,4-methylenedioxymethamphetamine (MDMA) in animals trained to discriminate hallucinogens from saline. Soc. Neurosci. Abstr. 13 (part 3):1720(#476.2).
106. Gold, L.H. and Koob, G.F., 1988. Methysergide potentiates the hyperactivity produced by MDMA in rats. Pharmacol. Biochem. Behav. 29:645–648.
107. Glennon, R.A., Young, R., Rosecrans, J.A., and Anderson, G.M., 1982. Discriminative stimulus properties of MDA analogs. Biol. Psychiat. 17:807–814.
108. Glennon, R.A. and Young, R., 1984. Further investigation of the discriminative stimulus properties of MDA. Pharmacol. Biochem. Behav. 20:501–505.
109. Schechter, M.D., 1986. Discriminative profile of MDMA. Pharmcol. Biochem. Behav. 24:1533–1537.
110. Oberlender, R. and Nichols, D.E., 1988. Drug discrimination studies with MDMA and amphetamine. Psychopharmacology, 95:71–76.
111. Schechter, M.D., 1987. MDMA as a discriminative stimulus: Isomeric comparisons. Pharmacol. Biochem. Behav. 27:41–44.
112. Evans, S.M. and Johanson, C.E., 1986. Discriminative stimulus properties of (+/−)-3,4-methylenedioxymethamphetamine and (+/−) methylenedioxyamphetamine in pigeons. Drug and Alc. Depend. 18:159–164.
113. Kamien, J.B., Johanson, C.E., Schuster, C.R., and Woolverton, W.L., 1986. The effects of (+/−) methylenedioxymethamphetamine and (+/−) methylenedioxyamphetamine in monkeys trained to discriminate (+) amphetamine from saline. Drug and Alc. Depend. 18:139–147.
114. Woolverton, W.L., Virus, R.M., Kamien, J.B., Nencini, P., Johanson, C.E., Seiden, L.S., and Schuster, C.R., 1985. Amer. Coll. Neuropsychopharm. Absts., p. 173.
115. Gibb, J.W., Hanson, G.R., and Johnson, M., 1986. Effects of (+) 3,4-methylenedioxymethamphetamine [(+)MDMA] and (−) 3,4-methylenedioxymethamphetamine [(−)MDMA] on brain dopamine, serotonin, and their biosynthetic enzymes. Soc. Neurosci. Absts. 12:169.2.
116. Stone, D.M., Stahl, D.C., Hanson, G.R., and Gibb, J.W., 1986. The effects of 3,4-methylenedioxymethamphetamine (MDMA) and 3,4-methylenedioxyamphetamine (MDA) on monoaminergic systems in the rat brain. Europ. J. Pharmacol. 128:41–48.
117. Stone, D.M., Hanson, G.R., and Gibb, J.W., 1987. GABA-transaminase inhibitor protects against methylenedioxymethamphetamine (MDMA) induced neurotoxicity. Soc. Neurosci. Absts. 13(part 3):#251.3.
118. Johnson, M., Bush, L.G., Stone, D.M., Hanson, G.R., and Gibb, J.W. 1987. Effects of adrenalectomy on the 3,4-methylenedioxymethamphetamine-induced decrease of tryptophan hydroxylase activity in the frontal cortex and hippocampus. Soc. Neurosci. Absts. 13(Part 3):#464.6.
119. Brady, J.F., Di Stephano, E.W., and Cho, A.K., 1986. Spectral and inhibitory interactions of (+/−) 3,4-methylenedioxyamphetamine (MDA) and (+/−) 3,4-methylenedioxymethamphetamine (MDMA) with rat hepatic microsomes. Life Sci. 39:1457–1464.
120. Lim, H.K. and Foltz, R.L., 1988. Metabolism of 3,4-methylenedioxymethamphetamine (MDMA) in rat. FASEB Abst. 2(5):A-1060 (#4440).
121. Nichols, D.E., Lloyd, D.H., Hoffman, A.J., Nichols, M.B., and Yim, G.K.W., 1982. Effects of certain hallucinogic amphetamine analogues on the release of [3H] serotonin from rat brain synaptosomes. J. Med. Chem. 25:530–535.
122. Levin, J.A., Schmidt, C.J., and Lovenberg, W., 1986. Release of [3H] monoamines from superfused rat striatal slices by methylenedioxymethamphetamine (MDMA). Fed. Proc. 45:1059 (#5265).
123. Lyon, R.A., Glennon, R.A., and Titeler, M., 1986. 3,4-Methylenedioxymethamphetamine (MDMA): Stereoselective interactions at brain 5-HT1 and 5-HT2 receptors. Psychophar-

macology 88:525–526.
124. Battaglia, G., Brooks, B.P., Kulsakdinum, C., and De Souza, E.B., 1988. Pharmacologic profile of MDMA (3,4-methylenedioxymethamphetamine) at various brain recognition sites. Europ. J. Pharmacol. 149:159–163.
125. Kalix, P., 1986. A comparison of the effects of some phenethylamines on the release of radioactivity from isolated rat caudate nucleus prelabelled with 3H-dopamine. Arzneim. Forsch. 36:1019–1021.
126. Kalix, P., Yousif, M.Y., and Glennon, R.A., 1988. Differential effects of the enantiomers of methylenedioxymethamphetamine (MDMA) on the release of radioactivity from (3H) dopamine prelabelled rat striatum. Res. Commun. Subst. Abuse 9:45–52.
127. Battaglia, G., Kuhar, M.J., and De Souza, E.B., 1986. MDA and MDMA (Ecstasy) interactions with brain serotonin receptors and uptake sites. *In vitro* studies. Soc. Neurosci. Absts. 12:336.4.
128. Johnson, M.P., Hoffman, A.J., and Nichols, D.E., 1986. Effects of the enantiomers of MDA, MDMA, and related analogues on [3H] serotonin and [3H] dopamine release from superfused rat brain slices. Eur. J. Pharmacol. 132:269–276.
129. Wang, S.S., Ricaurte, G.A., and Peroutka, S.J., 1987. [3H] 3,4-methylenedioxymethamphetamine (MDMA) interactions with brain membranes and glass fiber filter paper. Eur. J. Pharmacol. 138:439–443.
130. Frye, G. and Matthews, R., 1986. Effect of 3,4-methylenedioxymethamphetamine (MDMA) on contractile responses in the guinea pig ilium. Pharmacologist 28:149 (#318).
131. Ricaurte, G.A., Bryan, G., Strauss, L., Seiden, L., and Schuster, C., 1985. Hallucinogenic amphetamine selectively destroys brain serotonin nerve terminals. Science 229:986–988.
132. Ingrasci, R., 1988. Personal communication.
133. Thornton, M., 1985. DEA will ban hallucinogen known to users as "Ecstasy." Washington Post, June 1, p. A1.
134. Stone, D.M., Hanson, G.R., and Gibb, J.W., 1987. Differences in the central serotonergic effects of methylenedioxymethamphetamine (MDMA) in mice and rats. Neuropharmacology 26:1657–1661.
135. Battaglia, G., Yeh, S.Y., and De Souza, E.B., 1988. MDMA-induced neurotoxicity parameters of degeneration and recovery of brain serotonin neurons. Pharmacol. Biochem. Behav. 29:269–274.
136. Logan, B.J., Laverty, R., Sanderson, W.D., and Yee, Y.B., 1988. Differences between rats and mice in MDMA (methylenedioxymethylamphetamine) neurotoxicity. Eur. J. Pharmacol. 152:227–234.
137. Ali, S.F., Scallet, A.C., Holson, R.R., Newport, G.D., and Slikker Jr, W., 1987. Acute administration of MDMA (Ecstasy): Neurochemical changes persist up to 120 days in rat brain. Soc. Neurosci. Absts. 13(Part 3):904 (#251.1).
138. Battaglia, G., Yeh, S.Y., O'Hearn, E., Molliver, M.E., Kuhar, M.J., and De Souza, E.B., 1987. 3,4-Methylenedioxymethamphetamine and 3,4-methylenedioxyamphetamine destroy serotonin terminals in rat brain: Quantification of neuro-degeneration by measurements of [3H] paroxetine-labelled serotonin uptake sites. J. Pharmacol. Exp. Ther. 249:911–916.
139. Champney, T.H., Golden, P.T., and Matthews, R.T., 1986. Reduction in hypothalamic serotonin levels after acute MDMA administration. Soc. Neurosci. Absts. 12:101.6.
140. Commins, D.L., Vosmer, G., Virus, R.M., Woolverton, C.R., Schuster, C.R., and Seiden, L.S., 1987. Biochemical and histological evidence that methylenedioxymethamphetamine (MDMA) is toxic to neurons in rat brain. J. Pharmacol. Exp. Ther. 241:338–345.
141. De Souza, E.B., Battaglia, H., Yeh, S.Y., and Kuhar, M.J., 1986. In vitro and in vivo effects of MDA and MDMA (Ecstasy) on brain receptors and uptake sites: Evidence for selective neurotoxic actions on serotonin terminals. Am. Coll. Neuropsychopharmacol., Dec. 8–12, p. 207.
142. Finnigan, K.T., Ricaurte, G.A., Ritchie, L.D., Irwin, I., Peroutka, S.J., and Langston, J.W., 1988. Orally administered MDMA causes a long-term depletion of serotonin in rat brain. Brain Res. 447:141–144.
143. Molker, D.J., Robinson, S.E., and Rosecrans, J.A., 1987. (+/−) 3,4-methylenedioxymethamphetamine (MDMA) produces long-term reductions in brain 5-hydroxytryptamine in rats. Eur. J. Pharmacol. 138:265–268.
144. Yeh, S.Y., Battaglia, G., O'Hearn, E., Mollivar, M.E., Kuhar, M.J., and De Souza, E.B.,

1986. Effects of MDA and MDMA (Ecstasy) on brain monoaminergic systems: In vivo studies. Soc. Neurosci. Absts. 12:#336.5.
145. Ricaurte, G.A., Finnigan, K.F., Nichols, D.E., DeLanney, L.E., Irwin, I., and Langston, J.W., 1987. 3,4-Methylenedioxyethylamphetamine (MDE), a novel analogue of MDMA, produces long-lasting depletion of serotonin in the rat brain. Eur. J. Pharmacol. 137: 265–268.
146. Schmidt, C.J., 1987. Acute administration of methylenedioxymethamphetamine: Comparison with the neurochemical effects on its N-desmethyl and N-ethyl analogs. Eur. J. Pharmacol. 136:81–88.
147. Molliver, M.E., O'Hearn, E., Battaglia, G., and De Souza, E.B., 1986. Direct intracerebral administration of MDA and MDMA does not produce serotonin neurotoxicity. Soc. Neurosci. Absts. 12:336.3.
148. Yeh, S.Y. and Hsu, F.L., 1987. Neurotoxicity of metabolites of MDA and MDMA (Ecstasy) in the rat. Soc. Neurosci. Absts. 13(Part 3):906 (#251.11).
149. Bird, M.P., Svensen, C.N., Knapp, C., Hrbek, C.C., Bird, E.D., and Kornetsky, C., 1987. Evidence for dopinergic and not serotonergic mediation of the threshold lowering effects of MDMA on rewarding brain stimulation. Soc. Neurosci. Absts. 13(Part 3):1323 (#365.13).
150. Gold, L.H., Hubner, C.B., and Koob, G.F., 1987. The role of mesolimbic dopamine in the stimulant action of MDMA. Soc. Neurosci. Absts. 13(Part 3):833 (#234.13).
151. Trulson, T.J. and Trulson, M.E., 1987. 3,4-Methylenedioxymethamphetamine (MDMA) suppresses serotonergic dorsal raphe neuronal activity, in freely moving cats and in midbrain slices in vitro. Soc. Neurosci. Absts. 13(Part 3):905 (#251.7).
152. Wilson, M.A., Ricaurte, G.A., and Molliver, M.E., 1987. The psychotropic drug 3,4-methylenedioxymethamphetamine (MDMA) destroys serotonergic axons in primate forebrain: Regional and laminar differences in vulnerability. Soc. Neurosci. Absts. 13(Part 3):905 (#251.8).
153. Ricaurte, G.A., Forno, L.S., Wilson, M.A., DeLanney, L.E., Irwin, I., Moliver, M.E., and Langston, J.W., 1988. (+/−) Methylenedioxymethamphetamine selectively damages central serotonergic neurons in nonhuman primates. JAMA 260:51–55.
154. Ricaurte, G.A., DeLanney, L.E., Irwin, I., and Langston, J.W., 1988. Toxic effects of MDMA on central serotonergic neurons in the primate: Importance of route and frequency of drug application. Brain Res. 446:165–169.

2. THE THERAPEUTIC USE OF MDMA

GEORGE R. GREER AND REQUA TOLBERT

1. INTRODUCTION

This chapter describes a method for the therapeutic administration of MDMA ((+/−), 3,4-methylenedioxymethamphetamine) to humans and includes five case reports. Comparisons are made to the approach of "Twelve Step programs" for substance abuse treatment and to sacred rites of passage. The importance of the mental set of the patient and therapist and the psychological preparation of both are emphasized. Screening criteria and informed consent information are also discussed. Results from 80 patients indicate that MDMA seems to decrease the fear response to a perceived threat to a patient's emotional integrity, leading to a corrective emotional experience that probably diminishes the pathological effects of previous traumatic experiences. The acquisition of effective skills for communicating feelings to family members also occurs. Psychological benefits were lasting up to a two-year follow-up for many patients, and relief from chronic pain and premenstrual symptoms occurred for one patient each. Double-blind controlled experiments utilizing the method presented are not feasible because the mental set is affected and the MDMA effect is easily perceived by patient and therapist. Suggestions for potential applications include the prevention and treatment of dysfunctional family relationships and of substance abuse.

We supervised MDMA-assisted therapy sessions for patients from 1981 until 1985, when MDMA was placed in Schedule I by the Drug Enforcement Administration. An outline of our method and a detailed summary of the results reported by the first 29 people administered MDMA have been published elsewhere [1].

2. THE ROLE OF MENTAL SET

The term "mental set" refers to the overall belief of both patient and therapist as to what the goal of the session is; how the therapist, the session procedure and the drug will help the patient achieve the goal; and what results are to be expected. From our own experiences and from the reports of other therapists, we found that the goal of developing a more compassionate attitude toward oneself and others was easily achieved by people undergoing MDMA-assisted therapy. Also, relief from chronic symptoms and behavior problems seemed greater when such a change in attitude occurred. Based on the success of other methods utilizing altered states of awareness to achieve this healing attitude, we approached sessions more as sacred rites of passage than as conventional therapy sessions [2–4]. We also viewed the effect of MDMA as secondary to the effect of the therapeutic ritual: assisting more than causing the patient to achieve the desired outcome from the session.

After the choice of the session goal, we found that the quality of the relationship between patient and therapist was the next most important variable in predicting the outcome for an MDMA session — more important even than the dose taken. In the absence of a healing-oriented relationship in which the patients felt safe enough in the therapists' presence to open themselves fully to new and challenging experiences, one was apt to have a more superficial experience. An essential factor in establishing a therapeutic relationship was the patient's knowing that the therapists had undergone MDMA sessions themselves and so would be able to understand the kind of experience the patient would be having. Hearing the stories of the therapists' experiences and seeing that no harmful effects had occurred served to provide hope and reassurance that the session would go well. Special preparation of the therapists before they began to supervise MDMA sessions was crucial for both enhancing the therapeutic relationship and for understanding the effect of MDMA on the therapy process.

3. PREPARATION AND MENTAL SET OF THE THERAPISTS

We first learning about MDMA-assisted therapy from a clinical psychologist, Leo Zeff, Ph.D., in 1980. Zeff had conducted LSD-assisted therapy sessions in the early 1960s and was the first psychotherapist to use MDMA extensively, beginning in 1976.

Zeff's approach was based on the concepts and techniques derived from the LSD research of psychiatrist Stanislav Grof [2], the peyote rituals of the Huichol tribe in Mexico [3], Buddhist Vipassana (Insight) meditation [4], the mystical traditions of both East and West, as well as his training and experience in traditional Western psychology and psychotherapy. He saw that Using MDMA had similarities to the Twelve Step programs that have proven to be so effective in the treatment of addicitive behaviors: the surrender of the patient's individual will to a Higher Power for personal guidance in order to achieve healthy self-control. The surrendering attitude is seen as essential for achieving a profound release from addictive attachment to relationships, beliefs, and behaviors that have been destructive in the person's past.

At the time we learned of Zeff's work, he had conducted hundreds of MDMA sessions and had achieved dramatic results without complications. Because he had done his work away from the public eye and had written nothing about it, we saw a need both to offer sessions and to document the results so that the research community would learn about the potential of MDMA as a pharmacological catalyst for psychotherapy. We began conducting sessions and recording information about patients both before and after their sessions. This information gathering was more in the spirit of a descriptive "medical anthropology" study than a rigorously controlled experiment designed to determine the efficacy of MDMA-assisted therapy [1].

Regarding our own preparation, one of us [GG] was a Board-Certified psychiatrist, had studied the same subjects as Zeff, had undergone long-term insight-oriented psychotherapy, and had practiced Vipassana meditation. The other [RT] practiced Vipassana and was a Master's level psychiatric nurse. Our training for conducting MDMA sessions began with our own experience taking it together. We were most impressed with the ease we had in communicating our feelings and thoughts about each other that previously had been too emotionally charged to be discussed, as well as with the effortless forgiveness we experienced for times we felt we had been hurt by the other, all with a clear sensorium and cognitive faculties. In the eight years since, we have continued to utilize the skill of intimate communication that developed spontaneously during that first session.

Daily practice of meditation helped us to develop the skill of observing the details of inner experience during the MDMA state of consciousness and, because we learned to achieve similar states of mind without the drug, prevented our seduction into the belief that MDMA was an exclusive panacea. In addition, having a few MDMA sessions supervised by a therapist experienced in its use made us familiar with the range of its effects. The experience of fearless communication and spontaneous forgiveness, or letting go of resentments, was particularly important in understanding how MDMA can be used effectively.

4. SCREENING AND PREPARATION OF PATIENTS

To foster development of the optimal mental set for patients, we essentially followed Zeff's method for screening, preparing, and conducting sessions, and we added pre- and post-session questionnaires and written informed consent. The entire procedure was also reviewed and approved by a peer review panel of psychiatrists and a psychologist who were experienced with the use of drug-assisted psychotherapy and with the effects of MDMA. When prospective patients requested to be considered for a session, we always asked them what they already had heard of our work, to help us assess their expectations. We then sent them a screening questionnaire, informed consent information, and an essay that addressed our philosophy of the use of psychoactive medicines in therapy. The questionnaire elicited a personal, medical, and psychiatric history and information about their use of other drugs. It also asked questions designed to orient them toward having the session, eg., "What is your purpose in having a session with MDMA?" and "What are your expectations and/or fears of what will happen?"

Screening of candidates was very important and involved several issues. Medically, we excluded those who were hypertensive or had cardiovascular disease; were taking psychotropic medication; were hyperthyroid, epileptic, diabetic, or hypoglycemic; or had liver disease, actual or possible pregnancy, or any other medical condition that would have placed the person at risk for significant morbidity or mortality. We also excluded those who, due to a mental or emotional disturbance, had been unable to function at work or socially for more than a day or so. For those who were in psychotherapy, we obtained clearance from their therapists to give the session. Although sessions with MDMA were useful to individuals who were at times unable to take care of themselves due to psychiatric problems, we only worked with functional, relatively well-adjusted people [5].

If, after reviewing the questionnaire, there were no reasons for not having the session, we arranged to have a screening interview. This interview was usually held in our home, as were most of the sessions. Opening our home to the patients allowed them both greater physical comfort and greater trust from knowing us better, than if we had used an office setting. We began by asking if they had any questions based on what they might have heard or read about MDMA. We then reviewed the questionnaire with the patient, having him or her clarify or elaborate on issues that interested or concerned us. We pursued any areas of past difficulty, and reviewed their medical history, paying special attention to any history of significant losses, their attitudes and beliefs about death, and their general spiritual orientation.

The most important information we elicited from the patient was a clear statement of the purpose for having the session. If the stated purpose was in clear opposition to our own philosophy (eg., if they *only* wanted an enjoyable experience, wished to avoid issues of current or past pain, or wished *only* to focus on their spouse's problems), further interviews would have been scheduled, or the applicant would have been excluded. In addition, we did not administer MDMA to those who aroused any feeling of uneasiness in either of us. We had learned that giving MDMA in the presence of an ill-defined misgiving in the therapist almost always resulted in complications in managing the session. We also refused sessions to those whose spouse or therapist was not supportive of the plan.

After going over the questionnaire, we always told people of our own backgrounds and how we came to work with MDMA. We asked that this information be held in confidence, just as we held information about them in confidence. This mutual sharing established a context of equal status in collaboration, intimacy, confidentiality, and trust. It also discouraged the development of transference projections, distinguishing our approach from that of traditional analytically oriented psychotherapy. We preferred to serve only as "sitters" or assistants to patients who were exploring themselves, rather than to involve ourselves in a long-term relationship in order to allow a classical transference to emerge and to be worked-through. If transference phenomena emerged, we helped the person understand and use them in a clinically appropriate manner and scheduled follow-up therapy sessions with or without MDMA, as indicated. (This occurred only once: with the single patient who was in psychotherapy with one of us [GG] before having MDMA sessions.)

To establish an attitude of safety and security and to further screen out inappropriate patients, we required patients to make an explicit contract of four agreements. These served as the core structure of our relationship with them: 1) therapists and patients all agreed to remain on the premises until all agreed that the sessions was over and that it was safe to leave; 2) the patients agreed to refrain from any activity that could have been destructive to themselves, to others, or to any property; 3) there would be no sexual activity between the patients and the therapists; and 4) patients agreed to follow any instructions given to them by a therapist, when it was explicitly given as part of the structure of the session. This last agreement did not include various therapeutic suggestions we made.

Through the agreements, patients were asked to allow us to manage issues of physical safety during the course of the MDMA session. We believed that if there were some distrust of us, it would have been brought out during the discussion of the contract. If patients were uncomfortable with any of these requests, more time could have been spent in preparation until agreement occurred. It was never necessary to exclude patients due to their inability to accept the ground rules, and all were able to respond appropriately at the rare times when these rules were invoked.

With the agreements in place, we encouraged patients to ask for anything they wanted during the sessions, in order to encourage their becoming conscious of repressed desires, knowing that they would not be allowed to act them out destructively. For example, with an explicitly stated agreement of "no sex," one could feel, express, and even fulfill an infantile desire to be held or comforted without fear of a therapist taking sexual advantage. Within the context of safely defined external boundaries, patients could devote full attention and concern toward introspection.

5. INFORMED CONSENT

A major consideration in our using MDMA was informed consent. After a discussion of personal histories, the informed consent information was reviewed. In addition to going over all the known possible benefits and risks, the form listed the members of our peer review committee, stated the above-mentioned agreements and the protocol for the session, and listed alternative procedures for achieving similar results.

Benefits were briefly and generally described and included improved communication, personal insights, and elevated mood. Physiological side effects were primarily those that came from stimulation of the sympathetic nervous system: muscle tightness, restlessness, nausea, increased pulse, and increased blood pressure. If we were still conducting sessions at this time, we would also inform patients of the reports of human deaths associated with recreational MDMA use and the reports of serotonin depletion and neurotoxicity in rats and primates, as well as any other risks that might be known at the time of our obtaining informed consent [6–8]. The translation of human mortality data from use in uncontrolled situations and animal toxicology data into risk factors for humans under medical supervision is highly controversial and a matter to be decided by peer review and human experimentation review panels.

The issue of unwanted, or "negative," psychological effects or emotions was a special one to consider. With MDMA, as with any other drug that can compromise psychological defense mechanisms, it was common to see the pain of unfinished grief or earlier traumatic experience arise both psychologically and somatically. Physical symptoms such as headache, shortness of breath, pain, or other discomforts sometimes occurred and often were felt by the patient, to be associated with previously forgotten memories or repressed feelings. Depression and/or anxiety occasionally were felt during the session or in the days that followed until the person felt a sense of completion with the pertinent issues.

Rarely did unwanted reactions last more than a day or two, and usually the person found those experiences quite useful, although difficult. Even at the time of this writing (1988), we have not heard of any long-lasting problems following MDMA sessions supervised by professional psychotherapists. Because of this fact, we have not been overly concerned by the reports of neurotoxicity in animals [7, 8]. We currently believe that, for all but extremely rare cases, there is a significant gap between the highest therapeutic doses of 200 mg taken monthly and clinically significant toxic doses [9]. Further support for this view comes from the fact that fenfluramine, an appetite suppressant approved for daily use by the Food and Drug Administration, elicits a neurotoxicity pattern in animals that is very similar to that of MDMA (Molliver, M., personal communication) [10, 11].

Because we could not predict all of the specific elements of a difficult experience, patients were required to be willing to experience anything that might arise during or after the session, including the worst experience they had ever had in the past. If there was at least a conscious desire to open oneself to pain without resisting, then when painful experiences did occur, they could be worked through more quickly.

Hearing the details of the many unpleasant physical symptoms that we described in giving informed consent could have added an unnecessary element of anxiety. In spite of these considerations, a thorough process of informing people of what they might experience was both ethical and practical. If individuals were so frightened by our process of giving informed consent that they chose not to have a session (and this happened several times), then we believed that it was not a good time for them to have the experience in the first place.

As much as possible, everything we did or said in preparing people to take this compound attempted to give this implicit message: "You are consciously taking a medicine to open yourself to whatever teachings you may need at this time. Neither you nor we know what these teachings are or how they may occur. We will provide a safe place for your explorations and be available to assist you with any difficulties, but all that you learn that is real comes from yourself or from the Divine within you — not from us or from the medicine itself."

We found that the more attention patients placed on their preparation, the more meaning and value was achieved from the session, and the more the person claimed responsibility for it. It was useful for them to have a clear notion of what their expectations were, not so much to be able to fulfill them, but to facilitate a letting go of them beforehand. Meditation, keeping a journal, or other practices could all potentiate the effect of the session.

On the practical side, regular consumption of alcohol or other psychoactive drugs seemed to decrease the effects of MDMA; so abstaining from any use of these compounds was advised for the few days before the session. Food, especially milk products, seemed to decrease absorption of MDMA in the stomach and to predispose patients to nausea or vomiting. For this reason, fasting overnight or for at least six hours before ingesting MDMA was advised. Additionally, in planning when to have the session, we instructed patients to refrain from making any work or social obligations the day after. Frequently, there was much psychological material for the person to consciously integrate, as well as a tendency to feel tired.

With regard to alternative procedures, we knew of no other drug or procedure that produced the characteristic effects of MDMA. However, we informed patients of the many ways to achieve similar results with varying degrees of success. These included other techniques using MDMA or other mind-altering compounds, special deep breathing techniques, practices of meditation and prayer, hypnosis, psychotherapy, prescribed psychotropic medications, and certain massage and bodywork techniques. We felt the procedures that did not involve the use of drugs, when supervised by a skilled practitioner, were generally safer than those that did. Before giving someone a session, we made sure that the probable benefits significantly outweighed the risks when compared to the alternative procedures. We believed that for a person who was fully committed to a goal of honesty, psychological growth, and well-being, there was no one method that was necessary to make progress toward a therapeutic goal. The commitment and the willingness to encounter the difficulties that arose were all that were really required.

6. CONDUCTING THE SESSION

When patients arrived for their sessions, they were first given time to bring us up to date on their lives. Decisions about exact dosages of MDMA were then made. For men, the range was usually from 100 to 150 mg. Women took 75 to 125 mg. We did not know if there was a sex difference or a difference based solely on weight, but women seemed to be more sensitive to MDMA than were men. If the session was for an individual who wished primarily to focus his or her attention internally, a larger dose was suggested. For couples who wanted to spend time together, a smaller dose was more useful. Often the general intensity of effects and side effects was described for the dosage ranges, the person indicated his or her wish for a "low, medium, or high" dose, and we translated that into an actual amount. Especially in an initial session, we believed this ability to have some control over the situation would be comforting.

Time was sometimes spent in silence, prayer, or meditation before taking the MDMA. After ingestion, the patient sat quietly waiting to feel the effects, or lay down, donning eyeshades to decrease outside distractions. Music was played, usually via headphones, and was always instrumental, except for vocal pieces sung in foreign languages. The genre was classical, ethnic, or modern. Typical composers included Mahler, Beethoven, Wagner, Fauré, and Deuter. The decision to play a given piece of music at any given time was usually made intuitively by one of us. Patients could ask to change a piece of music or have silence.

Couples were encouraged to begin their experiences in separate rooms. This allowed them to attend to individual issues in the MDMA state and to notice fully the initial physical effects. After a couple of hours, partners usually had much to talk about with each other and so came together when they both felt ready.

We rarely initiated psychotherapeutic interaction with people during their sessions. We were, of course, available and supportive if difficult or painful experiences occurred. After conducting the first few sessions, we found that talking about or "reporting" one's experience and thoughts during the session was often done with *our* benefit in mind and only diluted the inner process. If this sort of "monologue conversation" with us occurred, we suggested that the person either talk into a tape recorder for future reference or simply focus his or her attention inside rather than toward us. We could hear about it when it was all over. The main thing for us to do was to be available to provide for physical needs and comfort and to help give perspective when requested.

After one-and-a-half to two hours, patients were offered an additional dose of MDMA (usually 50 mg) to extend the peak part of the experience another hour and to make the wearing off of the drug more gradual. Since dehydration was a common effect, water was offered periodically. After patients felt that the MDMA state had mostly passed, they usually set up and began talking to us about what had happened. We usually spent one to three hours discussing the session, to assist in the integration of the experience into daily life. In all, either or both of us usually spent a total of six to eight hours with the patients on the day of their session. We did not routinely offer interpretations of the meaning of the experiences, but tried to facilitate a smooth transition back to the usual state of consciousness.

We made sure that patients were alert and able to function normally, before they were allowed to leave. Blurred vision due to pupillary dilatation, and the visual "trails" that were rarely seen behind moving objects, had to be absent before we allowed anyone to drive. To gather follow-up information, a questionnaire was given, to be answered after one or two weeks. The Peak Experience Profile (Pahnke, W., Grof, S., and Dileo, F., 1981, unpublished manuscript) was also given to patients during the latter years of our work, to be completed as soon as possible. All patients were encouraged to call us whenever they wanted to discuss any problems or to relate their thoughts about the experience.

Roughly 90% of the people we saw in this context had powerful and generally positive and useful experiences, according to their follow-up reports [1]. About one third returned to have a single subsequent session, and another third had more than two sessions. The following are the stories of five people who had more dramatically beneficial sessions than most, though the quality of the sessions was typical for the other seventy-five people who had sessions with us:

Case 1: A married man in his early seventies with two grown children

A retired geophysicist and farmer, he had always been a successful man in charge of his own life. At the time of his sessions, he had been told that he was among the longest-living survivors to date with multiple myeloma, which had been diagnosed in 1975. He had undergone group therapy for two years (predating his cancer diagnosis) to help with depression over family problems. On being diagnosed with cancer, he began therapy in a group format, where he learned deep relaxation, meditation, and visualization to combat his cancer and to assist in pain control. He did, in fact, learn to achieve states where his pain was as reduced as it was with narcotics, but he still endured much pain.

At the time of our first meeting, his main complaint was "movement pain" from four collapsing vertebrae, secondary to the myeloma. Over the preceding months, the pain had increased, decreasing his physical and sexual activity and his ability to go fishing or to fly his plane. He was also troubled by the depression that usually followed the numerous fractures of his spine, which necessitated confinement to bed. The goal for his session with MDMA, which he wished to take with his wife, was to cope with his pain in a better way and to receive help in adjusting to his current life changes.

He took 125 mg, his wife took 100 mg, and they remained in separate rooms listening to music, with eyeshades and headphones. He hummed along with the classical music being played. Shortly after his second dose of 50 mg of MDMA, two hours after the first, he announced ecstatically that he was free of pain and began singing aloud with the music and repeatedly proclaiming his love for his wife and family. He spent several hours in this rapturous state. Afterwards he said it was the first time he had really been pain free in the four years since the current relapse of his myeloma had begun. He described his experience of being inside his vertebrae, straightening out the nerves, and "gluing" fractured splinters back together.

In a letter written two weeks after his session, he stated that his pain had returned, but that his ability to hypnotically "re-anchor" his pain-free experience greatly assisted him in reducing the pain by himself. He had four MDMA sessions spaced over the course of nine months; each time he achieved relief from his physical pain, and he had greater success in controlling painful episodes in the interims by returning himself to an approximation of the MDMA state. He noted in particular that the feelings of "cosmic love" and especially forgiveness of himself and others would usually precede the relief of physical pain. He described an episode from his second session:

> As I was finishing the meditation, time ceased to exist, my ego fell away, and I became one with the cosmos. I then started my visualization of my body's immune system fighting my cancer, of the chemo[therapy] joining with my immune system to kill the cancer cells in my vertebrae, and of positive forces coming from the cosmos to fight my cancer. Gradually I went deeper in to where the feeling of love, peace, and joy were overwhelming. Although I had heard the new age music before, many details of the music became clear and more beautiful.

The series of sessions stopped because MDMA was placed in Schedule I by the DEA. The FDA denied us permission to continue the treatment, pending further animal studies. He remained quite functional and mostly pain free for a few months after the last session, but eventually his pain began to return and he died very peacefully in his wife's presence soon afterward.

Case 2: A single man in his mid-30's and administrator of a small inpatient substance abuse treatment facility

He had taken LSD in Vietnam and was a little concerned that he might have flashbacks to those times during the session. However, he had no significant psychological problems when he came to us, was curious about MDMA, and wanted a session to find out new things about himself. He was a smoker and was surprised to find he had no desire for a cigarette for the few hours during the session. He was given 125 mg of MDMA with diazapam (5 mg) to reduce muscle tension, followed by another 50 mg of MDMA after an hour. One of us [GG] took the same combination for the purpose of learning how it would affect the relationship. (This procedure was followed in a few cases where more of a research goal than a specific therapeutic goal was the purpose of the session [12].) He listened to music with headphones for about an hour and then spent the rest of the time in conversation with us.

Three days later he said that he felt none of the physical tensions he feared he would feel from memories of his LSD experiences. Two days later, at work, he noticed he felt more relaxed on the job than ever before. Two years later he was sent the follow-up questionnaire and reported that, "It was a very enjoyable experience. I experienced a state, while under the MDMA influence, in which I found it difficult to concentrate on negative subjects (thoughts or feelings)." He did not expect to feel as close to us as he did: "I felt as if they were able to understand how I was feeling and thinking." The only unpleasant aspect was that the MDMA "wore off," because it had felt so good. His curiosity had been satisfied, but he did not believe he learned anything new about himself. He concluded his report by saying, "I believe the most beneficial aspect of how I felt during the session was that I felt very little defensiveness.... I thought about things in myself I didn't like. I was able to accomplish this without feeling guilty or defensive." He reported no long term benefit from the session.

Case 3: A real estate agent in her mid-thirties, married and mother of two daughters

She is the child of two Jewish Holocaust survivors from Poland and was born in a displaced persons' camp after the war. Her parents live in her community, and she had always been close to her father, who had been in a concentration camp, but she had a fairly difficult relationship with her mother. She had experienced some "anxiety attacks" in graduate school and had dropped out for some time. Subsequent to psychotherapy and re-entering school, she completed a Master's Degree in counselling. Her only significant medical history was a complaint of premenstrual syndrome — she would become quite irritable and emotionally labile during the premenstrual period every month. Her expressed purpose in having an experience with MDMA, which she wished to take with her husband, was to achieve "increased awareness and personal expansion."

She took 100 mg for her first session with no second dose. During the initial phase of the experience, she felt that she was "in Eternity" and was among the clouds (her eyes were closed). Then, gradually, disturbing thoughts intruded, and each one heralded a wave of nausea. Various fears and associations to a concentration camp were prominent. She tried to vomit several times but could not. Her nausea subsided as she released much of her "concentration camp consciousness" and the associated emotions. She felt she had taken on those feelings and attitudes from her parents, who had lived through the "Holocaust nightmare" where so many in their families had died. She noted that the pain of those years and, indeed, of the entire Holocaust had subtly colored her emotions and her life. It was after her "decision" to vomit during her session that her fears subsided, "moved through" her, and left. She felt a new appreciation and love for her parents for enabling her to be living in the world. The rest of her experience was generally positive.

The next day she was intensely angry for a short period of time and had her "worst fight in thirteen years" with her husband, as both continued to release old tensions and negative feelings. For the next two days, although she continued to have some nausea and her digestion was retarded, she felt well emotionally and more grounded than usual: "I was a different person."

She subsequently had eight MDMA sessions over the course of a year; four of those times she took only 50 mg during her premenstrual periods for the relief of tension and irritability, which she unexpectedly had discovered it offered. Her marijuana intake decreased from several times a week to occasional use, and cocaine ceased to have any appeal. Generally, she felt that the release of negative and painful material gave her more energy and creativity. She has observed that she argues less with her mother and feels closer to her. At the same time, she is less concerned with her parents' inevitable deaths, having a newly reinforced belief in the eternity of the soul — that "we are *not* our bodies."

Almost three years after her first session she said:

> I still am a different person. I'm not prone to getting caught up in the negative dark influences that are present in my character. I have more choice over how I feel. I can handle my emotions and I understand how they work more.

Cases 4 and 5: A married couple in their early 30's with no children

The husband was teaching creative writing and writing a novel, and the wife was a graduate student in foreign literature. She had undergone an abortion a few months before because their lives simply did not have room for a child, though they both wanted to have children later.

In stating her purpose for the session, the wife said, "I hope to achieve a new level of communication with [my husband] — one we can remember and continue to draw on in the future." He wrote, "I hope to clarify my thinking about myself, my work, and my short-term goals, and to share a visionary and intensely conscious experience with [my wife]."

One of us [GG] and the husband took 75 mg initially, the other [RT] and the wife took 50 mg. All four of us took three more doses of 50 mg each at 45 minutes, 1 ½ hours, and 4 hours after the initial dose. They spent their session talking with us, alternating with time to themselves. Two weeks later, the wife wrote the following:

> I wish I could be writing to tell you that the exhilaration both [my husband] and I felt two weeks ago is still alive. . . . but with a return to the daily world of responsibilities, the feeling has diminished. Not that it's left completely: what has remained is the memory of that [day] and the clarity of thought and emotion it left me with. And that is very precious indeed. . . .

I fell in love with [my husband] all over again, and I seemed to see how the anxieties of this year have taken their toll on him.... But when I saw his face released from cares, it was a great insight to me — and this was the face I first loved. So we've had some long talks and a lot of things that had been only superficially resolved now seem completed. We vow to work always to be more open with each other.

Perhaps the most obvious and delightful effect of the drug was that it freed me from feeling trapped inside my body. These past few months following the abortion have been excruciating, apart from the emotional pain. [My husband] and I have always enjoyed each other tremendously — physically — and somehow I was so shaken by what our bodies had done, that I developed a kind of fear or reluctance to take any more chances. This was exaggerated by the complications I had, but even once I got back onto a normal cycle, I could hardly believe that simple pills could prevent pregnancy. None of this was deep-rooted in me, because I had never felt it before and was consciously trying to overcome it. But the MDMA did the trick, like a miracle. I was able to put everything into perspective and realize that one accident does not necessarily mean another, and that in the meantime there is a lot of enjoying to do.

In her follow-up questionnaire much later, she wrote, "There was a great sense of communality — that we're in this life together — and we are still drawing on this shared realization now, after 1 ½ years."

He wrote the following after ten days:

The positive effects of the drug — calmness, fearlessness, renewed love for [my wife], a sensation of personal intensity or power, re-alignment of one's proper place in the universe — all these have been wearing thinner over the past week and a half.

Still, the effects haven't entirely worn off, and I'm happy that it's the feeling of renewed love which has held up the best. The sensation was (and still is) as if I were seeing [my wife] through new eyes, not unlike the eyes I saw her with when we first fell in love, but not quite the same ones either. Wider ones, I think; less wary ones, for sure.

We heard many similar stories from other therapists who used MDMA differently from us, though their basic attitudes and purposes were the same.

7. CONCLUSION

From our own observations and those of others, we believe that, in the right circumstances, MDMA reduces or somehow eliminates the neurophysiological fear response to a perceived threat to one's emotional integrity. Though we do not understand how MDMA reduces the experience of feeling threatened, it does consistently reduce the primary somatic symptom of fear: the tightness and nervous feeling in the throat, chest, abdomen, and skeletal musculature. There is also a moderate anesthesia to pain (but not to touch) in the skin during the acute effect, which may parallel the anesthesia to emotional pain or fear without reducing emotional sensitivity. With this barrier of fear removed, a loving and forgiving awareness seemed to occur quite naturally and spontaneously. People found it unusually comfortable to be aware of, to communicate, and to remember thoughts and feelings that are usually accompanied by fear and anxiety. Alcohol can reduce the same kind of fear, but cognitive clarity and conscious recovery of repressed feelings are not possible. Anxiolytic drugs and beta sympathetic blockers also reduce anxiety but do not facilitate the access of repressed memories or feelings.

Presumably both common and unique childhood traumas had caused the formation of conditioned fear responses, which made it desirable for patients to avoid having certain feelings or thoughts symbolically associated with the traumas. Without the conditioned fear inhibiting access to the information contained in these thoughts, feelings, or memories, patients' value judgements about their past, their relationships, and their self-worth could be based upon more accurate information. They could reassess any aspect of their lives and relationships that they chose, from the broader perspective of security and love, rather than from one of vulnerability and fear. With the fear removed, a corrective emotional experience could occur, and it seemed natural and easy for most people to begin to trust the validity of their own unfearful feelings, as well as those of a significant other who was experiencing the same state with them.

Because MDMA did not distort perception, thinking, or memory (except in doses well over 100 to 150 mg), the learning that took place during the session often became consolidated and applied to patients' everyday lives long after the session had ended. Couples who had a session together frequently began to base their relationships much more on love and trust than on fear and suspicion. Some of our patients said that under the influence of MDMA, and for days to years afterward, they "feel more loving," "can easily forgive pain of the past," or "let go of grudges or misunderstandings." We believe these results were not caused by MDMA, but were achieved by the patients making decisions based on what they learned during their MDMA sessions, and by their remembering and applying those decisions for as long as they were able to and willing to after the session was over. We believe this occurred because taking MDMA with an intention to learn, with an attitude of acceptance, and in a safely structured setting enabled people to experience their true nature, which is essentially loving and forgiving. About 75 of the 80 patients we treated reported significant benefit from their session(s).

Unfortunately, a double-blind controlled experiment testing the efficacy of our method is impossible because the optimum mental set requires that the patient and therapist know that MDMA is being taken and because the MDMA altered state is so obvious to both. Motivation would be severely compromised if therapists and patients thought there was only a 50% chance that they were really taking MDMA and that the primary goal of the session would be to study the effects of the drug itself rather than for the patients to learn something for themselves.

One potential application of MDMA therapy could be in the prevention and treatment of addictive behaviors. Pathological childrearing, with its traumas and deprivations, is a major cause of the development of both addictive behaviors and the co-dependency of family members, which helps sustain the addiction. If those at risk could acquire the skills of becoming aware of their deepest feelings and communicating these to family members, it could prevent the transmission of dysfunctional family relationships from one generation to the next.

Traditional cultures often used consciousness-altering drugs in a ritual context as a rite of passage into adulthood, while such powerful rites are virtually absent in modern Western culture. As an example, the Native American Church has successfully used peyote rituals within a Christian context to treat alcohol abuse among its members [13]. A number of our patients spontaneously reduced their intake of cocaine and marijuana and noticed a decreased desire to consume them, even though that was not a goal for having a session [1]. Such potential benefits of the careful use of MDMA should be considered when evaluating the potential risks of toxicity from therapeutic doses.

ACKNOWLEDGEMENT

The authors wish to acknowledge the assistance of Rick Strassman, M. D., in the preparation of the manuscript.

REFERENCES

1. Greer, G. and Tolbert, R., 1986. Subjective reports of the effects of MDMA in a clinical setting. J. Psychoactive Drugs 18(4):319–327.
2. Grof, S., 1980. *LSD Psychotherapy*. Pomona, CA: Hunter House.
3. Myerhoff, B., 1978: Peyote and the mystic vision. In *Art of the Huichol Indians*. Berrin, K., ed. New York: Harry N. Abrams, pp. 56–70.
4. Goldstein, J., 1983. *The Experience of Insight*. Boulder, CO: Shambala.
5. Wolfson, P.E., 1986. Meetings at the edge with Adam: A man for all seasons? J. Psychoactive Drugs 18(4):329–333.
6. Downing, G.P., et al., 1987. "Eve" and "ecstasy": A report of five deaths associated with the use of MDEA and MDMA. JAMA 257:1615–1617.
7. Commins, D.L., et al., 1987. Biochemical and histological evidence that methylenedioxymethylamphetamine (MDMA) is toxic to neurons in the rat brain. J. Pharmacol. Exp. Ther. 241(1):338–345.
8. Ricaurte, G.A., et al., 1988. (+/−)3,4-methylenedioxymethamphetamine selectively destroys central serotonergic neurons in nonhuman primates. JAMA 260(1):51–55.
9. Hayner, G.N. and McKinney, H.E., 1986. MDMA: The dark side of ecstasy. J. Psychoactive Drugs 18(4):341–347.
10. Molliver, M.E., 1987. Serotonergic neuronal systems: What their anatomic organization tells us about function. J. Clin. Psychopharmacol. 7(6):17S.
11. Molliver, D.C. and Molliver, M.E., 1988. Selective neurotoxic effects of (+/−) fenfluramine upon 5-HT axons in rat brain: Immunocytochemical evidence. Abstract, Society for Neuroscience Annual Meeting.
12. Shulgin, A.T., et al., 1986. A protocol for the evaluation of new psychoactive drugs in man. Meth. Find. Exptl. Clin. Pharmacol. 8(5):313–320.
13. Grinspoon, L. and Bakalar, J. 1979. *Psychedelic Drugs Reconsidered*. New York: Basic Books, p. 222.

3. TESTING PSYCHOTHERAPIES AND DRUG THERAPIES: THE CASE OF PSYCHEDELIC DRUGS

JAMES B. BAKALAR AND LESTER GRINSPOON

The drug revolution that began 30 years ago has transformed psychiatry, but it has left little imprint on psychotherapeutic procedures themselves. Little attention has been given to the possibility of using drugs directly to enhance the process of psychotherapy — fortifying the therapeutic alliance and facilitating the production of memories, fantasies, and insights. A change may now be coming; for example, a psychiatrist known for his research on the therapeutic alliance has proposed that a "pharmacotherapy of interpersonal processes" might be considered both to study and to improve the alliance [1]. The wait has been long partly because the research involved is complex and hard to perform. The theoretical bases for the two types of therapy are vastly different; these differences are reflected in the way experiments are conducted and the results are evaluated. Reconciliation and unification will not be easy to achieve. One of the best ways to see why that is so is to examine the different significance assigned to placebo effects in drug experiments and psychotherapy studies.

1. INTRODUCTION

Everyone now takes it for granted that the correct psychiatric and medical procedure for determining the effectiveness of drugs is the controlled double-blind trial with random assignment of otherwise matched patients to the experimental drug or a placebo. In medicine the controlled trial is, of course, a standard way to establish causal relationships; it is one form of Mill's method of agreement and difference. The method is now a legal rule as well as a scientific one: the FDA usually does not allow a drug to be marketed for therapeutic purposes unless it has been proved effective in double-blind experiments.

It is in the last twenty years that researchers have become interested in applying this standard to psychotherapy. Psychotherapists are beginning to think that they must obtain scientific credentials in order to get government support and justify insurance payments. They have also been influenced by the success of lithium, phenothiazines, tricyclic antidepressants, and other psychiatric drugs. The experimental literature in the field is already large. But unlike controlled drug experiments, which usually come to straightforward conclusions that produce a fairly quick consensus in the psychiatric community, psychotherapy research is plagued by empirical doubts and conceptual confusions. Even the experts have found it hard to agree on what the results mean. For example:

> I believe that it is fair to say that psychotherapy produces modestly or moderately positive results with certain types of cases, but there is tremendous variability among studies, including the types of control group used [2].

> All that has been demonstrated in the literature so far is that psychotherapies seem more efficacious than nothing for relatively minor conditions . . . what remains to be shown is that any psychotherapy is more efficacious than simple helping relationships that reduce demoralization [3].

> The effects of psychotherapy are equivalent to the effects of a relatively minimal placebo, which is essentially equivalent to knowledge that one is in treatment [4].

> As I have attempted to show, research is not likely to adduce precise data on such issues as the "safety and efficacy" of psychotherapeutic modalities or their "cost effectiveness" [5].

The main reason for the lack of consensus is the problem of placebo effects. In testing drugs, it is relatively simple to isolate them, because we have a theory that explains why psychiatric drugs do something that sugar pills do not do: they act directly on the brain, altering the synthesis, release, or breakdown of neurotransmitters by virtue of their chemical structure. This deceptive simplicity has produced some popular but very inadequate definitions of placebo; for example, placebos are said to be inert or non-specific and placebo effects are said to consist of the features that all effective treatments have in common.

To define placebos as non-specific is to confuse effects on neurotransmitters with effects on the disorder or symptom being studied. Each type of drug has a specific effect on neurotransmitters and a sugar pill does not. But a sugar pill acts just as specifically as an aspirin pill when it relieves a headache. Placebos might be regarded as less specific in another sense: they relieve a great many different symptoms, whereas the neurochemical action of a drug affects only a few. But non-specificity in the sense of wide usefulness cannot be the *definition* of a placebo. If we discovered a single drug that reliably relieved the symptoms of most psychiatric disorders, it would be regarded as a genuine panacea, not a placebo. For similar reasons, the placebo effect cannot be defined as the factors that all useful treatments for a disorder have in common. If we had a theory that explained all those common factors and allowed us to exploit them for the benefit of patients, we would no longer call the factors placebos or part of a placebo response. They would simply be the effects predicted by a true theory of extraordinarily wide applicability.

Distinguishing a placebo by its alleged inertness also means confusing activity with respect to neurotransmitters and activity with respect to the disorder being treated. When a psychiatrist gives a patient a sugar pill and relieves depression, the treatment is certainly active, although the pill itself is neurochemically inert.

The misleading but attractive definitions of placebos as inert, non-specific, or common factors are not even plausible in the case of psychotherapy. Psychotherapy could easily be claimed to be non-specific, inert, or made up only of "common factors." It has no direct effect on the brain and no well-defined neurochemical action. It is supposed to relieve a great many symptoms and conditions rather than a few specific ones. All of the procedures called psychotherapies apparently make use of mechanisms that they have in common with other human relationships and ways of helping. The problem of isolating placebo effects in psychotherapy is so complicated in practice and confusing in theory that some authorities have even recommended abandoning the concept, or at least the word, entirely [6].

The hunt for the elusive placebo in psychotherapy brings to light ambiguities in the term "placebo" as it is used in modern medical and psychiatric research. Its traditional meaning is, very roughly, the "psychological" aspects of treatment — the therapeutic relationship and the attitudes and expectancies of patients and therapists. The second meaning, more formal and technical, has always been implicit in modern drug experiments and has now been fully explicated by the philosopher of science, Adolf Grunbaum [7, 8]. He defines a placebo effect as an improvement in mental or physical health that occurs with no help from the mechanism that is the source of relief predicted by the theory on which the experiment is based. In drug testing the two definitions of placebo more or less coincide, but in psychotherapy research the ambiguity creates serious problems.

Using Grunbaum's definition, a medical treatment is a placebo only in relation to the disorder that is being treated and the theory that purports to explain the effectiveness of the therapy being tested. In other words, what is regarded as a placebo depends on what the experimenter is trying to prove. The therapeutic theory describes certain characteristic constituents of the treatment. Any result of what happens in the therapist's office or clinic is regarded as a placebo response for the purpose of the experiment if it is incidental from the perspective of the theory. Different theories have different characteristic mechanisms. In controlled experiments testing these theories, various responses not predicted by the theories will occur: the incidental or placebo effects. Since no effect is incidental from the perspective of all possible theories, nothing is everywhere and forever a placebo. We can always construct a theory that explains what first appears as a placebo response and then test that theory in another experiment.

Grunbaum's definition of a placebo is strictly methodological; it is neutral with respect to content. By his account, any therapeutic response that occurs in a control group not receiving the experimental treatment — any incidental or placebo effect — has a mechanism that is indeterminate until further experiments are done. In practice, however, physicians have always had some notions about the kinds of factors that produce most of the incidental responses both in clinical practice and in experiments. These are roughly described by the traditional conception of the placebo as "psychological" elements in treatment — the professional and charismatic authority of the healer, the healer's contagious confidence in the therapeutic procedure, the patient's expectations and suggestibility, the attention devoted to the patient, and so on. Although we understand these influences poorly, we can easily correct for them in a double-blind drug experiment because we have background theories that explain the effects of drugs. In psychotherapy, there is no accepted background theory and therefore no easy way to distinguish specific psychotherapy effects from placebo effects.

Grunbaum's definition of placebo allows for the validation of psychotherapy, because nothing is an incidental mechanism in all possible theories. But separating the characteristic mechanisms of specific psychotherapies from the incidental effects of healing or helping in general seems to require an experimental control treatment that is not a form of psychotherapy but employs, to an equal degree, the "psychological" mechanisms associated with most professional healing and with most other forms of helping as well — expectancies, beliefs, authority, status, social approval, confidence, enthusiasm, emotional warmth, and so on.

Jerome Frank [9] has devised a theory to circumvent this problem. He says that psychotherapy is a healing technique that works by creating an expectation of improvement through inducing the patient to follow certain rules within a relationship with a trusted authority in a place identified as a place of healing. The authority provides a coherent, plausible explanation of why the procedure works; the patient's hope and sense of mastery are mobilized to combat demoralization and accelerate natural recovery. All these things are placebo mechanisms (incidental factors) according to the theories that justify specific types of psychotherapy; they become characteristic mechanisms in Frank's theory.

The theory is pleasantly ecumenical and describes fairly well much of what goes on in psychotherapy and medicine. If the therapist's enthusiasm and the patient's hopes are aroused and the patient is made to feel secure through symbols of authority, the patient is likely to start feeling better. Even rats run mazes better when the experimenters in charge are confident because they have been led to expect an unusually good performance. There is little doubt that the transactions between psychotherapists and their patients often help the patients, but probably so did the transactions between 18th century doctors and their patients; we now attribute that to a placebo response. A strong application of healing rituals, almost in the pre-modern style, is still a recommended treatment for hypochondriacs, who refuse to talk to psychotherapists at all because they will not admit that their physical symptoms have emotional causes. Hypochondriasis is considered a psychiatric disorder, but many psychiatrists recommend that it be treated by a physician not necessarily trained in psychiatry, who should wear a white jacket, listen sympathetically to the patient's complaints, and offer reassurance along with harmless "medical" comforts, like stethoscope examinations, blood pressure measurements, heating pads, and vitamin pills.

Frank calls psychotherapeutic theories "healing myths." His account of psychotherapy fits very well the traditional conception of a placebo. A sugar pill administered by a sympathetic doctor also mobilizes healing symbols and associations. But even Frank concedes that this sort of charismatic healing without religious faith is hardly subject to experimental test. A control group in an experiment testing this theory would have to avoid using its extremely broadly defined characteristic mechanisms — the healing symbols and rituals, the office or clinic identified as a place of healing, attempts to encourage the patient's hopes and confidence in the therapist. But this would not demonstrate anything of interest.

How can it be shown that psychotherapy is more effective than commonsense helpfulness or charismatic healing? Perhaps the best-known study on the efficacy of psychotherapy is the work of Smith, Glass, and Miller [10]. They performed a meta-analysis, compiling and re-analyzing the results of hundreds of experiments covering more than 25,000 patients with a variety of disorders, who were given many different kinds of psychotherapy. Psychotherapy was defined as follows: the application of techniques derived from established psychological principles by persons qualified through training and experience to understand the principles and apply the techniques to assist patients in modifying behavior, values, and attitudes judged by the therapist to be maladaptive. They say they excluded no form of psychotherapy if the following conditions were fulfilled: clients were identified as having emotional or behavioral problems; they sought or were referred for treatment for the problem; the treatment was psychological or behavioral; the person delivering it was identified as a psychotherapist by virtue of training or professional affiliation. They go on to name some legitimate forms of psychotherapy — cognitive, behavioral, psychodynamic, rational-emotive, and so on.

Smith, Glass, and Miller estimated that the average patient who completes psychotherapy will be better off than similar patients with similar conditions who are not treated. The conclusion was based on elaborate statistical analysis of a vast number of experiments and presumably should be welcome to psychotherapists. Yet it was widely criticized, and the details of their study suggest some reasons why. First, they found that no psychotherapy has been shown to be better than any other psychotherapy for any particular emotional problem or psychiatric condition. It is as if a study concluded that drugs in general are better than pill placebos for a wide range of physical illnesses, but no drug has been shown to be better than any other drug for any particular illness. Almost all that is left is the bald conclusion that psychotherapy works; but this is hardly more useful or revealing than would be a study concluding that medicine works for physical illness. For example, it suggests that one form of psychotherapy cannot be used as control treatment in testing another form of psychotherapy.

They also found that the training and experience of the therapist and the duration of the treatment made no difference. Yet professional training and experience were included in their definition of psychotherapy. An essential feature of psychotherapy therefore seems to be irrelevant in practice.

The placebo control treatment included waiting lists, interview and assessment without further treatment or with periodic phone calls, or counseling and conversation of various kinds. In most of the experiments these apparently lacked credibility; that is, they lacked an equal dose of the features that define the traditional conception of placebo effects. For example, in some of the "counseling" sessions the mock therapist deliberately avoided discussing personal problems. Finally, the experiments were inevitably not double-blind.

The limitations of both Frank's work and the Smith, Glass, and Miller study suggest that there is no scientific test for the efficacy of psychotherapy in general. This is now commonly accepted, and the underlying reason is the lack of a general theory or model of psychotherapy analogous to the (admittedly rudimentary) theory of neurochemical activity that accounts for the characteristic effects of drugs. No explanatory mechanism is common to all psychotherapies, as distinct from all other agencies and procedures used in medicine and psychiatry. Without such a theoretical background, it is impossible to identify effects that are merely incidental and therefore should be regarded as placebo responses. Smith, Glass, and Miller include the application of "established psychological principles" in their definition of psychotherapy, but there are few if any scientifically established psychological principles and probably none that are relevant to psychotherapy.

Certain principles are socially established, in the sense that they are promulgated by various respected psychotherapeutic schools and professional guilds. But these schools and guilds have in some ways the same scientific status as those of 19th century medicine. Their ideas are not only divergent but often incompatible. To take the most obvious example, psychodynamic theory and behavioral theory apparently contradict each other at almost every point. What could it mean, in experimental terms, to say that both psychodynamic and behavioral therapies are effective as treatments for the same symptoms (psychological problems)? From the perspective of behavior therapy, psychoanalysis and psychoanalytic psychotherapy produce at most a placebo response, and from the psychoanalytic perspective, behavior therapy produces only placebo responses. The mechanisms each therapy regards as characteristic are incidental according to the other. It is not clear that they have any ingredient in common that can be contrasted with a placebo effect in the same way we distinguish the neurochemical action of drugs from the placebo response to the act of prescribing a pill.

Taking recourse to pluralism means evading the issue of incompatible principles. Suppose that in an experiment two drugs are found to be equally good treatments for a physical illness; a well-defined theory is available to account for each drug's presumed activity and the two theories are incompatible. The only proper conclusion would be that this experiment produced only placebo or incidental effects. The improvement could not be shown to result from the supposedly characteristic activity of either drug. The same conclusion is necessary when two psychotherapies with incompatible principles both prove equally effective in treating some psychiatric disorder or symptom.

It has become clear that psychotherapy research requires experiments in which patients are chosen for specific symptoms, standardized training and treatment manuals are used to apply uniformly the theory and technique of specific psychotherapies, therapists are monitored to make sure they observe the rules, and measures of outcome are carefully defined. In what has been properly greeted as the best research so far, the National Institute of Mental Health recently tested cognitive behavioral therapy and interpersonal therapy, two well-defined techniques for treating depression [11, 12]. At three university medical centers, these psychotherapies were compared with an antidepressant drug and with a placebo control consisting of sugar pills and weekly consultations with a psychiatrist. Preliminary results indicated that the drug began to work somewhat sooner, but at the end of 16 weeks the psychotherapies were as effective as the drug and considerably more effective than the placebo (50%-60% versus 29% substantially improved).

This experiment, careful as it is, still leaves many questions open. Cognitive therapy relies on correcting patients' faulty ideas about themselves and the world to improve their mood. Interpersonal therapy concentrates on altering present relationships with other people. It is unclear to what extent the theoretical principles on which these techniques rely are compatible. To the extent that they are not, as we have shown, the experiment would only prove that some common element in the situation, not necessarily an ingredient characteristic of either treatment, had a therapeutic effect. If the theories are compatible, or partially compatible, their characteristic mechanisms may not have produced the results. In fact, the researchers have emphasized that their main purpose was not to decide whether the psychotherapies were better than a placebo. They were more interested in comparing the psychotherapies against each other, and even in doing so their main purpose was not to decide which of the two was better. Instead, they wanted to find out which patients would do better with each treatment and why (11). Although the evidence is preliminary, they seem to have been largely disappointed. The outcome was about the same in both treatments, and no particular type of patient or symptom seems to have been improved significantly more by one therapy than by the other. Neither treatment was better than the placebo for the less severely depressed patients. The results of psychotherapy also varied greatly depending on the medical center where the treatment took place (which was not true of drug therapy).

In these circumstances it may prove hard to find any characteristic elements of the therapies that were effective. One obvious alternative explanation for the improvement is that the psychotherapy patients had the benefit of enthusiasm and concern from therapists who knew that they were conducting an important experiment. Several months of devoted attention conferred by a respected and authoritative person is a classically powerful placebo. Furthermore, the experiment was, of course, not blind. A double-blind experiment is almost impossible in psychotherapy research, because no one has found a good way to make therapists unaware of whether the treatment they are administering is supposed to be a control. The psychiatrists providing what they knew was meant to be a mere placebo treatment must have found it hard to preserve as much interest and enthusiasm as the ones providing psychotherapy.

The psychiatrists in the control group were also hampered by rules that prevented them from doing anything that overlapped with the psychotherapies tested or with any other forms of psychotherapy. The authors of the manual directing the administration of the control conditions (antidepressant drug and sugar pill) admit that "protocol demands may inhibit the application of the full range of usual and customary therapeutic techniques." They found that in training sessions "a self-consciousness seemed to evolve in some of the therapists resulting in a rigidity that diminished their usual and customary therapeutic responsiveness. Our observations provided us with a number of examples where an inflexible adherence to a rigidly interpreted protocol led to the abandonment of supportive interventions that the pharmacotherapist might ordinarily have made under practice conditions" [13].

Time is also an important element. The cognitive therapy patients had twenty 50-minute sessions, and the interpersonal therapy patients had sixteen to twenty 50-minute sessions. The patients who received either drugs or sugar pills combined with weekly consultations (described by the researchers as "minimal supportive therapy") spent about half as much time with the psychiatrist — one 50-minute session and sixteen to twenty 30-minute sessions.

The appropriate control for time is obvious. A control for the therapist's earnest conviction of effectiveness and devoted attention might be the use of counselors (or even friends) who are convinced that they can help and who talk about personal problems without artificial restrictions, but are not specifically trained in any type of psychotherapy. Smith, Glass, and Miller's conclusion that experience and training make no difference in the outcome of psychotherapy suggests the need for such a control.

A series of experiments of this kind, controlling for various aspects of the treatment, might distinguish the effective characteristic elements, if any exist, of psychotherapies and at the same time separate out characteristic ingredients of the placebo response in the traditional sense. A placebo effect, in the experimental sense elaborated by Grunbaum, is a therapeutic response without an explanation. When indeterminate placebo responses occur, researchers can develop a theory that explains their mechanisms and then devise another experiment to test the theory. Ingredients of what was once seen as a placebo response are employed as a new therapeutic technique, and the new control is some other agent or procedure that is not supposed to have the effects characterized by the new theory as therapeutic. In this way, experimental placebo responses might be analyzed and their constituents successively incorporated into new theories according to which they are no longer incidental effects but characteristic ones.

For example, experiments can test the physiological theory that some or most placebo effects involve the release of endogenous opioids in the body. Experimenters can also try to vary systematically certain characteristics of the patient or the therapist to see if they are associated with improvement. The most important element in any form of psychotherapy may be the way in which the psychotherapist communicates with the patient or the kind of working relationship that develops: the therapeutic alliance. Acknowledging this, some psychotherapists are now interested in process research, which might be seen as an attempt to examine in detail some of the mechanisms of what has been regarded as the placebo response [14]. There is no evidence that trained therapists are better at establishing an effective therapeutic alliance than sympathetic, intelligent lay people [15].

Unfortunately, the explanation and analysis of placebo responses in medicine is not far advanced. In any field, difficulty in determining what conditions do and do not influence the outcome indicates an early stage of science in which theories are inadequate. It is a sign of how little progress has been made in this field that some researchers can still write as though there is a generic placebo effect or as though there are agents or procedures that can serve as placebos in any experiment.

2. THE TESTING OF PSYCHOTHERAPY: FOCUS ON MDMA

In pre-industrial cultures there was an ancient tradition of using drugs to enhance psychotherapeutic healing; and from 1950 to the mid-1960s, there were 15 years of experimentation with drugs in Europe and the United States — an episode in the history of psychiatry that is now almost forgotten. The drugs used in this research were psychedelic or hallucinogenic substances, both natural and synthetic. It might now be possible to revive this tradition with a synthetic drug that has some of the virtues of the familiar psychedelics without their disadvantages.

Ever since experimentation with psychedelic plants began, users have maintained that the experience could be useful for self-exploration, religious insight, or relief of neurotic and somatic symptoms. The plants have been used for thousands of years in rites conducted by shamans and other professional healers. This religious and therapeutic use of psychedelic plants continues in the Amazon, in southwestern Mexico (where psychedelic mushrooms are used in healing rites), and in the Native American Church services of Indians in the western United States, which make use of the peyote cactus. Psychiatrists have proposed the peyote ritual as an adjunct to the treatment of alcoholism among American Indians [16].

Psychedelics were also used extensively in psychotherapy as experimental drugs, in Europe and the United States for almost two decades. A large number of clinical papers and several dozen books on the subject were published. The drugs were employed for a wide variety of problems, including alcoholism, obsessional neurosis, and sociopathy; they were also used to ease the process of dying [17]. With proper screening, preparation, and supervision, it was possible to minimize the danger of adverse reactions [18].

The literature contains impressive case histories, which can be questioned because they do not allow for spontaneous recovery, the effects of a therapist's special and prolonged devotion, or the therapist's and patient's biases in judging improvement. Most psychedelic drug studies also lacked controls and adequate follow-up. Beginning in the early 1960s, as illicit use of LSD and other psychedelic drugs increased, it became difficult to obtain the drugs legally or get funding for research, and professional interest declined. A generation of physicians and scientists has grown up without the opportunity to pursue human research on these drugs, and the financial and administrative obstacles remain serious. But psychedelic drugs should not be treated as entirely worthless and extraordinarily dangerous. The complexity of their effects may explain the inconsistent therapeutic results and the difficulty in sorting out their best uses.

We now have an opportunity to revive this research. Dozens of psychedelic drugs are known, and some have effects that are different from those of LSD and other familiar substances. In particular, some do not produce the same degree of perceptual change or emotional lability as LSD. MDMA is a relatively mild, short-acting drug. Both users and therapists have said that it heightens the capacity for introspection and intimacy and temporarily frees the user from anxiety and depression. There are no distracting changes in perception, body image, and the sense of self. As compared with the more familiar psychedelic drugs, it evokes a gentler, subtler, highly controllable experience that invites rather than compels intensification of feelings and self-exploration [19, 20, 21].

Patients in MDMA-assisted therapy report that they become less defensive and more emotionally open, able to get in touch with feelings and thoughts not ordinarily available to them. One patient put it this way: "I found it to be uncanny how easy it was to speak freely...about feelings. I'm generally not very good at that, but the MDMA apparently enabled me to let down the defenses and open up the offenses — but all in a gentle, matter-of-fact sort of way." Many patients find that they feel much closer to the therapist after one such session. If, as many believe, the strength of the therapeutic alliance is the best indicator of a good outcome in therapy, this characteristic of MDMA could be of very general usefulness [22].

But there are serious social and scientific obstacles to therapeutic research on such drugs as LSD and MDMA. The problem goes beyond legal restrictions or social attitudes towards altered states of consciousness and drugs used for non-medical purposes. An even deeper issue is the nature of the scientific validation required.

Probably the most influential controlled experiment in psychedelic drug therapy was the work of Ludwig, Levine, and Stark on alcoholism, which won a prize from the American Psychiatric Association in 1970 [22]. They chose 195 patients in an alcoholism clinic and separated them into four groups. One group received a standard residential milieu therapy. A second group was also given a single dose of LSD, a third group received LSD with psychotherapy, and the fourth group had LSD with psychotherapy and hypnosis. After three, six, nine, and twelve months, the results in all four groups were the same — about 75 percent improved. Some psychedelic drug researchers considered this experiment inadequate, on the grounds that the LSD sessions were not properly prepared and conducted. In some controlled experiments, the group treated with psychedelic drugs was doing better after several months. [24, 25]. (Psychotherapy has never been shown to have more durable effects.) But psychiatrists who used LSD never agreed on standards for preparing and conducting drug sessions, and most controlled experiments, including those performed by advocates of psychedelic drug therapy, were disappointing [26, 27, 28].

Some advocates of the therapeutic use of psychedelic drugs are impatient with this conclusion and with controlled studies in general. They say that psychedelics cannot be evaluated in the same way as ordinary psychiatric drugs with a simple consistent effect. The drug is given to create certain possibilities, not because it reliably relieves any symptoms. Its effects are overwhelmingly dependent on the set, the setting, and the relationship between therapist and patient.

These complaints amount to saying that the use of psychedelic drugs is not a drug treatment but a form of psychotherapy — or at least inseparable from the psychotherapeutic process it is alleged to promote. It has been observed that most studies on psychedelic drug therapy do not even describe the set, setting, and intentions of the experimenters well enough for replication. They leave out so much that the literature is "rich in the variability of results and premature in its conclusions" [29]. In this respect, it is much like the literature on other forms of psychotherapy. But standards for psychotherapeutic techniques are not the same as standards for drug therapies. Controlled experimental proof of efficacy is not required legally or even demanded for public and professional acceptance. Although the study by Ludwig and his colleagues seemed to show that psychotherapy was no more effective than LSD, alcoholics are still given psychotherapy.

Because no one has yet determined precisely what a placebo is in psychotherapy, controlled research has hardly affected practice in this field. Psychotherapists rarely stop using a technique because of experiments that seem to put its efficacy in doubt. The few exceptions only confirm the difficulty of distinguishing placebo from psychotherapy effects. For example, many professionals gave up psychotherapy with schizophrenic patients after controlled studies showed that drugs and psychotherapy were no better than drugs alone [30, 31]. But psychotherapy of schizophrenics had always been frustrating and unrewarding, perhaps precisely because they have such a poor placebo response (in the traditional sense rather than Grunbaum's technical sense): they are bad at reading feelings, adapting themselves to social situations, developing expectations of their own, and sensing the expectations of others.

Psychiatrists take the results of experimental drug trials much more seriously than they take the results of psychotherapy experiments, and that leaves them much less freedom. Psychedelic drug therapy, as a hybrid between pharmacotherapy and psychotherapy, is in a peculiarly uncomfortable position. Its claims are no easier to establish than those of any other form of psychotherapy, and all the same problems involving placebo effects arise, but the accepted standards, both legal and professional, are the same as those of drug therapy.

In the literature on psychedelic drug therapy, these problems are presented mainly in the form of complaints about the difficulty in conducting a double-blind experiment. In this respect, as in many others, psychedelic drug therapy resembles psychotherapies much more than it resembles standard drug therapies. Unlike psychotherapy researchers, psychedelic drug researchers have felt obliged to try to satisfy the demand for double-blindness. One approved method is to introduce an "active placebo." Since any effective placebo is active, this phrase is misleading. A better term might be "mimicking placebo": a drug that is neurochemically active and mimics certain immediate side effects of the experimental drug but is not thought to have any direct neurochemical effect on the disorder being treated. For example, a proposed antidepressant drug might be too easy to identify because it inhibits salivation, causing dry mouth; the experimenter employs as a placebo another drug that is also known to inhibit salivation but is not believed to have any effects on depression. This method is hard to apply in practice, since no drug has exactly the same immediate side effects as any other, and the drug used as a mimicking placebo usually has short- and long-term effects of its own that complicate the experiment. In most drug trials, mimicking placebos are not used; experimenters simply take the chance that side effects may be recognized.

Because the immediate effects of psychedelic drugs are often so overwhelmingly obvious, the demand for a mimicking placebo seems both highly justified and almost impossible to fulfill. Experimenters have reported using amphetamine, amphetamine-barbiturate combinations, or a small dose of LSD; they rarely say whether the subjects or therapists recognized the placebo. But concern about the technical issue of double-blind experiments may be misleading here. The deeper problem is that a psychedelic drug, like psychotherapy itself, it given precisely to create a certain experience, not to relieve symptoms unawares, like a standard psychiatric drug. In psychedelic drug therapy as in any other form of psychotherapy, a true mimicking placebo — one that produced the same immediate "experience" as the treatment — would be the same, for all practical purposes, as the treatment itself. Separating immediate effects from the desired therapeutic activity seems impossible, when precisely this immediate conscious response is the basis for treatment. If a double-blind experiment seems impossible in both psychedelic drug therapy and psychotherapy, it also seems undesirable.

Nevertheless, advocates of psychedelic drug therapy must meet the standards of drug therapy rather than those of psychotherapy. Even if double-blind experiments prove impossible, they will have to accept the burden of showing that, say, various forms of psychotherapy for depression along with MDMA give better results than various forms of psychotherapy along with a dextroamphetamine or sugar placebo. In other words, it seems that psychedelic drug therapy must be proved unequivocally superior for at least some patients and some conditions before it will be accepted or even legalized. We have already seen how hard it is to do this for any psychotherapy.

Conceptual and empirical improvements in psychotherapy research may provide some hope. For example, much recent process research finds that good therapeutic results are highly correlated with the early establishment of a solid therapeutic alliance [32]. But this result is subject to the same doubts as the research on outcome in psychotherapy. Can the therapeutic alliance be defined in a way that distinguishes it from other elements of friendship, expectancy, and authority that are part of the traditional placebo response? Is the alliance really a mechanism of improvement, or is it the result of the therapist's and patient's personalities and other conditions that precede therapy? Experimenters may have to find ways of manipulating and improving the alliance systematically, allowing therapeutic ingredients to emerge from the placebo background — the procedure described earlier as turning placebo responses into characteristic mechanisms. If certain drugs proved to enhance or alter the alliance, they might be valuable in such experiments. The phrase "pharmacotherapy of interpersonal processes" [1] is one that some therapists who have used MDMA could probably adopt as their own.

The placebo problem is associated with another important distinction between drug therapy and psychotherapy. Psychotherapy is accepted despite the lack of scientific proof, partly because it is regarded as something more than a medical procedure designed to bring a patient back to an accepted standard of health. One authority even suggests that the idea of a placebo is inapplicable in psychotherapy because it serves to "subordinate psychotherapy research to scientific and political considerations that are relevant to medical research. This is self-defeating" [6]. The standard or ideal of normal mental health is much vaguer than the standard of physical health. Psychotherapy is often used by people who simply feel dissatisfied with some aspect of their lives and want to improve it. What is regarded as improvement often depends on the patient's own judgment and the judgment of the therapist, rather than a scientific consensus about what is emotionally normal or healthy. (Smith, Glass, and Miller even include the therapist's judgment about what is adaptive in their definition of psychotherapy.) It is not acceptable to use drugs in this way, to enhance rather than to normalize. The history of psychedelic research shows that a mixture of therapeutic claims involving drugs with talk of mysticism, transcendence, and consciousness-expansion is regarded with great suspicion. But psychedelic drugs are not the only ones affected; consider the public outrage about the use of drugs by athletes to improve performance. Here again psychedelic drug therapy will be subject to different standards from those applied to other forms of psychotherapy.

Members of the Native American Church, who use peyote (mescaline) in their rituals, are the only exception to the rule; they have been given the right to use a drug for a kind of enrichment or enhancement in the name of the religious freedom guaranteed by the Constitution. To grant to non-Indians the same right in the name of psychotherapy, we would have to drastically modify our social understandings about appropriate drug use.

3. CONCLUSIONS

This social obstacle to the acceptance of psychedelic drug therapy is closely related to the scientific obstacles. The standards of drug research are different from those of psychotherapy research, and the differences are deeply rooted in our science and society. The social standards for life-enhancement with drugs are also different from those applied to life-enhancement by other means, including psychotherapy. These differences too have deep cultural roots. Therefore any possibility of acceptance for psychedelic clinical research will have to wait on improved scientific standards for psychotherapy research in general.

Psychotherapy survives as a craft with aspirations to the status of science or a way of providing a new experience for people who feel dissatisfied and want to change their lives. Scientific standards demand more: a series of experiments asking a series of questions to distinguish characteristic from incidental effects, where each new question is suggested by the results of a previous experiment. Progress in this kind of testing will require much more careful and creative thinking about placebo effects, and it is still not clear whether psychotherapy can realize its scientific aspirations. But even without scientific status, most forms of psychotherapy will continue to be socially accepted practices. The dominant cultural and scientific understandings about drugs make social acceptance impossible for psychedelic drug tharapy, whether the agent is MDMA or some other drug. Therefore this kind of therapy will probably never become available until psychotherapy succeeds in the difficult task of obtaining scientific credentials of the same quality as those now granted to antidepressant or antipsychotic drugs.

REFERENCES

1. Docherty, J.P., 1985. Introduction to Section V: The Therapeutic Alliance and Treatment Outcome. In *The American Psychiatric Association Annual Review, Vol. IV*, (Hales R.E. and Frances A.J., eds). New York: American Psychiatric Press.
2. Garfield, S., 1984. Psychotherapy: Efficacy, generality, and specificity. In *Psychotherapy Research: Where are We and Where Should We Go?* (Williams J.B. and Spitzer R.L., eds.) New York: Guilford Press.
3. Klein, D.F., 1983. Talking often helps: The efficacy, generality, and specificity of psychotherapy. Biol. Psychiat., 18:1101–1105.
4. Prioleau, L., Murdock, M., and Brody, N., 1983. An analysis of psychotherapy versus placebo studies. Behav. Brain Sci. 6:275–310.
5. Strupp, H.H., 1986. Psychotherapy: Research, practice, and public policy. (How to avoid dead ends). Am. Psychol. 41:120–130.
6. Wilkins, W., 1986. Placebo problems in psychotherapy research: Social-psychological alternatives to chemotherapy concepts. Am. Psychol. 41:551–556.
7. Grunbaum, A., 1981. The placebo concept. Behav. Res. Ther. 19:157–167.
8. Grunbaum, A., 1985. Explication and implications of the placebo concept. In *Placebo: Theory, Research, and Mechanisms* (White, L., Tursky, B., and Schwartz, G.E., eds). New York: Guilford Press.
9. Frank, J.D., 1973. *Persuasion and Healing: A Comparative Study of Psychotherapy.* Second Edition. Baltimore, MD: Johns Hopkins University Press.
10. Smith, M.L., Glass, G.V., and Miller, T.I., 1980. *The Benefits of Psychotherapy.* Baltimore, MD: Johns Hopkins University Press.
11. Elkin, I., Parloff, M.B., Hadley, S.W., et al., 1985. NIMH treatment of depression collaborative research program: Background and research plan. Arch. Gen. Psychiat. 42:305–316.
12. Elkin, I., Shea, T., Imber, S., et al., 1986. NIMH treatment of depression collaborative research program: Initial outcome findings abstract. American Association for the Advancement of Science, May.
13. Epstein, P. and Fawcett, J. Treatment of Depression Collaborative Research Program, Pharmacotherapy Training Site. Addendum to Clinical-Management-Imipramine-Placebo Administration Manual. Rush Presbyterian, St. Luke's Medical Center, Dept. of Psychiatry.
14. Gomes–Schwartz, B., 1978. Effective ingredients in psychotherapy: Prediction of outcome from process variables. J. Consult. Clin. Psychol. 46:1023–1035.
15. Moras, K. and Strupp, H.H., 1982. Pretherapy interpersonal relations, patients' alliance, and outcome in brief therapy. Arch. Gen. Psychiat. 39:405–412.
16. Albaugh, B.J. and Anderson, P.O., 1974. Peyote in the treatment of alcoholism among American Indians. Am. J. Psychiat. 131:1247–1251.
17. Grinspoon, L. and Bakalar, J.B., 1979. *Psychedelic Drugs Reconsidered.* New York: Basic Books.

18. Strassman, R.J., 1984. Adverse reactions to psychedelic drugs: A review of the literature. J. Nerv. Ment. Dis. 172:577–595.
19. Shulgin, A.T., 1983. Twenty years on an ever-changing quest. In *Psychedelic Reflections* (Grinspoon, L. and Bakalar, J., eds.). New York: Human Sciences Press.
20. Riedlinger, J.E., 1985. A pharmacist's perspective in the matter of MDMA scheduling. J. Psychoactive Drugs 17:167–172.
21. Smith, D.E., Wesson, D.R., and Buffum, J., 1985. MDMA: "Ecstasy" as an adjunct to psychotherapy and a street drug of abuse. California Society for the Treatment of Alcoholism and Other drug Dependencies News 12:2:1–3.
22. Grinspoon, L. and Bakalar, J.B., 1986. Can drugs be used to enhance the psychotherapeutic process? Am. J. Psychother. 40:393–404.
23. Ludwig, A.M., Levine, J., and Stark, L.H., 1970. LSD and *Alcoholism: A Clinical Study of Treatment Efficacy*. Springfield, IL: Charles C. Thomas.
24. Hollister, L.E., Shelton, J., and Krieger, G., 1969. A controlled comparison of lysergic acid diathylamide (LSD) and dextroamphetamine in alcoholics. Am. J. Psychiat. 125:1352–1357.
25. Kurland, A., Savage, C., Pahuke, W., Grof, S., and Olsson, J.E., 1971. LSD in the treatment of alcoholics. Pharmakopsychiat. Neuropharm. 4:83–94.
26. Cheek, F.E., Osmond, H., Sarett, M., et al., 1966. Observations regarding the use of LSD-25 in the treatment of alcoholism. J. Psychopharm. 1:56–74.
27. Soskin, R.A., 1973. The use of LSD in time-limited psychotheray. J. Nerv. Men. Dis. 157:410–419.
28. Abuzzahab Sr, F.S. and Anderson, B.J., 1971. A review of LSD treatment in alcoholism. Int. Pharmacopsy 6:223–235.
29. Yensen, R., 1984. From mysteries to paradigms: Humanity's journey from sacred plants to psychedelic drugs. In *Proceedings of the Association for the Responsible Use of Psychoactive Agents*, Esalen Institute, Big Sur, California, 1984. Albany, NT: State University of New York Press, in press.
30. May, P.R.A., 1968. *Treatment of Schizophrenia: A Comparative Study of Five Treatment Methods*. New York: Science House.
31. Grinspoon, L., Ewalt, J.R., and Shader, R.I., 1972. *Schizophrenia: Pharmacotherapy and Psychotherapy*. Baltimore, MD: Williams and Wilkins.
32. Hartly, D.E., 1985. Research on the therapeutic alliance in psychotherapy. In *The American Psychiatric Association Annual Review, Vol. IV* (Hales, R.E. and Frances, A.J., eds.). Washington, D.C.: American Psychiatric Press.

4. RECREATIONAL USE OF MDMA

STEPHEN J. PEROUTKA

1. INTRODUCTION

The recreational use of MDMA in the United States has never been documented adequately. According to some reports in the popular press, it is possible that MDMA may be one of the most widely used recreational drugs of the late 1980s. Alternatively, MDMA use may be restricted to a few locations and may be relatively inconsequential in comparison to drugs such as cocaine and marijuana. There are simply no good data on which to adequately define and assess the extent of recreational MDMA use in the United States.

As best as can be determined, however, the recreational use of MDMA does appear to have increased significantly in the mid-1980s. For example, it has been reported that approximately 10,000 doses of MDMA were being distributed monthly by a California laboratory in 1976 (1). By 1984, the monthly production increased to 30,000 doses per month at the same laboratory. By mid-1985, this single laboratory was reportedly producing nearly 500,000 doses of MDMA per month.

The manufacturers of MDMA appear to have responded to a surge in demand for the drug largely among recreational users. Prior to 1985, "Ecstasy" was popular as a recreational drug in two main areas of the United States: Texas and California. In Texas, the drug was sold openly in bars in many student areas of Dallas and Houston (2, see chapter by Beck, this volume]. MDMA was usually sold as a yellow tablet and cost approximately $10 to $35 per tablet. According to one anecdotal report, MDMA tablets were actually given away for free in at least one Houston bar on June 30, 1985, the day

Table 1. Myths about MDMA.

–	MDMA was developed by the CIA to be the ultimate "truth serum"
–	MDMA can damage the kidney and should always be taken with large amounts of fluids.
–	MDMA drains the spinal fluid
–	If a male takes MDMA more than 3 times, he will be rendered permanently impotent. However, MDMA is an incredible aphrodisiac during the first 3 doses.

before the compound was placed on Schedule I by the FDA. The relatively extensive open use of MDMA may account for the fact that the majority of documented toxic reactions and deaths due to MDMA have been reported in Texas [3, see chapter by Dowling, this volume]. The use of MDMA as a "legal" euphoriant in Texas may also have played a large role in the decision by the FDA to place this compound on Schedule I in 1985.

The increase in popularity that occurred with MDMA in the mid-1980s has many possible causes. First, at all, "word of mouth" information concerning the supposed unique psychoactive effects of MDMA seems to have spread among recreational drug users and undergraduate students in the United States during the mid-1980s. At the same time, the potential dangers of drugs such as cocaine were beginning to be reported and discussed widely in the national media. Cocaine was also becoming an increasingly expensive recreational drug. Finally, a number of popular magazines published stories in mid-1985 concerning the purported psychotherapeutic benefits of MDMA. Articles on MDMA appeared in *Newsweek*, *Time*, *Life*, and *New York Magazine* and seem to have stimulated interest in this novel psychoactive compound among recreational drug users. The *Newsweek* article, for example, stated that the effects of a single dose of MDMA were equivalent to "a year of therapy in two hours" [4]. The well-publicized controversies surrounding the placement of MDMA on Schedule I by the Food and Drug Administration on July 1, 1985, also contributed to an increased awareness of its potential use as a recreational drug.

A variety of myths have developed about MDMA. Some of these claims and beliefs, which are derived from my interviews and discussions with approximately 200 MDMA users, are summarized in Table 1. MDMA, for example, was said to have been developed by the CIA in the 1950s as the "ultimate truth serum", despite the fact that it was patented in 1914. Supposed dangers of MDMA have included a propensity to damage the kidneys, a statement that may relate to the presumed dehydrating effect of the drug. Many undergraduates have been told that MDMA can "drain the spinal fluid," perhaps in reference to the frequent myalgic complaints of recreational users on the day following MDMA use. One of the more interesting comments concerns the ability of MDMA to induce sterility in males after four or more uses. However, in the words of one student, "it's an incredidible aphrodisiac during the first three doses!" In fact, MDMA has no apparent effect on sexual behavior, according to a study published in 1986 [5].

Although MDMA has been available on certain college campuses since the mid-1970s, recreational MDMA use among college students in the United States appears to have become much more common since 1985. In many ways, MDMA might be considered a "seductive" drug to students. The proponents of its use frequently cite its ability to make the user feel "warm and friendly" and claim that the drug causes them to have an increased sense of "closeness" with other people. This type of information is provided in an information sheet that is frequently distributed among recreational MDMA users [1]. Student users have also been told that "doctors" use MDMA because of its psychotherapeutic applications, and that the drug is "legal" (or at least that MDMA *was* "legal" until 1985). As a result, it is clear that recreational MDMA use may have appealed to many novice recreational drug users prior to the recent wave of publicity concerning the neurotoxic effects of the drug in laboratory animals.

2. RECREATIONAL MDMA USE ON A SINGLE UNDERGRADUATE CAMPUS

As noted above, the exact incidence of recreational MDMA use during the past few years is impossible to determine accurately. In the Fall of 1986, my laboratory began working on the pharmacological and neurotoxicological effects of MDMA in laboratory animals. At that point in time, reports from several laboratories clearly indicated that MDMA could produce long-term neurochemical changes in the brains of laboratory animals. However, these scientific data had not yet been communicated to recreational MDMA users.

In March, 1987, the Student Health Center at Stanford University published a study (that was performed in early 1986) that indicated that 8% of undergraduate students reported having used MDMA as a recreational drug [6]. The undergraduate students in my laboratory were surprised by this finding and felt that the use of the drug was actually higher, especially during the 1986–87 academic year. The interest of the students derived from their awareness of the significant controversies that existed concerning the legal status of MDMA and the possibility that it may cause irreversible neurotoxicity in human users. Importantly, the students in my laboratory felt that most of these issues were simply unknown to the vast majority of their undergraduate colleagues who were recreational users of MDMA. As a result, two students volunteered to do an informal survey of their colleagues in an attempt to obtain some data on the recreational use of this compound on a single undergraduate campus [7, 8]. The results of their survey are summarized below.

3. SUBJECTIVE EFFECTS OF MDMA IN RECREATIONAL USERS

3.1. Subjects and methods

Undergraduate students at a major university were polled anonymously by two student colleagues concerning possible recreational use of MDMA between May 4, 1987 and June 3, 1987. Subjects were selected and interviewed (in approximately equal numbers) at the student union, at an undergraduate library, and at three dormitories containing all four classes of students. The subjects were first asked whether or not they were undergraduate students at the school where the informal poll was conducted. If they responded positively, the subjects were then asked whether they had ever taken "Ecstasy" or "MDMA." A total of 369 subjects were interviewed, as previously reported in preliminary form [7]. If the subject admitted having used the drug, he or she was asked to complete a questionnaire concerning the subjective effects of the drug. The questionnaire was based on previous reports of subjective MDMA effects [9–11]. A copy of the questionnaire is available upon request.

The subjects were asked whether or not they experienced a variety of both psychological and physiological effects on both the day of drug usage and on the day following MDMA use. Subjects were also asked whether the effects of the drug changed with successive doses and whether the drug was felt to produce any permanent change in their behavior or personality. The use of this questionnaire was formally reviewed and approved for use in this study by the Human Subjects Committee at Stanford University.

3.2. Frequency of use

Of the 369 subjects initially interviewed, 143 (39%) reported that they had used MDMA at least once. Importantly, this unexpectedly high usage figure was not unique to this single institution, since a similar study at the University of Colorado [12] in 1987 reported that 20% of undergraduates had taken MDMA. Significant recreational MDMA use at other universities has also been reported anecdotally.

A totally of 100 individuals (70% of positive responders) agreed to complete a questionnaire concerning their subjective experiences while using the drug, the results of which have been reported recently [8]. The age of the respondents ranged from 18 to 25. The frequency of use by the subjects ranged from one to 38 doses of the drug (Figure 1). Among these 100 individuals, the median amount of MDMA usage was four doses while the mean number of doses taken was 5.4. The amount of drug taken in a single dosage ranged from 60 to 250 mg (approximately 1–4 mg/kg). Of note is the fact that the majority (55%) of the recreational users reported taking less than six doses of the drug. Only 33% reported taking six to ten doses, and only 12% took greater than ten individual doses.

3.3. Acute effects of MDMA

A total of 90% of the individuals reported that they had an increased sense of "closeness" with other people in the first few hours after taking MDMA (Table 2). The subjects thought that they were more verbal during this time and were able to interact better with other people. This subjective sense of enhanced awareness is in keeping with numerous anecdotal reports, as well as with the reported effects of the drug under psychotherapeutic conditions [9–11].

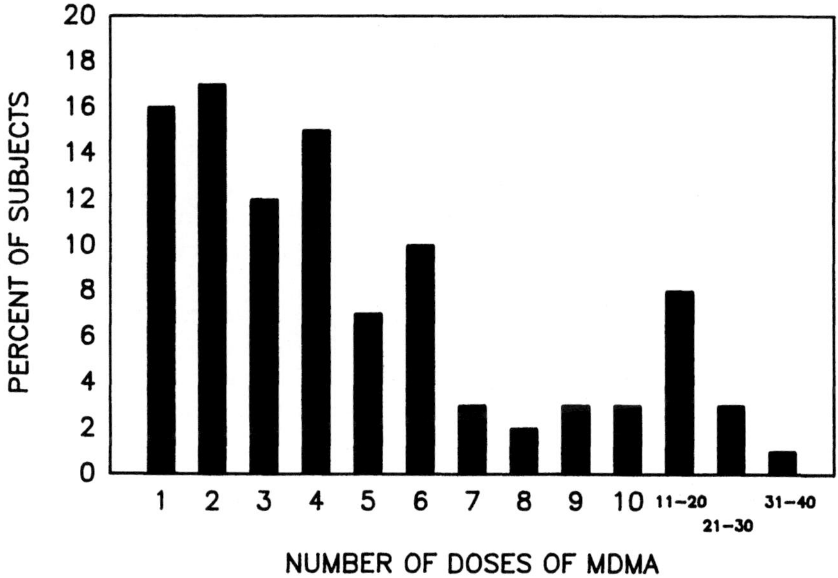

Figure 1. Frequency of MDMA use by recreational users (n = 100). (Taken from reference 8).

Table 2. Acute effects of MDMA (n = 100).

Subjective sensation	Number of subjects
Sense of "closeness" with other people	90
Trismus	75
Tachycardia	72
Bruxism	65
Dry mouth	61
Increased alertness	50
Luminescence of objects	42
Tremor	42
Palpitations	41
Diaphoresis	38
Difficulty concentrating	38
Paresthesias	35
Insomnia	33
Hot or cold flashes	31
Increased sensitivity to cold	27
Dizziness or vertigo	24
Visual hallucinations	20
Blurred vision	20

Taken from reference 8.

In addition, a variety of physiological effects were reported that were indicative of the sympathomimetic effects of the drugs. Tachycardia (72%), dry mouth (61%), tremor (42%), palpitations (41%), diaphoresis (38%), and parasthesias (35%) were the most frequently reported sympathomimetic effects. Trismus ("tight jaw muscles") and bruxism ("grinding of the teeth") were reported by 75% and 65% of the subjects, respectively. Similar sympathomimetic effects are observed with high-dose amphetamine use [13]. Although increased alertness was reported by 50% of the subjects, 38% reported that they had difficulty concentrating during the acute phase of MDMA effects.

Visual hallucinations were noted by 20% of the respondents. However, these visual sensations rarely consisted of well-formed hallucinations. Rather, the subjects reported sensing that a flash of light or an object was in their peripheral visual field but, when looking in that direction, they saw no objects. There were no reports of auditory hallucinations.

3.4. Effects of MDMA twenty-four hours after ingestion

The subjects were also asked to report any residual effects of MDMA that were experienced on the day following the ingestion of the drug (Table 3). The most commonly reported effect was drowsiness (36%), which was attributed to the insomnia that was acutely reported by 33% of the respondents. Diffuse muscle aches and general fatiguability were reported by 32% of the subjects. Depression (21%), difficulty concentrating (21%), and headache (17%) were considered to be "negative" aspects of MDMA use occurring on the day after ingestion. Both the sense of "closeness" with other people and trismus continued into the second day for 22% and 21% of the respondents, respectively. Finally, a general sense of anxiety, worry, or fear, as well as irritability, on the day following MDMA use, were reported by 12% of the subjects. Because of these frequent "negative" side effects, the respondents tended to prefer to use the drug on either a Friday or Saturday evening, so

Table 3. Subacute effects of MDMA (n = 100).

Subjective sensation	Number of subjects
Drowsiness	36
Muscle aches or fatigability	32
Sense of "closeness" with other people	22
Depression	21
Tight jaw muscles	21
Difficulty concentrating	21
Headache	17
Dry mouth	14
Anxiety, worry, or fear	12
Irritability	12

Taken from reference 8.

that these drugs effects would not interfere with their school and/or work performance.

The subjects were also asked whether the beneficial effects of MDMA decreased with usage. In the 43 subjects who had taken two to five separate doses of the drugs, 21 (49%) reported that the effects of the drug decreased with subsequent doses. In subjects who had taken six or more doses of MDMA, 67% reported a decrease in beneficial effects over time. In general, the subjects reported that the "positive" effects of the drug decreased while the "negative" effects increased with successive doses. An increase in the size of a single dose of MDMA was found to increase the "negative" effects of the drug while decreasing the "positive" effects.

3.5. Long-term effects of recreational MDMA use

Only two of 100 subjects reported any long-term effect of the drug. One claimed a tendency to clench the teeth when anxious for a period of months following two separate doses of MDMA. A second subject attributed increased emotionality to the effects of three separate doses of MDMA.

4. ANECDOTAL CASE REPORTS

The lack of adequate epidemiological and clinical data on recreational MDMA use is a serious impediment to the analysis of this drug. Moreover, because of its placement on Schedule I and the increasing public awareness of its neurotoxicological potential, the human effects of MDMA will be extremely difficult, if not impossible, to study adequately in the future. At the present time, anecdotal reports must be used in an attempt to determine the effects of this drug in recreational users. The following "Case reports" are derived from my interviews and discussions with approximately 200 recreational users of MDMA. These personal observations are an attempt to provide a reasonable, although nonscientific, overview of this complex issue.

4.1. "Ecstasy"

A frequently mentioned fact among recreational MDMA users is that the first ingestion of the drug is "the best." One individual reported that MDMA was taken for the first time before a day of skiing.

"The air was crisper, the snow was whiter, and the sky was more brilliant than I've been seen. I also skied better than I ever had. At the end of the day, I felt like I had experienced the most incredible day of my life. Moreover, when I think back on that day, the memories continue to be extremely vivid. I remember it as one of the best experiences that I ever had."

This report is typical of many first-time MDMA users. I once asked an undergraduate how MDMA users could be identified after having taken the drug. The individual said that if I saw a group of students walking together, holding hands, and laughing or singing, then they may have ingested MDMA. This observation is in keeping with the fact that MDMA is most commonly used by undergraduates in groups before attending social functions on weekend nights. The perceived increase in verbal behavior and decrease in defensiveness, both induced by the drug, is thought by recreational users to facilitiate social interactions. MDMA is rarely reported to be taken by individuals who remain alone during the first few hours after ingestion.

4.2. Panic attacks

During the past two years, I have been contacted independently by four individuals who shared an atypical but similar experience following MDMA use. Each individual took a relatively large dose of MDMA over a relatively short period of time. For example, one individual took a total of four doses of 125 mg (total of 500 mg MDMA) over a 24 hour period, while a second individual reported taking three separate doses of approximately 150 mg MDMA (total of 450 mg MDMA) over a six hour period. Each of the four individuals reported the onset of "panic attacks" (as defined by the Diagnostic and Statistical Manual III) within approximately seven days of MDMA use. None of the four individuals reported a previous history of neurological or psychiatric disease. The number of "panic attacks" per individual ranged from two to eight. The "panic attacks" slowly became less frequent and ceased in all four individuals within eight weeks of the last ingestion of MDMA. Similar anecdotal reports of "panic attacks" have been published by other investigators [2, 14, 15].

4.3. Death

In the summer of 1988, a 37 year-old woman was found dead in Palo Alto, California, after supposedly ingesting an unknown quantity of MDMA [16]. According to police reports, she died approximately 90 minutes after MDMA ingestion. The individual who administered the drug to the woman was arrested on suspicion of possessing a controlled substance and administering a controlled drug to another person. As clearly documented in the chapter by Dowling (this volume), MDMA does possess the potential for lethality. This fact appears to be unknown to the vast majority of recreational MDMA users. The actual risk of MDMA, unfortunately, will remain unknown until adequate epidemiological and pathological data can be developed.

5. CONCLUSIONS

Although the observations described in this chapter do not constitute a formal epidemiological study, they do represent the first and, to date, the only analysis of the subjective effects of MDMA in recreational users. Previous descriptions of MDMA effects have focused on patients who used the drug as an adjunct to psychotherapy [10, 11]. As noted above, unconfirmed reports from various university campuses have suggested that the recreational use of this compound has rapidly gained popularity since 1985. Indeed, two independent surveys on a single university campus, taken a year apart, found an 8% to 39% incidence of undergraduate MDMA use [6, 7], and a third study, at another major university, documented a 20% undergraduate exposure rate [12]. These data clearly indicate that a significant number of students in the United States have ingested this compound for recreational purposes. These observations are significant, since MDMA has been shown to produce neurotoxicity in animals [see Chapters 7–13, this volume]. MDMA has also been associated with acute toxicity and death in human users [see Dowling, Chapter 5, this volume].

A serious concern is the observation that the majority of multiple time MDMA users state that the "positive" effects of the drug decrease over time. This finding has now been reported in both recreational MDMA users and individuals who used MDMA in a therapeutic setting. For example, Greer and Tolbert [10] found that frequent or high doses of MDMA diminished the pleasurable effects of the drug while increasing its side effects. Conceivably, this finding may suggest subtle long-term effects of the drug on the human central nervous system, since the primary psychoactive effects of MDMA last only three to five hours [9]. These observations suggest strongly that the recreational use of MDMA should be avoided at the present time.

ACKNOWLEDGEMENTS

I thank Bruce G. McCarthy for his helpful comments. This work was supported in part by the McKnight Foundation.

REFERENCES

1. Kirsch, M.M., 1986. "Ecstasy". *In* Designer Drugs. Minneapolis, MN: Complare Publications, pp. 74–97.
2. Beck, J. and Morgan, P.A., 1982. Designer Drug Confusion: A focus on MDMA. J. Drug Educ. 16: 287–302.
3. Dowling, G.P., McDonough, E.T., and Bost, R.O., 1986. "Eve" and "Ecstacy": A report of five deaths associated with the use of MDEA and MDMA. JAMA 257:1615–1617.
4. Adler, J., 1985. Getting high on "Ecstasy." Newsweek, April 15, p. 96.
5. Buffum, J. and Moser, C., 1986. MDMA and human sexual function. J. Psychoactive Drugs 18: 355–360.
6. Calvert, C., 1987. Psychedelic drug use up on Farm. Stanford Daily, March 3.
7. Peroutka, S.J., 1987. Incidence of recreational use of 3, 4-methylenedioxymethamphetamine (MDMA; "Ecstasy") on an undergraduate campus. N. Engl. J. Med. 317:1542–1543.
8. Peroutka, S.J., Newman, H., and Harris, H., 1988. Subjective effects of 3, 4-methylenedioxymethamphetamine in recreational users. Neuropsychopharmacology 1:273–277.
9. Shulgin, A.T., 1986. The background and chemistry of MDMA. J. Psychoactive Drugs 19:291–304.
10. Greer, G. and Tolbert, R., 1986. Subjective reports of the effects of MDMA in a clinical setting. J. Psychoactive Drugs 18:319–328.
11. Downing, J., 1986. The psychological and physiological effects of MDMA on normal volunteers. J. Psychoactive Drugs 18:335–340.
12. Accola, J., 1988. MDMA: Studies of popular illicit drug raise questions about effects. Rocky Mountain News, March 4, p. 72.

13. Khantzian, E.J. and McKenna, G.J., 1979. Acute toxic and withdrawal reactions associated with drug use and abuse. Ann. Int. Med. 90:361–372.
14. Seymour, R.B., 1986. *MDMA*. San Francisco, CA: Haight-Ashbury Publications.
15. Whitaker–Azmitia, P.A. and Aronson, T., 1989. Panic attacks associated with MDMA (Ecstasy). Am. J. Psychiat., in press.
16. Brazil, J., 1988. Controversy continues to surround the drug "ecstasy." Peninsula Times Tribune, July 26, p. A1.

5. HUMAN DEATHS AND TOXIC REACTIONS ATTRIBUTED TO MDMA AND MDEA

GRAEME P. DOWLING

1. INTRODUCTION

3,4-Methylenedioxymethamphetamine (MDMA) and 3,4-methylenedioxyethamphetamine (MDEA) are synthetic amphetamine analogues that have received considerable media attention as recreational drugs popular among college students and young professionals. MDMA, more commonly known as "Ecstasy," has been available on the illicit drug market since 1968 [1], with increasing popularity in the late 1970s and early 1980s. On the other hand, MDEA, also known as "Eve," has only started to gain prominence since the placement of MDMA on Schedule I of the Controlled Substance Act by the Drug Enforcement Administration (DEA) on July 1, 1985. MDMA has been investigated by a small number of psychiatrists for its potential use as a psychotherapeutic agent. Uncontrolled trials of MDMA in clinical settings seem to indicate that it helps to facilitate therapeutic communication and increase patient insight and self-esteem [2, 3]. MDMA in the hands of psychiatrists and both MDMA and MDEA among those who use them recreationally have generally been regarded as safe drugs with some minor short-term side effects [3–5]. Indeed, from 1977 to 1985 the Drug Abuse Warning Network (DAWN) reported only eight admissions to emergency rooms, across the United States, for treatment of individuals who claimed they had taken MDMA [6]. When one considers that the prevalence of MDMA use has been estimated at 10,000 doses nationally in 1976 to more current estimates of 30,000 doses nationally per month in 1985 [7] (and perhaps as high as 30,000 doses per month in one city, as reported by the DEA [8]), then the low number of emergency room admissions is remarkable, to say the least. Likewise, well-documented deaths related to the use of these two drugs are exceptionally rare [9, 10].

Several chapters in this text deal with the potential neurotoxic effects of MDMA and the implications this may have in the long run for individuals who use it either therapeutically or recreationally. This chapter discusses those rare instances of acute toxic reactions and sudden death that have been caused solely by MDMA or MDEA, or where MDMA or MDEA are thought to have contributed significantly to a toxic reaction or death.

2. TOXIC REACTIONS TO MDMA

When MDMA is taken in a therapeutic or recreational setting in the usual dosage range of 75–150 mg orally, a wide variety of minor side effects are often noted. These include increased heart rate, tremor, tightening of the jaw muscles, bruxism, sweating, nauseau, and occasionally vomiting [1, 3, 11, 12]. The side effects are generally short-term in nature, although some individuals complain of residual jaw muscle tension, fatigue, depression, anxiety, nauseau, and occasionally flashbacks, which persist for up to two weeks after a single dose [3, 11–13]. Physiologic side effects that have been documented in uncontrolled studies and that are related to MDMA's sympathomimetic effects include tachycardia and a fairly consistent rise in blood pressure of approximately 15 mm Hg above baseline levels [11, 14]. Generally, as the dose is increased, these side effects become more disturbing, and this has often been cited as a feature that distinguishes MDMA from other drugs of abuse [3, 15, 16]. Although several individuals have noted numbness and tingling in the extremities, increased color acuity, or luminescence of objects, true auditory, visual, or tactile hallucinations are usually not experienced until very high doses are takan [1]. Impairment of judgement, coordination, and the ability to concentrate have all been documented in a clinical setting. For this reason, it is recommended that tasks requiring coordination, concentration, and problem solving skills not be performed by individuals under the influence of MDMA [11].

For the most part, the documentation of severe toxic reactions to MDMA or MDEA has been scanty and primarily anecdotal in nature [17]. As mentioned earlier, only eight admissions to emergency rooms for MDMA-related toxic reactions were reported to DAWN between 1977 and 1985. Perhaps as a reflection of the increasing popularity of MDMA, this number jumped to 28 in 1985, but this still represents only 0.03% of the drugs mentioned most frequently at the time of admission to emergency rooms [18]. The Haight–Ashbury Clinic has reported that these individuals usually present with anxiety and tachycardia. Reassurance is the only treatment necessary; however, on occasion, counselling sessions may be needed for those in whom symptoms of anxiety or paranoia persist [12, 19]. Siegel very briefly described more severe adverse reactions to MDMA in three users [1]. One woman apparently became mute and semicatatonic for three days after ingesting her monthly dose of 130 mg, and two males developed hallucinations and paranoia that lasted 24 and 48 hours (the latter after a dose of 700 mg!). Wolfson states that there are anecdotal reports of seizures occurring with MDMA use, but provides no further details [20]. All of these reports must be tempered by the fact that only rarely are clinical diagnoses of MDMA toxicity confirmed with toxicology. Analysis Anonymous®, a confidential drug testing service that began operating in 1972, found that 58% of 101 samples submitted as MDMA contained only MDMA, 24% contained MDMA plus other substances, and 16% contained drugs other than MDMA [21]. Although these figures are quite good when compared to other illicit drugs, such as cocaine, it is easy to see that individuals who present with a toxic reaction to MDMA may, in fact, be exhibiting symptoms produced by drugs other than MDMA.

There are only three well-documented cases of serious toxic reactions to MDMA and four clinical cases, from Dallas, where the role of MDMA or MDEA in producing toxic reactions is much less clear. These cases are outlined in Table 1.

2.1. Case 1

Case 1 represents an example of toxic psychosis that is supposed to have been caused by MDMA. Unfortunately, the clinical diagnosis could not be confirmed by toxicology. This 34-year-old male presented to a clinic during the course of receiving therapy for opiate and benzodiazepine dependency. He was agitated and hypertensive and admitted to the intravenous administrations of two syringefuls and the ingestion of ½ teaspoonful of MDMA over a period of 48 hours ending two days prior to his presentation. A diagnosis of MDMA toxic psychosis was made, and he received haloperidol and phenobarbital. Five days later he was found to be increasingly agitated, and the following day he was exhibiting a severe reaction with paraonia and auditory hallucinations that required hospitalization. Hayner and McKinney report that generally toxic psychosis, such as that displayed by this patient, only occurs with doses of MDMA in excess of 200 mg, although it is possible that lower doses could trigger a similar response in susceptible individuals [12].

2.2. Case 2

Case 2 is extremely interesting in that a 32-year-old woman presented to hospital one half hour after ingesting only 60–65 mg of MDMA. She was hallucinating and was agitated, diaphoretic, tachycardiac, hyperthermic, and hypotensive. This had apparently been her second experience with MDMA. The first one had been described as "the best time of her life" [12]. Her five day course in hospital was complicated by continuing hallucinations, rhabdomyolysis, and coagulopathy. Serum MDMA levels were 6.5 mg/L (33.7 umol/L) one hour after admission and 7.0 mg/L (36.3 umol/L) three hours

Table 1. Clinical cases of acute toxic reactions involving MDMA and MDEA.

Case no.	Age/sex	Summary of presentation of clinical course	Toxicology	
1 [12]	34/M	Agitation and hypertension following intravenous and oral MDMA two days prior to presentation. Five days later developed paranoia and auditory hallucinations requiring hospital admission.	Not obtained	
2 [12, 22]	32/F	Agitation, hallucinations, hypotension, hyperthermia, diaphoresis, wheezing 1/2 hour after taking 60–65 mg MDMA p.o. Five day course complicated by continuing hallucinations, rhabdomyolysis, coagulopathy.	Serum:	MDMA 6.5 mg/L (33.7 umol/L)
3 [17]	50/M	Hypertension, palpitations, difficulty controlling movements, and unintelligible speech after taking one MDMA tablet and 15 mg phenelzine p.o. Recovered in 15 hours with supportive care.	Blood:	Ethanol 0.01 gm% (2.2 mmol/L)
			Urine:	Benzodiazepines, meprobamate, cimetidine, MDMA present
4 [10]	30/M	Unconscious following amitriptyline overdose. Recovered.	Blood:	Ethanol 0.16 gm% (34.8 mmol/L) amitriptyline 2.1 mg/L (7.6 umol/L) nortriptyline 0.4 mg/L (1.5 umol/L)
			Urine:	MDEA, amitriptyline, nortriptyline, lidocaine present.
5 [10]	38/M	Bizarre behviour, weakness, dizziness after ingesting MDEA. Released following psychiatric evaluation.	Blood:	Ethanol 0.24 gm% (52.2 umol/L) MDEA present.
6 [10]	26/M	Hallucinations and agitation after taking cocaine.	Blood:	Cocaine 1.6 mg/L (5.3 umol/L)
			Urine:	MDMA, amphetamine, cocaine present.
7 [10]	32/F	Unconscious after taking ethanol, cocaine, MDMA, and valium. Hospitalized 40 days.	Blood:	Diazepam and demethyldiazepam present.
			Urine:	MDMA, cocaine, lidocaine present.

later. These are the highest human blood levels of MDMA reported to date. Interestingly enough, a friend of the patient ingested a similar quantity of MDMA from the same packet as the patient and suffered no ill effects. The clinical course of this individual is remarkably similar to the hospital course described in individuals with toxic reactions to large doses of amphetamines [23, 24] and 3,4-methylenedioxyamphetamine (MDA) [24, 25]. The complications are also similar to those seen in cases of hyperthermia due to heat exposure, and it has been postulated that the hyperthermia seen in all of these instances may be the underlying mechanism by which rhabdomyolysis and coagulopathy come into play [23]. If one looks only at the blood levels of MDMA reported in this case, it seems to be an example of an extreme overdose [26]. Yet, judging from the alleged dose taken and the absence of toxic effects on the part of the patient's friend, this case has the appearance of an idiosyncratic response to MDMA's sympathomimetic effects. Thus, it is not entirely clear whether this patient's complications were the result of an overdose or an idiosyncratic reaction to MDMA.

2.3. Case 3

The last example of a serious toxic reaction to MDMA (Case 3) actually represents an interaction between MDMA and a monoamine oxidase inhibitor (MAOI). This individual presented to hospital with hypertension, diaphoresis, altered mental status, and hypertonicity 3½ hours after taking his usual 15 mg dose of phenelzine and 4½ hours after ingesting one tablet of MDMA. His symptoms resolved with supportive care in 15 hours. The presence of MDMA was confirmed in urine samples. The clinical picture was that of a fairly typical interaction between an MAOI and a sympathomimetic agent, which results in the exaggerated release of monoamine oxidase substrates, such as epinephrine and norepinephrine. Such reactions have been documented with MAOIs and amphetamine, methamphetamine, and several other sympathomimetic agents [17]. Thus it appears that the use of MDMA should be avoided by those on MAOIs.

2.4. Other cases

The four clinical cases reported from Dallas by Bost [10] are difficult to interpret as toxic reactions caused by MDMA or MDEA. In each instance, either ethanol or some other drug was present in sufficient quantities to account for the patient's symptoms. Thus it is impossible to sort out what role, if any, MDMA or MDEA played in their symptomatology.

In summary, it would appear that although minor side effects produced by MDMA are quite common, serious toxic reactions are exceptionally rare. These can take the form of toxic psychosis, drug interactions, and complications of MDMA's sympathomimetic effects, either in an overdose situation or possibly as an idiosyncratic reaction. These toxic reactions are generally managed with supportive care only, but they can be life-threatening.

3. HUMAN DEATHS AND MDMA

As with toxic reactions, the topic of human deaths related to MDMA and MDEA use has, until recently, been anecdotal and poorly documented. For example, in 1985 it was commonly believed that at least two and possibly three deaths had been caused by MDMA. Later, it was found that one of the cases was, in fact, an MDA overdose, and in the second instance the presence of MDMA by toxicology was not confirmed [14]. Likewise, in 1985 the DEA reported that there were 22 Medical Examiner cases where the presence of MDMA was cited [27], but the details of these cases have not been published. Five deaths related to the use of MDMA and MDEA were reported by Dowling et al. in 1987 [9]. These five deaths occurred in Dallas, Texas, and its surrounding counties between July, 1985 and March, 1986. Table 2 summarizes these cases together with the details of 11 other deaths in which MDMA or MDEA was detected by postmortem toxicology. These cases were compiled by review of the available literature, by obtaining the most recent information on human deaths related to MDMA and MDEA use from DAWN, and by review of records at the Office of the Chief Medical Examiner in Dallas County. It should be stated at the outset that this list is likely to be incomplete, as most of the cases are from Dallas, and other major centers where MDMA and MDEA use have been reported, including Florida, New York, California, and New England [1], are underrepresented. It is probable that deaths have occurred in these states, but they have not been reported in the literature or to DAWN. The 16 cases are divided in Table 2 into those cases where MDMA or MDEA were the underlying cause of death, those cases where MDMA or MDEA are thought to have contributed to death, and those cases where the drugs were simply incidental findings at autopsy or where their role in contributing to death is uncertain. Not included in Table 2 is the recent death of a young woman in Santa Clara County, California, wherein MDMA is thought to have caused death; however, this has not yet been confirmed by toxicology [28].

At the present time, there are only two documented cases where the acute toxic effects of MDMA were thought to be the underlying cause of death. No deaths solely related to MDEA use have been reported. The first case (Case 8) involved an 18-year-old previously healthy female who allegedly ingested 1 ½ "hits" (approximately 150 mg) of MDMA at a Dallas bar. A short time later, she collapsed and was found to be in ventricular fibrillation upon the arrival of paramedics. No natural disease processes were identified at autopsy that would account for death, and the only finding of significance was an MDMA level of 1.0 mg/L (5.2 umol/L) in postmortem blood, together with a small amount of ethanol. The second case (Case 9) was that of a 34-year-old male who appeared at his physician's home claiming that he had ingested LSD, MDMA, and valium. He suddenly collapsed, and all attempts at resuscitation were unsuccessful. Again, there was no anatomical cause of death identified at autopsy, but 1.46 mg/L (7.6 umol/L) of MDMA was found in postmortem blood. Unfortunately, the detection and quantitation of LSD is exceptionally difficult and was not attempted in this case. Therefore, it is not possible to clarify the role, if any, that LSD played in this individual's death.

Both of these deaths occurred very suddenly in previously healthy individuals, and it would appear that in both instances the mechanism of death was a cardiac arrhythmia induced by MDMA. Such arrhythmias are well-recognized mechanisms of sudden deaths attributed to amphetamine overdoses [24, 29]. What is not clear in both cases is whether these deaths were the result of true overdoses or of an idiosyncratic reaction to MDMA. The estimated dose of MDMA taken in Case 8, 150 mg, is well within the range of this drug taken therapeutically and recreationally. However, this estimate was derived from witnesses at the scene and, therefore, may be misleading. There is a small but growing body of evidence, to be discussed later in this chapter, that indicates that the levels of MDMA found in these cases are quite high. Thus these may, in fact, be examples of overdoses.

As can be seen in Table 2, there are six cases where MDMA or MDEA are thought to have contributed significantly to death. Cases 10–13 have been described in detail by Dowling et al. [9] and only their salient features will be repeated here. In three of these four cases, MDMA or MDEA was found in individuals whose deaths could be accounted for by underlying natural disease processes. These diseases include coronary atherosclerosis (Case 11), an idiopathic cardiomyopathy (Case 13), and asthma (Case 12). The sympathomimetic effects of MDMA have been documented in an uncontrolled clinical trial and include mild tachycardia together with a consistent elevation of blood pressure [11]. Therefore, in individuals like those in Cases 11 and 13 with significant underlying cardiovascular disease, it is not unreasonable to suggest that MDMA or MDEA could precipitate a fatal arrhythmia. In fact, as a result of its documented sympathomimetic effects, Greer has recommended that MDMA not be administered to those suffering from "hypertension, heart disease, hypothyroidism, diabetes mellitus, hypoglycemia, seizure disorder, glaucoma, diminished liver function, and actual or possible pregnancy" [3]. These would seem to be reasonable guidelines in view of the postulated role played by MDEA in these two deaths. The contributory role of MDMA in the death of the asthmatic (Case 12) is perhaps more tenuous. This individual's asthma was poorly controlled, which is one finding noted in those who die suddenly of asthma [30]. One cannot rule out the possibility, however, that MDMA potentiated a cardiac arrhythmia in an individual who was already compromised as a result of his asthma. It is interesting to note that the young asthmatic woman who suffered a severe toxic reaction to MDMA (Case 2) developed wheezes and respiratory distress requiring treatment with methylprednisolone and isoetharine [12, 22]. Thus MDMA may play some as yet undefined role in the exacerbation of asthma.

Of the remaining three cases in Section B of Table 2, only Case 10 has been described previously in the literature. This 22-year-old male was apparently

Table 2. Deaths associated with MDMA and MDEA use.

Case no.	Age/sex/ city reported	History	Cause of death	Toxicology
		A. Cases where MDMA or MDEA are sole cause of death		
8 [9]	18/F Dallas	Sudden collapse after ingesting 1½ "hits" MDMA with ethanol.	Acute MDMA intoxication	Blood: Ethanol 0.04 gm% (8.7 mmol/L) MDMA 1.0 mg/L (5.2 umol/L)
9 [10]	35/M Berkeley	Sudden collapse after allegedly ingesting LSD, valium, MDMA.	Acute MDMA intoxication	Blood: MDMA 1.46 mg/L (7.6 umol/L) MDA 0.03 mg/L (0.2 umol/L) LSD not determined.
		B. Cases where mDMA or MDEA contributed to death		
10 [9]	22/M Dallas	Climbed electrical utility tower at 0130 hours. Electrocuted and fell.	Electrocution and multiple injuries.	Blood: MDMA present.
11 [9]	25/M Dallas	Collapsed while driving vehicle.	Atherosclerotic cardiovascular disease	Blood: MDEA 0.95 mg/L (4.6 umol/L) Butalbital 0.8 mg/L (3.6 umol/L)
12 [9]	32/M Dallas	Found dead beside car after evening of bar-hopping.	Asthma	Blood: MDMA 1.1 mg/L (5.7 umol/L) Ethanol not present.
13 [9]	21/M Dallas	Found dead in shower after ingesting 3 "MDMA" capsules, 65 mg propoxyphene, and ethanol.	Idiopathic cardiomyopathy	Blood: MDEA 2.0 mg/L (9.7 umol/L) MDMA not present. Propoxyphene 0.26 mg/L (0.8 umol/L) Norpropoxyphene 1.0 mg/L (3.1 umol/L)
14[a]	49/M Oklahoma City	Found dead at home with car running in garage and exhaust fumes in house. Had been snorting cocaine and taking MDMA.	Carbon monoxide intoxication	Blood: MDMA 1.4 mg/L (7.3 umol/L) Cocaine — trace Carboxyhemoglobin 43%
15[b]	25/M Dallas	Involved in motor vehicle collision after running red light.	Multiple injuries	Blood: Ethanol 0.23 gm% (50 mmol/L) MDEA 0.33 mg/L (1.6 umol/L)

C. Cases where MDMA or MDEA not related to cause of death

16[b]	34/M Dallas	History of depression and drug abuse. Found dead at home.	Gunshot wound of head	Blood:	Ethanol 0.17 gm% (37.0 mmol/L) MDMA present. Acetaminophen 430 mg/L (2.8 mmol/L) Cocaine 0.4 mg/L (1.3 umol/L)
17[b]	22/M Dallas	Struck by car while crossing freeway.	Multiple injuries	Blood:	Ethanol 0.24 gm% (52.2 mmol/L) MDMA present.
18[b]	34/F Dallas	Lost control of vehicle, rolled, ejected.	Multiple injuries	Blood:	Ethanol 0.05 gm% (10.9 mmol/L) MDMA present. Cocaine 0.11 mg/L (0.3 umol/L) Dextromethorphan 0.11 mg/L (0.4 umol/L)
19[b]	21/F Dallas	Found dead in garage with car running. Suicide note present.	Carbon monoxide intoxication	Blood:	MDMA 0.26 mg/L (1.35 uol/L) Carboxyhemoglobin 90%
20[b]	20/M Dallas	Found dead in bed with plastic bag over head and cylinder of nitrous oxide nearby.	Smothering and inhalation of nitrous oxide	Blood:	Nitrous oxide detected. MDMA 0.08 mg/L (0.41 umol/L)
21[b]	18/F Dallas	Girlfriend of Case 20, found at same scene.	Smothering and inhalation of nitrous oxide	Blood:	Nitrous oxide detected. MDMA 0.39 mg/L (2.0 umol/L)
22[b]	20/M Dallas	Unwitnessed hit-and-run.	Multiple injuries	Blood:	Ethanol 0.22 gm% (47.9 mmol/L) MDMA 0.38 mg/L (2.0 umol/L)
23[b]	34/M Dallas	Suspect in a theft, shot by police.	Multiple gunshot wounds	Blood:	Cocaine 0.03 mg/L (0.1 umol/L) MDMA 0.27 mg/L (1.4 umol/L)

[a] Personal communication, L. Balding, M.D., Office of the Chief Medical Examiner, Oklahoma City, 1988.
[b] Personal communication, C.S. Petty, M.D. and R. Bost, Ph.D., Office of the Chief Medical Examiner, Dallas, 1988.

climbing an electrical utility tower at 1:30 a.m., when he was electrocuted and fell to his death. In view of the fact that there was no evidence that his actions were suicidal in nature, it was concluded that MDMA was the cause of his bizarre behavior. Unfortunately, the MDMA was not quantitated in postmortem blood samples. In Case 15, a 25-year-old male died of multiple injuries received when he drove through a red light and collided with another vehicle. Although a high level of ethanol was detected in his blood, the Medical Examiner was of the opinion that MDEA contributed to his state of intoxication and thus was a factor in causation of the accident. Finally, Case 14 is interesting in that this 49-year-old male was found dead in his home, with a car running in the attached garage. He had apparently been snorting cocaine and taking MDMA several hours prior to the time his body was found. The circumstances surrounding this death were such that the manner of death (i.e., suicide or accident) could not be determined. The cause of death was thought to be carbon monoxide intoxication. However, the carboxyhemoglobin level of 43% was substantially below the level of 60%–70%, which causes death in healthy individuals [31]. Although this may be partially accounted for by the fact that the patient was actively resuscitated, it is also likely that the relatively high blood level of MDMA was a significant factor in this man's death.

The final eight cases listed in Table 2 represent instances where the presence of MDMA or MDEA was an incidental finding during postmortem drug screening or cases where the role that these drugs played in contributing towards death is unclear. It might reasonably be argued that MDMA contributed to the state of intoxication of the individuals in Cases 17, 18, and 22, and that this, in turn, may have been a factor in the motor vehicle and pedestrian deaths. However, the MDMA was not quantitated in Cases 17 and 18, and the circumstances surrounding the hit-and-run incident in Case 22 were unknown. Therefore, these cases are considered to be instances where the role that MDMA played in contributing towards death is not known.

When the level of MDMA and MDEA found in postmortem blood samples from those cases in Sections A and B of Table 2 is compared with those of Section C, an interesting difference becomes apparent. With the exception of case 10, in which MDMA was not quantitated, and Case 15, in which MDMA was only thought to have contributed to a general state of intoxication, all of the cases in which MDMA is thought to have caused or contributed significantly towards death have shown levels of this drug in excess of 1.0 mg/L (5.2 umol/L), with a mean of 1.2 mg/L (6.4 umol/L) (N = 4). In contrast, those cases in which MDMA was simply an incidental finding have shown a maximum MDMA level of 0.39 mg/L (2.0 umol/L), with a mean of 0.28 mg/L (1.4 umol/L) (N = 5). Likewise, the MDEA levels in Cases 11 and 13, where MDEA contributed significantly towards death, are 0.95 mg/L (4.6 umol/L) and 2.0 mg/L (9.7 umol/L), respectively. Although there have been no deaths described in which MDEA was simply an incidental finding, Bost has reported the blood levels of MDEA found in four living individuals who were arrested for driving while under the influence of drugs [10]. The maximum MDEA level found was 0.59 mg/L (2.9 umol/L), with a mean of 0.31 mg/L (1.5 umol/L) (N = 4). Thus, although the number of cases is relatively small, it is apparent that a blood level of MDMA or MDEA in excess of 1.0 mg/L has the potential to cause or contribute significantly towards death, whereas levels below approximately 0.6 mg/L appear to be consistent with a state of MDMA or MDEA intoxication. In fact, one is left to speculate whether or not the high levels of MDMA or MDEA found in Cases 11, 12, and 13 represent the actual cause of death, with the natural disease processes being contributory factors.

If one assumes that the high levels of MDMA and MDEA found in Cases 11, 12, and 11–14 are the result of overdoses of these drugs, then the question remains: What amount of MDMA or MDEA taken orally represents an overdose? The cases themselves are not enlightening in this regard. Only Cases 8 and 13 provide any information as to the amount of MDMA or MDEA taken, but there is no way to be certain of the actual dose and purity of the drugs ingested. Taken at face value, the estimated doses of 150 mg of MDMA in Case 8 and 300 mg of MDEA in Case 13 are not particularly large, when one considers that some individuals have allegedly taken 700 mg of MDMA orally in one session and survived [1]. Thus individual susceptibility is probably a major factor in determining whether any given dose of MDMA or MDEA is potentially lethal.

Regrettably, virtually nothing is known about the pharmacokinetics of MDMA or MDEA. Only one study has been reported in humans [26], in which a healthy 74 kg 40-year-old male ingested 50 mg (0.68 mg/kg) of MDMA. The peak plasma MDMA level was 0.106 mg/L (0.55 umol/L), measured two hours after administration of the dose. Sixty-five percent of the 50 mg dose was recovered from the urine as unchanged MDMA and 7% as MDA. Although it is not possible to draw any definitive conclusions from a single case study, the peak serum MDMA level of 0.106 mg/L (0.55 umol/L) following a 50 mg dose does tend to support the idea that blood levels of MDMA in excess of 1.0 mg/L (5.2 umol/L) result from the ingestion of large quantities of this drug.

Turning to animal studies of MDMA toxicity, Hardman et al. determined the mean lethal dose (LD50) of MDMA in several animal species [32]. The LD50 was 97 mg/kg i.p. in mice, 49 mg/kg i.p. in rats, 14 mg/kg i.v. in dogs, and 22 mg/kg i.v. in monkeys. Orally administered, MDMA appears to be much less toxic, with an LD50 of 325 mg/kg p.o. reported in rats [33]. Although it is difficult to extrapolate animal data to humans, the orally effective dose of MDMA in humans is 1.5 mg/kg, which is less than one-tenth of the parenteral LD50 reported in animals [14]. Thus it would appear that there should be a high margin of safety between therapeutic and lethal doses of MDMA. It is interesting to note, however, that Frith et al. described the sudden death of one experimental dog following a single oral dose of 15 mg/kg MDMA [34]. This same dose was tolerated once daily for 28 days by five other dogs in the same study, thus supporting the concept that individual susceptibility is an important factor to consider when trying to establish what a so-called lethal dose of MDMA is in humans.

4. CONCLUSIONS

Human deaths that can be attributed to MDMA or MDEA appear to be exceedingly rare, especially when one considers the widespread use of these drugs in the United States. There is evidence to suggest that deaths can occur either as a result of the direct toxic effects of high doses of MDMA or MDEA (especially in those with underlying cardiovascular diseases) or as a result of the intoxicating effects of these drugs upon individuals who are engaged in activities requiring intact concentration, judgment, and coordination (e.g., driving an automobile). The possibility that some deaths arise as an idiosyncratic reaction to low doses of MDMA or MDEA cannot be ruled out at this time. Although the rarity of serious toxic reactions and deaths may indicate that these drugs are safe to use, one must always exercise caution in making such a judgment. For one thing, the long-term sequelae of their use is still unknown, and this is a matter of considerable debate, given the present findings which suggest that MDMA is neurotoxic [35, 36]. Secondly, one must always keep in mind the lessons taught to us by drugs such as cocaine, which only ten years ago was generally considered to be safe [37]. Clearly, more research is needed into the pharmacokinetics and toxicity of MDMA and MDEA, together with the continued documentation of toxic reactions and deaths related to their use, before we can be assured of their safety.

REFERENCES

1. Siegel, R.K., 1986. MDMA: Nonmedical use and intoxication. J. Psychoactive Drugs 18:349–354.
2. Greer, G. and Strassman, R.J., 1985. Information on "Ecstasy." Am. J. Psychiat. 142:1391.
3. Greer, G. and Tolbert, R., 1986. Subjective reports of the effects of MDMA in a clinical setting. J. Psychoactive Drugs 18:319–327.
4. Baum, R.M., 1985. New variety of street drugs poses growing problem. Chem. Eng. News 63(36):7–16.
5. Adler, J., 1985. Getting high on "Ecstasy." Newsweek April 15, p. 96.
6. Eisner, B., 1988. Ecstasy: The MDMA story (Part One). High Times, August, pp. 32–35, 73.
7. Klein, J., 1985. The new drug they call "Ecstasy." New York, May 20, pp. 38–43.
8. DEA, 1985. Temporary placement of 3,4-Methylenedioxymethamphetamine (MDMA) into Schedule I. 21 CFR Part 13013.
9. Dowling, G.P., McDonough, E.T., and Bost, R.O., 1987. "Eve" and "Ecstasy": A report of five deaths associated with the use of MDEA and MDMA. JAMA 257:1615–1617.
10. Bost, R.O., 1988. 3,4-Methylenedioxymethamphetamine (MDMA) and other amphetamine derivatives. J. Forensic Sci. 33:576–587.
11. Downing, J., 1986. The psychological and physiological side effects of MDMA on normal volunteers. J. Psychoactive Drugs 18:335–340.
12. Hayner, G.N., and McKinney, H., 1986. MDMA: The dark side of Ecstasy. J. Psychoactive Drugs 18:341–347.
13. Shafer, J., 1985. MDMA: Psychedelic drug faces regulation. Psychol. Today 19(5):68–69.

14. Shulgin, A.T., 1985. What is MDMA? Pharmchem. Newsletter 14(3):3–5, 10–11.
15. Riedlinger, J.E., 1985. The scheduling of MDMA: A pharmacist's pespective. J. Psychoactive Drugs 17:167–171.
16. Dowling, C.G., Barnes, E., Peters, S., and Zich, J., 1985. The trouble with Ecstasy. Life 8(9):88–94.
17. Smilkstein, M.J., Smolinske, S.C., and Rumack, B.H., 1987. A case of MAO inhibitor/MDMA interaction: Agony after Ecstasy. Clin. Toxicol. 25:149–159.
18. Data from the Drug Abuse Warning Network, 1985. Series 1, No. 5. Rockville, Md: National Institute on Drug Abuse, pp. 24–25.
19. Seymour, R.B., 1985. MDMA: Another view of Ecstasy. Pharmchem. Newsletter 14(3):1–2, 8–9.
20. Wolfson, P.E., Meetings at the edge with Adam: A man for all seasons? J. Psychoactive Drugs 18:329–333.
21. Renfroe, C.L., 1986. MDMA on the street: Analysis Anonymous®. J. Psychoactive Drugs 18:363–369.
22. Brown, C. and Osterloh, J., 1987. Multiple severe complications from recreational ingestion of MDMA ("Ecstasy"). (Letter) JAMA 258:780–781.
23. Ginsberg, M.D., Hertzman, M., and Schmidt–Nowara, W.W., 1970. Amphetamine intoxication with coagulopathy, hyperthermia, and reversible renal failure. Ann. Intern. Med. 73:81–85.
24. Buchanan, J.F. and Brown, C.R., 1988. "Designer drugs": A problem in clinical toxicology. Med. Toxicol. 3:1–17.
25. Simpson, D.L. and Rumack, B.H., 1981. Methylenedioxyamphetamine: Clinical description of overdose, death, and review of pharmacology. Arch. Intern. Med. 141:1507–1509.
26. Verebey, K., Alrazi, J., and Jaffe, J.H., 1988. The complications of "Ecstasy" (MDMA). (Letter) JAMA 259:1649–1650.
27. Climko, R.P., Roehrich, H., Sweeney, D.R., and Al–Razi, J., 1986–87. Ecstasy: A review of MDMA and MDA. Int. J. Psychiat. Med. 16:359–372.
28. Peroutka, S.J., 1988. Personal communication.
29. Benowitz, N.L., Rosenberg, J., and Becker, C.E., 1979. Cardiopulmonary catastrophes in drug-overdosed patients. Med. Clin. North Am. 63:267–296.
30. Benatar, S.R., 1986. Fatal asthma. N. Engl. J. Med. 314:423–429.
31. Finck, P.A., 1977. Exposure to carbon monoxide. In *Forensic Medicine* (Tedeschi, C.G., Eckert, W.G., and Tedeschi, L.G., eds.) Philadelphia PA: W.B. Saunders Co., pp. 840–849.
32. Hardman, H.F., Haavik, C.O., and Soevers, M.H., 1973. Relationship of the structure of mescaline and seven analogs to toxicity and behaviour in five species of laboratory animals. Toxicol. Appl. Pharmacol. 25:299–309.
33. Goad, P.T., 1985. Report: Acute and subacute oral toxicity study of Methylenedioxymethamphetamine in rats. Protocol No. EMD-AT-001. Redfield, AR: Intox Laboratory.
34. Frith, C.H., Chang, L.W., Lattin, D.L., Walls, R.C., Hamm, J., and Doblin, R., 1987. Toxicity of Methylenedioxymethamphetamine (MDMA) in the dog and the rat. Fundam. Appl. Toxicol. 9:110–119.
35. Schmidt, C.J., 1987. Neurotoxicity of the psychedelic amphetamine, Methylenedioxymethamphetamine. J. Pharmacol. Exp. Ther. 240:1–7.
36. Ricaurte, G.A., Forno, L.S., Wilson, M.A., Delanney, L.E., Irwin, I., Molliver, M.E., and Langston, J.W., 1988. (±) 3,4-Methylenedioxymethamphetamine selectively damages central serotonergic neurons in nonhuman primates. JAMA 260:51–55.
37. VanDyke, C. and Byck, R., 1982. Cocaine. Sci. Am. 246:128–141.

6. THE PUBLIC HEALTH IMPLICATIONS OF MDMA USE

JEROME BECK

1. INTRODUCTION

MDMA has been thrust upon the public awareness as a largely unknown drug which to some is a medical miracle and to others a social devil. . . . There have been the born-again protagonists who say that once you have tried it you will see the light and will defend it against any attack, and there have been the staunch antagonists who say this is nothing but LSD revisited and it will certainly destroy our youth [1].

The above appraisal by psychopharmacologist Alexander Shulgin aptly illustrates the public health controversy surrounding the emergence of this unique substance in American society. Frequently referred to as "Ecstasy," "XTC" or "Adam," MDMA suddenly became the object of extensive media coverage in 1985, highlighting what appeared to be dramatic increases in both therapeutic and recreational use. A controversy ensued providing widely divergent perspectives on the substance. Representing one faction were various psychiatrists and researchers who viewed MDMA as a valuable therapeutic adjunct and saw minimal harm associated with carefully monitored use [2–6]. The other side was largely composed of drug enforcement officials, who viewed MDMA as a dangerous "designer drug" possessing potentially harmful actions, with increasing abuse occurring outside of the therapeutic community [7–9].

The uniqueness of MDMA (3,4-methylenedioxymethamphetamine) is exemplified by the terminological confusion in adequately describing its actions.

As an N-methyl analogue of MDA, it is related to both the amphetamines and mescaline. Although MDMA has been most commonly labeled a psychedelic drug, it possesses stimulant properties as well. Moreover, it is rarely hallucinogenic and seldom produces the sensory phenomena or mental confusion associated with other psychedelics [10–12].

A public health appraisal should examine what is known about a particular substance and its users in assessing the potential benefits and harms associated with the drug. Much of the current research examining MDMA's therapeutic as well as toxic potential is amply explored elsewhere in this volume. In attempting to construct a comprehensive public health appraisal of MDMA, this chapter combines a brief overview of research findings with a review of epidemiological data and governmental policy. Complicating such an evaluation is the fact that our current knowledge about MDMA is almost entirely derived from anecdotal data and preliminary research. As a consequence, the most immediate public health implication concerning MDMA is how little we actually know about both the drug and its users.

An adequate public health appraisal should also assess the strengths and weaknesses of current governmental policies regarding a substance and suggest changes thought to maximize benefits and minimize dangers. As such, this chapter not only addresses ways to remedy gaps in knowledge but also examines current policies that possess significant public health implications. This task is best accomplished through separate examinations of the therapeutic and recreational use of MDMA. An assessment of the significant impact that illegality has had on both forms of use is included within each of these analyses.

2. THE SCHEDULING CONTROVERSY

The Drug Enforcement Administration (DEA) first encountered MDMA in 1972 through a street sample bought in the Chicago area [12]. However, such reports were infrequent and it was not until a decade later that the Drug Control Section of the DEA began soliciting information regarding MDMA. In 1982, an early article on MDMA quotes a DEA spokesman as stating, "If we can get enough evidence to be sure there's potential for abuse, we'll ban it" [13].

Satisfied that they had accumulated enough evidence regarding abuse potential, the DEA administrator recommended the placement of MDMA into Schedule I on July 27, 1984 [14]. In support of this action, a DEA chemist concluded that "MDMA has a high potential for abuse based on its chemical and pharmacological similarity to MDA, its self-administration without medical supervision, its clandestine synthesis, and its distribution in the illicit drug traffic" [8].

What appeared to be a routine scheduling process of a little-known substance was quickly challenged by a rather well-organized group of psychiatrists and researchers who strongly believed in MDMA's therapeutic potential

[2–6]. Citing LSD as a case example, therapists argued that a schedule I status would severely hinder any research into the drug's therapeutic potential. The government's surprise at the therapists' reaction was evidenced by a DEA pharmacologist's statement that they "had no idea psychiatrists were using it" [15].

In actuality, a number of psychiatrists and other therapists had been using MDMA since the late 1970s as an adjunct for various purposes, particularly in facilitating communication, acceptance, and fear reduction [10, 16–17]. Despite their belief in MDMA's efficacy, therapists were reluctant to publicize their preliminary findings for fear that any publicity would inevitably result in its illegality and removal from therapeutic research and use [18].

In response to MDMA proponents' challenges, federal administrative law hearings were held in three cities (Los Angeles, Kansas City, and Washington, D.C.) to determine the final scheduling of MDMA. The DEA (together with the FDA) clearly believed that MDMA belonged in Schedule I. Their attorneys set out to prove that MDMA fit all three criteria necessary for such a placement: a high potential for abuse, no currently accepted medical use, and a lack of safety for use under medical supervision [19].

Shortly before the first hearing, the DEA Administrator unexpectedly invoked the emergency scheduling powers granted by the Comprehensive Crime Control Act of 1984 [20]. As a result, MDMA was temporarily placed in Schedule I on July 1, 1985. The primary rationale behind this new federal law was an attempt at counteracting the sudden advent of so-called designer drugs (primarily synthetic opiate analogues) in the early 1980s [21]. This amendment provides the Attorney General authority to place any substance posing "an imminent hazard to public safety" into Schedule I for a period of one year (plus an additional six months, if necessary), while the final scheduling process is underway [20].

A number of rationales were provided for the necessity of this action. The primary justification centered on an as yet unpublished study associating high dosage, intravenous use of MDA in rats with suspected serotonergic neurotoxicity of unknown significance [22].

Perhaps an even more significant reason behind the emergency scheduling was the active promotion of "XTC" as a legally available euphoriant by a Texas-based operation. Beginning in the early 1980s, this mass-production and marketing scheme stood in sharp contrast to the typically smaller-scale, more clandestine distribution networks found in other parts of the country. [21, 23]. The blatantly open sales of MDMA in numerous bars and nightclubs in the Dallas area presented a very public and problematic drug use pattern to authorities [7, 8].

Although virtually everyone at the hearings argued that there should be at least some controls placed on MDMA (thus outlawing non-medical use), therapists and other proponents proposed that it remain available for clinical use and research. Their lawyer attempted to refute the DEA's contentions by arguing that MDMA has only a low to moderate abuse potential, is safe under medical supervision, and possesses significant therapeutic value. As a consequence, they argued that MDMA be placed into a lower schedule that would allow for the continuation of human research and therapy [2–6, 18].

Many researchers and therapists feared that a Schedule I status would make it almost impossible to continue using MDMA, even experimentally. Therapists argued that their quiescence in publicizing preliminary findings was justified in light of historical examples involving other psychedelics. Numerous LSD studies involving over 40,000 subjects were conducted throughout the 1950s and early 1960s. Major reviews of these studies concluded that, in general, LSD research had compiled a remarkable safety record with arguable efficacy in at least some studies [24–27]. Nevertheless, strict controls established in the late 1960s resulted in the almost total discontinuance of LSD research. Once-flourishing explorations into the therapeutic potential of other psychedelic substances (including MDA) also came to a virtual standstill upon their placement into Schedule I of the 1970 Controlled Substances Act [27].

These actions were not supported by much of the therapeutic community. According to Grinspoon and Bakalar, "Almost everyone who has worked with psychedelic drugs, and many who have not, think that their research potential is great; and many who have worked with them also think they have therapeutic potential" [27]. They reinforced their contentions by citing survey findings of LSD researchers as well as randomly selected American Medical Association members [27].

The DEA attorneys countered therapist concerns by arguing that a Schedule I status does not preclude appropriately conducted research into MDMA's therapeutic potential. Various government witnesses were called upon to testify that such substances could still be studied if the correct protocol was followed [28–30]. However, citing historical examples, MDMA proponents argued that stringent Schedule I requirements significantly discouraged research and claimed that no substance had ever been removed from Schedule I. An FDA official refuted this latter point by noting that sufentanil had been rescheduled from Schedule I to II in 1984 [31]. This appears to be one of only two exceptions to the rule, however, with almost all changes in scheduling having occurred in the direction of increased control [32].

Several psychiatrists and other researchers testified on behalf of MDMA's therapeutic potential at the administrative law hearings [2–6]. In general, they argued that a major advantage of MDMA over traditional psychedelics is that it produces far less distortion of sensory perception and fewer unpleasant emotional reactions. The MDMA experience is generally seen as both personal and familiar and seems to differ only in its degree of intensity from that of everyday experience. This stands in sharp contrast to the effects of most other psychedelics, where the experience is often perceived as unfamiliar and transpersonal [10]. As Grinspoon argued, "MDMA appears to have some of the advantages of LSD-like drugs without most of the corresponding disadvantages" [3].

In countering the optimism expressed by MDMA advocates, DEA attorneys called upon various research experts to critique the largely anecdotal nature of the therapist's testimony [29, 31, 33–34]. These government witnesses also gave little credence to the two preliminary studies conducted by proponents [17, 35]. In general, their critique could be summed up in Kleinman's conclusion that "although these reports make interesting reading, their lack of scientific design, methodology, and controls makes them scientifically unsound" [28].

The anecdotal evidence and preliminary research offered by MDMA advocates fell far short of meeting the FDA's exacting specifications regarding safety and accepted medical use. Nevertheless, proponents' testimony at the hearings provided strong arguments for MDMA's therapeutic promise as well as safety (considering the small number of doses that would be given to any one patient). After criticizing Greer's study on methodological grounds, one government witness, John Docherty, went on to urge that formal, well-controlled studies be conducted to assess what he viewed as a potentially valuable compound for psychotherapy [36]. Acknowledging the necessity of such research, Greer confidently stated that "because every therapist I know who has given MDMA to a patient has found it to be of significant value, I am convinced that it can be shown scientifically to be efficacious" [2].

The persuasiveness of proponents' arguments was evidenced by the DEA Administrative Law Judge's findings and recommendations, which largely concurred with their contentions [37]. Citing MDMA's therapeutic potential and safety and noting the lack of significant abuse, the judge recommended a Schedule III placement. This would have substantially eased research requirements and allowed the continued therapeutic use of MDMA by physicians.

The DEA attorneys took sharp exception to the judge's ruling, once again emphasizing the absence of well-controlled, double-blind studies, necessary in meeting the "currently accepted medical use" criteria required of drugs in Schedules II–V. In addition, they argued that the unresolved neurotoxicity question demonstrated a significant lack of "safety for use under medical supervision" — the other major criteria for Schedules II–V [38]. Relying on these arguments, the DEA Administrator subsequently rejected the judge's recommendation and attempted to permanently place MDMA in Schedule I on November 13, 1986 [39].

Four recent appellate court decisions challenged the validity of the DEA's attempts to temporarily and permanently schedule MDMA in 1985 and 1986. The major consequence of these rulings was the apparent invalidation of these placements on technical grounds [40].

Three of the appellate court decisions concerned the validity of prosecutions and convictions resulting from MDMA's temporary placement in Schedule I on July 1, 1985 [41–43]. Although differing somewhat as to the challenges and resulting decisions, all three rulings were in basic agreement that the emergency scheduling was invalid. As a consequence, the convictions were overturned on the basis of technicalities associated with faulty implementation of the emergency scheduling law [18, 40].

Grinspoon, M.D. vs. Drug Enforcement Administration differed from the other three cases in not challenging a previous conviction [44]. Instead, this appeal to the First Circuit Court in late 1986 was a continuation of the therapist's battle to get MDMA out of Schedule I. In September of 1987, the Court overturned the DEA Administrator's final scheduling and remanded the DEA to reanalyze its scheduling of MDMA. In so doing, the Court ruled that the DEA had erred in narrowly defining the criteria for "currently accepted medical use" as equivalent to an FDA New Drug Application (NDA) or Investigational New Drug (IND) protocol approval. The Court also found the DEA to have erred in equating "lack of accepted safety for use of the drug or other substance under medical supervision" with lack of FDA approval. To address these definitional problems, the Court directed the DEA to better define these two scheduling criteria and demonstrate how they apply to MDMA [44].

The First Circuit Court did concur, however, with the DEA's contention that MDMA possesses a "high potential for abuse." Although the Court agreed with proponents in noting the lack of sufficient evidence demonstrating significant abuse of MDMA, it nevertheless ruled that the DEA had succeeded in establishing the *potential* for such abuse. The Court acknowledged that a Schedule I placement might impede MDMA research, but concluded that this reason was not sufficient to require moving MDMA out of Schedule I [44].

Addressing the First Circuit Court's directives, the DEA Administrator concluded that MDMA fell far short of meeting the DEA's revised definitions of what constituted acceptable medical use. Although acknowledging that, "many witnesses in this proceeding, including those presented by the agency, indicated that MDMA may have a potential therapeutic use, such a potential use is not sufficient to establish accepted medical use" [45]. As a consequence, he once again ordered the placement of MDMA into Schedule I on March 23, 1988.

An unforeseen legal complication resulted from the First Circuit Court's decision. Although temporarily dismissing MDMA's placement in Schedule I, the Court's ruling appeared to have little effect on MDMA's legality, since it was generally assumed that it would continue to be illegal under the Controlled Substance Analogue Enforcement Act passed in late 1986. The Analogue or "Designer Drug" law makes illegal any substance that is similar in structure or psychological effect to any substance already scheduled, providing it is manufactured, posssessed, or sold with the intention that it will be consumed by humans [46].

It now appears, however, that anyone arrested for MDMA offenses prior to the second rescheduling cannot be prosecuted under the Analogue Act, as a result of an unusual technicality. Since MDMA had already been made a controlled substance it could no longer be considered an analogue of a controlled substance [40].

In summary, a major consequence of the various appellate court rulings is the likelihood that most, if not all, convictions for MDMA offenses before March 23, 1988, will ultimately be reversed if appealed. As a US Department of Justice lawyer recently concluded, "it appears that all federal prosecutions based upon MDMA's previous status as a Schedule I controlled substance will be subject to challenge and that such challenges are likely to be sustained" [40]. It should be noted, however, that although these rulings fault the DEA's actions in attempting to schedule MDMA, they lend little or no support to proponents' contentions. Barring unforeseen circumstances, it appears that Schedule I will remain MDMA's home for many years to come.

3. THERAPEUTIC IMPLICATIONS

It is important to examine the various obstacles that may impede or prevent an adequate assessment of the therapeutic potential of substances such as MDMA in our society. More to the point, let us assume for a moment that MDMA does indeed possess both significant therapeutic value and relative safety at prescribed doses. What would it take and how long would it take to convince the government of its efficacy and safety?

A number of formidable obstacles confront any attempt to generate the substantial funding necessary to finance research into MDMA's therapeutic potential. Aside from the neurotoxicity question, the most significant obstacle currently facing MDMA is its "orphan drug" status. Since MDMA was patented in 1914, it is now in the public domain, which means that any company could produce and market it for approved conditions. Little incentive exists for a pharmaceutical company to invest the millions of dollars on research necessary to possibly obtain FDA approval, only to have other firms market the same product with minimal investment [10, 47].

An additional problem inhibiting pharmaceutical interest concerns the probable lack of profit associated with the marketing of MDMA or similar substances employed as adjuncts to psychotherapy. Profits accruing from the few doses given a typical patient would be minimal compared with those garnered from currently prescribed psychotropic medications (e.g. tranquilizers or antidepressants) that are often intended for daily use.

Finally, the placement of MDMA into Schedule I probably completes the task of discouraging pharmaceutical interest in the substance. As the Second Triennial Report to Congress from the Secretary of Health and Human Services states, "it is unlikely that pharmaceutical companies will develop a drug, no matter how promising it is, that is in Schedule I of the Controlled Substances Act" [48].

Despite the above obstacles, efforts are still being made to generate the interest and funding necessary to investigate MDMA's therapeutic potential. A major goal in this regard is convincing the FDA to certify MDMA as an "orphan drug." This action would greatly increase MDMA's chances of being researched and marketed as a therapeutic adjunct. However, the possibility of an orphan drug designation is largely dependent on a successful resolution of the neurotoxicity question and subsequent FDA approval for human research [47].

Pharmaceutical disinterest is not the only obstacle facing MDMA or other substances utilized for their potential insight-enhancing properties. As Seymour, Wesson, and Smith point out: "No other medication is currently recognized by the Food and Drug Administration as an adjunct to psychotherapy, and even psychotherapy in its many variants is not accepted by mainstream medical practitioners as bona fide medical treatment" [49].

The above skepticism can be attributed in part to the empirical difficulties that plague any attempt at conducting well-controlled double-blind research with insight-oriented therapeutic techniques. As Sidney Cohen succinctly pointed out with regard to LSD research, "no method of using LSD therapeutically has as yet met rigid scientific requirements,...but, in truth, no other type of psychotherapy has been fully tested by these exacting methods either" [50]. Although psychotherapy research has improved since that observation was made over two decades ago, the methodological problems involved in conducting such studies remain daunting at best.

Given the obstacles, it is not surprising that Grinspoon and Bakalar conclude from their comprehensive review of psychedelic research that the therapeutic use of these substances "obviously has potentialities that are not being allowed to reveal themselves" [27]. Citing historical examples to reinforce his argument, Nichols argues that "as a result of this government-induced stagnation in the field of drug-assisted psychotherapy research, psychiatry has not been offered various types of novel psychoactive drugs for assessing their value to medicine" [51]. He goes on to declare that,

> the very nature of the organization of the FDA precludes it from taking any kind of risk — theoretical or actual. Yet risk is an essential part of drug discovery. The paternalistic idea has developed that consumers must be protected from any risk, of any kind, from the cradle to the grave [51].

The limits of the current psychiatric system are exemplified by the drugs commonly utilized in treating the gamut of mental problems. Practically all of these psychotropic medications are pharmacological depressants prescribed primarily for symptomatic relief. Nichols decries this lack of pharmaceutical options, declaring that, "It is a harsh reality indeed that tells patients with emotional pain to suffer quietly, that they will not be helped except with drugs that dull the mind" [51].

Frustrated by the intractability of the current system, a number of researchers have argued that revisions are necessary to adequately deal with substances such as MDMA. Nichols, Grinspoon, Bakalar, and others advocate the creation of adequately conducted informed consent procedures, allowing adults to voluntarily participate in research involving drugs with significant therapeutic promise [27, 51]. According to Smith and Seymour: "There is some movement currently to create a new category for experimental psychoactive drugs with low abuse potential, no established medical uses, but high therapeutic potential, so that these drugs may possibly be used in treatment-center research" [52].

Even assuming the creation of such a category, the neurotoxicity question remains the most formidable obstacle blocking the human research necessary to assess MDMA's therapeutic value. The majority of animal studies have found varied degrees of suspected serotonin nerve terminal degeneration in certain areas of the brain [53–56].

The significance of this alleged neurotoxicity, however, remains unknown. Also unknown is whether it occurs (and at what dosage levels) in humans and, if so, whether it is permanent or transient in nature. Finally, MDMA proponents point out that there have been no documented cases of MDMA-related neurological impairment among any of the hundreds of thousands of MDMA users [57].

Government officials and other researchers respond to these arguments by warning that disorders or problems associated with other neurotoxic substances (e.g., MPTP) were not always immediately apparent in users [53]. As Charles Schuster, Director of the National Institute on Drug Abuse (NIDA) and co-author of the original MDA neurotoxicity study, cautions, "What we don't know is whether twenty or thirty years from now, at the age of 45, they [MDMA users] may begin to be showing central nervous system degenerative signs that ordinarily would not be seen until they get to be 70 or 80 [58].

Proponents have countered this commonly expressed fear by noting that a number of other sympathomimetic drugs suspected of neurotoxicity continue to be medically prescribed, often for daily use [57, 59]. The most notable of these is fenfluramine (Pondamin®), which produces serotonergic changes similar to those of MDMA, at dosage levels scarcely above the effective therapeutic dose [60]. In summarizing the research of Schuster and colleagues at the University of Chicago, Johanson concludes that fenfluramine produces, "a long lasting depletion of serotonin in the striatum, hippocampus, and rest of brain at a dose only 1.25 times the ED50 dose for anorexia" [61].

Proponents also argue that any risk associated with the therapeutic use of MDMA would be minimal, considering the small number of doses given to any one patient. Nevertheless, the FDA has rejected all Investigational New Drug (IND) applications to date, in each case citing the neurotoxicity issue as its major rationale — even in proposals involving therapeutic research with terminally ill patients [47].

The government's well-publicized stand on MDMA stands in sharp contrast to its seeming lack of concern regarding fenfluramine. As the Adminstrative Law Judge concluded, the "FDA has approved the daily use of fenfluramine in humans on a chronic basis. Fenfluramine is a controlled substance, but this proven neurotoxic substance is only in Schedule IV [37]. An examination of the 1988 Physician's Desk Reference reveals that fenfluramine remains in Schedule IV, with no mention informing physicians or other readers of the neurotoxicity research [62].

Numerous problems complicate an adequate determination regarding the appropriate place in modern society for potentially valuable therapeutic adjuncts like MDMA. Nevertheless, the inadequacy of current efforts lead Grinspoon and Bakalar to conclude that "our legal and political institutions, like our natural science and psychiatry, are failing to supply the complex responses these complex drugs demand. We should show more confidence in our capacity to tolerate and make use of them" [27].

4. RECREATIONAL IMPLICATIONS

Although MDMA first appeared on the street in the early 1970s, use remained limited until the end of the decade. Recreational use increased at a somewhat faster pace during the early 1980s, with information about the drug disseminated largely through word of mouth and anonymously written "flight" guides providing detailed instructions regarding proper use [10, 63].

This relatively quiet popularization suddenly changed with the proposed scheduling and ensuing reaction by therapists, which brought MDMA to national attention in mid-1985. Within a few months, the print and electronic media had discovered "Ecstasy." Almost every major newspaper and magazine printed stories about MDMA, often sensationalizing its reputed euphoric, sensual, and therapeutic qualities [15, 23, 64–65].

The rise in publicity was accompanied by what appeared to be an exponential increase in street demand. During the administrative law hearings, UCLA's Ronald Siegel testified "that street use had escalated from an estimated 10,000 doses distributed in all of 1978 to 30,000 doses distributed per month in 1985" [66].

The DEA found evidence of increased use throughout much of the country, particularly in the Dallas area, where it was estimated that "30,000 dosage units of MDMA are distributed each month" [7]. As mentioned earlier, this mass-production and marketing scheme included blatantly open sales of MDMA in certain bars and nightclubs. The DEA also noted the promotion of MDMA as a legal euphoriant by means of fliers, circulars, and promotional parties [7].

Although the media blitz resulted in a dramatic increase in interest throughout the country, it appears that MDMA was already popular in certain areas. Shulgin estimated that two million doses had been consumed prior to the DEA's proposed scheduling [37]. Doblin's interviews with major dealers leads him to believe in even higher estimates of use. One group of distributors told him that they had dispensed approximately 500,000 doses over a seven year period up to 1984. Another group (the Texas operation) claimed to have already distributed two to three million doses by 1984. Nevertheless, the impact of media coverage is evidenced in the same group's claim to have sold two million doses in the month prior to the emergency scheduling [67].

The above estimates clearly attested to MDMA's increasing popularity, as well as the power of free publicity. Nevertheless, all of these figures were highly speculative and limited in what they told us about MDMA users.

Having first encountered MDMA as a drug educator at the University of Oregon in 1976, I found myself on the ground floor in researching the recreational use of this substance. Through my capacity as a drug educator, counselor, and researcher in Oregon and the San Francisco Bay Area, I was able to use informal ethnographic and qualitative interview strategies in sketching a profile of MDMA use in two areas where it enjoyed early and significant popularity. This observational analysis, combined with anecdotal accounts provided by various groups (e.g., media and therapists) and official statistical indicators (DEA and Drug Abuse Warning Network [DAWN] data), culminated in the publication of three articles [10, 21, 68].

In June of 1987, the National Institute on Drug Abuse (NIDA) approved a grant by Marsha Rosenbaum, Patricia Morgan, and myself to conduct a sociological exploration of MDMA users. MDMA's recent emergence, unique actions, and increasing popularity provide a rare opportunity to examine gradually evolving patterns of use among various groups. Unfortunately, our findings are still preliminary as of this writing allowing for only general observations to be reported here. This analysis is supplemented by my earlier research and by the only other known studies of recreational MDMA users: Siegel's exploratory study in Los Angeles [69] and Peroutka's informal survey of undergraduates [70, 71].

MDMA appears to be most popular in particular urban areas possessing established distribution networks for the drug. Its use has been associated most commonly with college students, gays, yuppies, and "New Age" seekers of psychological and/or spiritual growth. A typical dose ranges from 100 to 150 milligrams and costs between 10 and 25 dollars [10, 12].

Although many respondents in our study consider MDMA to be a "drug of choice," they offer radically different points of view regarding its perceived value. On the one hand are those who see "Adam" as a valuable therapeutic and spiritual tool. Many of these individuals pursue "New Age" spiritual directions and, with the exception of other psychedelic experiences, often report little use of other substances. On the other extreme are those who seek the acclaimed euphoria and sensuousness associated with "Ecstasy." These individuals tend to have substantial experience with a wide array of psychoactive drugs and find that MDMA provides many of the qualities previously sought in other substances (e.g., cocaine). Although extremists on either side often have great difficulty understanding the other, the vast majority of users fall somewhere in between, sensing and often pursuing both "therapeutic" and "recreational" benefits in their experiences.

Oral ingestion is by far the most common route of administration among current users, although inhalation is occasionally reported and, in rare cases, injection. Taking the drug orally is preferred because it produces the longest, smoothest high with the least amount of stimulant side effects. Briefly summarized, effects generally appear within 20 to 60 minutes, with the user often experiencing a brief "rush" of energy, most often described as mild but euphoric. After this rush, the high levels off to a comfortable plateau, which usually lasts two to three hours and is followed by a gradual "coming down" sensation, culminating in a feeling of fatigue.

MDMA, although milder and shorter-lasting than MDA, still exerts amphetamine-like effects on the body, including dilated pupils, dry mouth and throat, tension in the lower jaw, occasional grinding of the teeth, and overall stimulation. Nausea and dizziness are occasionally reported, most often during the initial onset of the high. Individuals become dehydrated and should be drinking water or juice throughout the experience. Unfortunately, some choose to drink alcoholic beverages, which increase dehydration and negative aftereffects. In general, the presence and/or severity of various side effects is greatly affected by the individual's frequency of use, the size of dose, and overall mental and physical health. As a consequence of its sympathomimetic actions, MDMA use would likely be contraindicated for individuals with the following medical conditions: diabetes, diminished liver function, epilepsy, glaucoma, heart disease, hypertension, hypoglycemia, hyperthyroidism, and pregnancy [10, 12, 17].

During the hearings, proponents presented the results of a later published research project evaluating the effects of a single MDMA exposure on 21 healthy individuals [35]. All of the subjects had used MDMA on previous occasions. Using blood chemistry, physiological measures, and neurological examinations, the researchers concluded that,

> This experimental situation produced no observed or reported psychological or physiological damage, either during the twenty-four hour study period, or during the three month follow-up period. From the information presented here, one can say only that MDMA, at the doses tested, has remarkably consistent and predictable psychological effects that are transient and free of clinically apparent major toxicity [35].

The research design of this experiment was heavily criticized by an FDA pharmacologist at the administrative law hearings [31]. He essentially agreed with the study's conclusion that "there is insufficient evidence to judge accurately either harm or benefit" [35].

A unique attraction that differentiates MDMA from other stimulants is its capacity to induce a strong paradoxical sense of relaxation. This effect often leads users to be almost totally oblivious to many of the stimulant side effects [10]. As with therapeutic accounts, most recreational users cite a dramatic drop in defense mechanisms or fear responses, while also feeling an increased empathy for others. Combined with the stimulant effects, this often produces an increase in intimate communication.

Users tend to be predominantly positive when describing their initial MDMA experiences. Nevertheless, many of the users in our study have significantly cut down or discontinued use as their perception of costs begin to outweigh benefits. Although positive effects are often described as continuing well beyond the experience (e.g., carryover of insight, lessened fear), it is the negative aftereffects that often lead to discontinuance or sharp reductions in use. Necessary allowances for next day recovery underlie the infrequent use reported by many of the respondents in our study (particularly professionals), who state that job, school, and family demands rarely allow for what they consider to be a "two-day experience" [72].

There is wide variability as to the perceived severity of aftereffects. Some individuals report frequent use with minimal problems, whereas others quickly discontinue use. As with other stimulants, individuals under the influence of MDMA are often capable of ingesting large quantities of alcohol with few immediately discernible effects. As a consequence, overuse of alcohol plays a significant role in many of the next day hangovers. What could be a potentially toxic interaction between MDMA and alcohol merits further investigation.

Factoring in a number of common culprits (e.g., taking too much too often, overindulging in alcohol or other drugs) offers a significant, yet incomplete, explanation for differences in perceived severity of negative aftereffects. Users are often aware of and attempt to control for a number of readily identifiable factors that contribute to the next day's hangover. Nevertheless, some users still complain of varying degrees of problematic aftereffects (fatigue, malaise, headaches) that often persist for a day or two (and in rare cases longer) after taking MDMA [10, 12, 70].

In earlier articles, I speculated from my observations of users that MDMA may have an adverse action on the immunological response of some individuals [10, 12]. This effect was most often (but not always) associated with repeated high dosage use, particularly in long-term users. Such individuals frequently complained of increasingly uncomfortable and prolonged "burnout" periods and reported an increased susceptibility to various ailments, particularly sore throats, colds, flus, herpes outbreaks, and bladder infections. Such reactions were rare among novice users and individuals in good physical and mental health. Since these problems have been noted by only a small number of respondents in our current study, the significance of such findings remains in question. Such problems are probably comparable to what might be expected from the overuse of other sympathomimetic substances (e.g., the amphetamines). Further study would be useful in addressing this concern.

In addition to valid concerns regarding potential neurotoxicity or negative aftereffects, we have seen the emergence of user mythologies, spawned by erroneous media reports and hard-to-trace rumors of the drug's toxic potential. A primary problem has been the frequent confusion of MDMA with various synthetic opiate analogues, commonly referred to as "designer drugs." This label has been applied to the intentional process of chemically engineering existing controlled substances to create legal substitutes that possess similar psychoactive properties [73]. Although MDMA is often referred to as a designer drug, such a designation is debatable, since it was first synthesized in 1912, before the passage of any national drug legislation. The primary designer drugs are synthetic opiate analogues employed as substitutes for heroin. The use of these substances has resulted in significant problems: MPTP has been associated with Parkinson's disease, while the extremely potent fentanyl analogues have been responsible for a large number of fatal overdoses. Unfortunately, MDMA has often been confused with these drugs, both in the media and on the street, resulting in a number of erroneous beliefs [21].

Another surprisingly common belief (particularly among college students on the West Coast) is that MDMA somehow "drains the spinal fluid" or "fuses the spinal cord." Neck and backaches that occasionally follow MDMA use probably contributed to the formation of this user mythology [12, 21].

Psychological problems associated with MDMA use appear to be rare but are potentially troublesome when they do occur. Although MDMA enjoys a reputation for producing an "easy-to-handle" experience, infrequent panic reactions and/or hyperventilation during the initial onset of the high (the "rush" phase) have been noted. I have found this to occur most often with novice users who become overwhelmed by the sudden power of the initial rush. Fearing that this is simply a portent of worse to come, they then succumb to a generalized panic reaction. Reassurance that this "peak" phase is transitory generally lessens the problem [10, 12, 66].

Very little data exists regarding prevalence and types of psychological problems associated with MDMA use. What little information is available regarding the treatment of MDMA-related psychological problems comes from the Haight–Ashbury Free Medical Clinic. In 1985, they reported that three to four individuals a month sought treatment for problems related to MDA, MDMA, or related drugs [12]. This number represented less than 1% of their entire case load and has significantly decreased since that time. Unfortunately, it is impossible to discern whether this decline can be attributed to lessened use or changing patterns of use associated with a more informed user subculture. Many of the clients appearing at the clinic present acute symptoms that include anxiety, rapid pulse, and in advanced cases, paranoia. Unlike abusers of other substances, these clients rarely returned for further treatment, leading to speculation that they were novice users who had learned their lesson [74]. According to Seymour:

With MDMA and the methoxylated amphetamines, as is the case with most stimulants and psychedelics, the acute toxicity symptoms that are usually seen in treatment are similar and result from taking too much of the drug. These dose-related symptoms usually dissipate as the drug wears off, and the patient can be discharged within a few hours [12].

More chronic psychological problems have recently been noted by the Haight–Ashbury Free Medical Clinic. A "delayed anxiety disorder" has been observed in a few individuals following an MDMA experience. The problem typically occurs among novice users of MDMA, with manifestations ranging from mild anxiety or depression to more full-blown symptomatology [12]. Although rare, these initial findings underscore a potential danger underlying unsuccessful attempts at "self-therapy" by individuals who run the risk of exacerbating emotional problems with unsupervised use.

During the administrative law hearings, the DEA attorneys attempted to establish that MDMA possesses the "high potential for abuse" criteria necessary for a Schedule I placement. In support of their contention, they cited two animal studies which found that most primates will self-administer MDMA at regular intervals (although less frequently than cocaine) [75–76].

The attractive qualities frequently ascribed to the "Ecstasy" experience certainly suggest a high abuse potential. Although Seymour [12] and others have stated that MDMA doesn't seem to pack a "euphoric punch" or "rush" comparable to other drugs, just the opposite appears to be true. Among individuals in our study who have tried both MDMA and cocaine, many express a clear preference for the longer, smoother euphoria provided by MDMA. One could assume that a significant number of MDMA users would eventually experience major problems from overuse.

Nevertheless, in sharp contrast to cocaine, this does not appear to be the case with MDMA, at least among current user groups. In testimony submitted on the DEA's behalf, Siegel noted from his exploratory study that the most common patterns of MDMA use were "experimental" (ten times or less in a lifetime) or "social-recreational" (one to four times per month). He went on to state that "compulsive patterns marked by escalating dose and frequency of use have not been reported with MDMA users" [66].

Our preliminary findings generally support Siegel's observations on use patterns. The most frequent use of MDMA tends to occur during the months following the initial experience. After first exposure, a small minority of individuals attempt to continually re-experience the positive aspects of the drug. This abusive cycle tends to be brief, however. Within a short period of time, the frequent use of MDMA almost invariably produces a strong dysphoric reaction that is only exacerbated with continued use. The increasing number of unpleasant side effects, coupled with an almost total loss of desired effects, appear to occur with greater rapidity and intensity than they do with other more commonly abused substances.

In agreement with Siegel's findings, respondents in our study who continue to use MDMA almost invariably do so in a controlled, infrequent manner. Nevertheless, a very small minority of our respondents describe "binging" with MDMA upon occasion, in a manner similar to their use of cocaine — sometimes individually going through a half gram or more of MDMA in an evening. In addition, approximately 10% of our respondents have reportedly taken MDMA over 100 times, usually over a time period of five or more years. This stands in contrast to Peroutka's sample of 100 undergraduates where 38 was the maximum number of experiences and the vast majority had taken it less than 10 times [70]. The major reason for this disparity in number of experiences was our intentional oversampling of frequent and long-term users, who tend to be much older than the undergraduate sample.

In attempting to establish MDMA's abuse potential, the DEA attorneys cited commonly accepted drug problem indicators (e.g., seizures, drug treatment, and emergency room admissions) to support their contention. However, proponents reversed this argument by nothing the remarkable dearth of reported problems considering the DEA's own estimates of MDMA use [10, 12]. Their contention was later supported by Newmeyer's epidemiological review of drug problem indicators, which concluded that MDMA "has given hardly any indication that it is a problem for Americans, either in terms of adverse reactions, treatment admissions, or police involvement" [77].

It appears that chronic binging or other problematic use patterns are rare among current MDMA users. However, since the popularity of MDMA is fairly recent, more time is needed to see how long-term patterns develop among current user groups. In addition, continued research is necessary to adequately assess MDMA's abuse potential for new user groups introduced to the drug. It could prove to be a self-limiting phenomenon among certain user groups, while others encounter significant problems related to overuse and/or more potent routes of administration (e.g., I.V. use). As Newmeyer cautions:

> It may well be that MDMA currently enjoys controlled, careful use by a number of cogniscenti, somewhat as LSD did around 1960. Perhaps in future years, a much larger number of less sophisticated individuals will be drawn into MDMA usage and will find ways to evince adverse reactions, police involvement, and other unpleasant consequences from use of the drug [77].

Finally, one must consider the public health implications surrounding the scheduling of MDMA. For a short time, the outlawing of MDMA led to increased interest in the development and sale of other, still legal, methoxylated amphetamines [21]. However, users generally found these substances to be lacking in comparison to MDMA. A small number of our respondents reported having tried the most popular of these — MDE ("Eve"). They gave it mixed reviews, with all preferring MDMA. Nevertheless, following

MDMA's scheduling, "Eve" quickly replaced "Adam" in some Dallas bars [23]. However, the "designer drug" legislation that passed in 1986 put a damper on this maneuver by outlawing the sale of controlled substance analogues [46].

The demise of street drug analysis in recent years adds to the difficulty of ascertaining what effect (if any) MDMA's newly controlled status has had on street quality and availability. Although analysis results have generally shown high levels of purity for MDMA samples submitted before scheduling, not enough information is available since scheduling [63]. Nevertheless, one can anticipate an increase in problems resulting from bathtub chemistry and increased adulteration, which are commonly associated with other illicit drugs. Considering MDMA's narrow dose range, these factors could present significant problems to users.

5. CONCLUSIONS AND RECOMMENDATIONS

The scheduling of any substance has obvious public health implications. As previously discussed, the placement of MDMA into Schedule I has significant ramifications regarding both therapeutic research and recreational use. As Smith, Wesson, and Buffum pessimistically conclude:

> Moving a drug to Schedule I does not stop illicit availability. The whole notion of controlled drugs is a misnomer. Nothing is so out of control as those drugs the DEA and FDA have appropriated to Schedule I. Moving a drug to Schedule I does, however, have consequences. Price generally increases, the quality control of licit manufacture is destroyed, and responsible research becomes almost impossible. MDMA, a drug with low abuse potential and possible therapeutic use, is the latest victim of our misguided drug control policies [78].

The MDMA controversy did succeed in challenging many of the basic precepts underlying the Controlled Substances Act. The Administrative Law Judge was not alone in his frustration regarding the limitations of the five available schedules and the vagaries of the criteria utilized in determining proper placement into them [79]. In describing the dilemma facing him, the judge noted that four of the five schedules are reserved exclusively for substances possessing demonstrated medical value and safety. These drugs are placed in differing schedules depending on their potential for abuse (II is the highest and V the lowest). Attempts to control substances of unproven medical value and safety have only one available option — Schedule I. However, the criteria necessary for inclusion in this schedule specifically state that the substance must possess a *high abuse potential*. Recognizing this gap in the Controlled Substances Act, the judge concluded that:

> The Acting Administrator should decide that a substance that has a potential for abuse less than a *high* potential, and no currently accepted medical use in treatment in the United States, cannot lawfully be placed in any of the five schedules established by

the Controlled Substances Act of 1970. The terms of the Act do not permit it. No amount of poring over legislative history empowers us to close the obvious gap left in the statutory scheme [79].

Since MDMA proponents fell far short of meeting the FDA's exacting specifications regarding safety and accepted medical use, the third criterion pertaining to abuse potential became even more significant. Of interest here is the process that determines whether a substances possesses the "high potential for abuse" necessary for inclusion into Schedule I. This was a key concern for both camps, as a result of the above-mentioned gaps in the Controlled Substances Act.

A number of ill-defined problems arise in any attempt to assess the abuse potential of various substances. Should any non-medical use of a psychoactive drug automatically be considered abuse? Even utilizing more exacting definitions of abuse (e.g., problematic use characterized by dysfunction or other negative consequences) still encounters problems regarding where to draw the line between low, moderate, or high abuse potential, since practically any psychoactive substance (licit or illicit) will be abused by at least some individuals.

The DEA's interpretation of abuse potential was repeatedly challenged by proponents, as well as by the Administrative Law Judge [2, 5, 37, 78, 80]. It appears that the DEA arbitrarily defines any recreational use of certain kinds of substances as evidence of abuse [19, 81–82]. However, what criteria are actually employed to differentiate drugs with low, medium, and high potentials of abuse?

It appears that three different standards are currently employed by the DEA in assessing abuse potential: one for such highly abused drugs as alcohol and tobacco, which are intentionally exempted from the scheduling process; another for medically accepted drugs found in Schedules II–V; and a final one reserved for those substances placed in Schedule I. An examination of Schedules II–V provides a fairly good ordering of drugs regarding their respective abuse potentials. Looking at opiates, for example, one can understand why morphine is in Schedule II (high abuse potential) while codeine cough syrups are in Schedule V (low abuse potential).

Comparison of abuse potential becomes increasingly problematic when examining those substances that have been placed in Schedule I. Although the inclusion of drugs such as PCP certainly makes sense, the rationale for other substances is less convincing. This is particularly true of the numerous psychedelic drugs found in Schedule I. How could an obscure drug such as ibogaine be determined to possess a high potential for abuse necessary for a Schedule I placement? This substance is used ritualistically among certain African cultures and remains virtually unknown in the United States [27].

Although LSD and psilocybin have certainly been abused by some individuals, one might question whether their abuse potential is substantially greater than that of diazepam (Valium®). Diazepam is a popular street drug possessing significant addiction potential and frequently cited in emergency room admissions [83–84]. Nevertheless, it is listed in Schedule IV, which is reserved for those substances possessing only a low potential for abuse. Assume for a moment that diazepam had no authorized medical uses. If that were the case, would it then be placed into Schedule I as a result of its street reputation and abuse potential? If so, Schedule I can then be seen as a catch-all for certain kinds of drugs lacking FDA approval regarding accepted and safe medical use, regardless of their actual abuse potential relative to other substances. Ultimately, as Seymour contends:

"What is being called into question is not just the control of one drug that may or may not have a high abuse potential. The core issue is one of scientific inquiry and medical progress and how these are to be balanced against public safety and integrity" [12].

In concluding that MDMA did not meet any of the three criteria necessary for a Schedule I placement (high abuse potential, no currently accepted medical use, and lack of safety of use under medical supervision), the Administrative Law Judge presented a significant challenge to his agency's interpretation of the scheduling process [37]. Although his findings were later rejected by the DEA Administrator [39] and largely overruled by the First Circuit Court [44], they nonetheless highlighted the troublesome dilemmas associated with attempting to apply the DEA's and FDA's understanding of these criteria to MDMA and other psychedelic substances.

The administrative law hearings also revealed an obvious lack of research in assessing both the potential benefits and harms of MDMA. The overall epidemiology of use was clearly a mystery as well. Consequently, both sides were limited to offering testimony based largely on anecdotal data or extrapolations from preliminary animal studies. The most significant point of agreement between the two camps was in recognizing the obvious need for more research to better determine the potential benefits and risks of a substance that was becoming a "drug of choice" for increasing numbers of Americans.

Since the hearings, MDMA research has primarily centered on various animal studies conducted to assess the neurotoxicity question. The importance of this research emphasis is highlighted by the fact that its eventual resolution will largely determine if and when needed human studies of MDMA's therapeutic potential are allowed to resume.

An additional research priority should focus on studying individuals who have taken or were prescribed various suspected neurotoxins, particularly fenfluramine. For obvious reasons, the invasiveness required of many physiological techniques necessitates extrapolations from animal data. Nevertheless, researchers have already begun to conduct spinal taps and other physiological and psychological measures on MDMA users [85]. Unfortunately, a number of methodological problems present significant challenges to the validity and reliability of these studies. There are always problems with attempts at collecting retrospective data from recreational users regarding total number of uses, dosage amounts, and purity of product, as well as frequently extensive use of other drugs. Even more importantly, most MDMA users have short use histories, frustrating attempts at addressing the long-term implications of MDMA use.

One option that would overcome many of the problems described above entails a study of individuals who were given MDA in research or therapeutic settings back in the 1960s and early 1970s [86–89]. Locating an adequate sample from this relatively small group would overcome many of the problems regarding dosage and purity, while also addressing the long-term implications of a substance suspected of being more neurotoxic than MDMA.

An even better alternative calls for studying users of fenfluramine, a substance that produces serotonergic changes similar to MDMA in roughly the same equivalent dose range. In contrast to MDMA users, it can be anticipated that individuals prescribed fenfluramine will have better recall of total doses (assisted by prescription records) and assurance of product purity. In addition, as a result of fenfluramine being prescribed for daily use (often for lengthy periods of time), users will frequently have longer exposure to greater amounts of a suspected neurotoxin than their MDMA counterparts. Perhaps most significantly, the fact that fenfluramine has been prescribed for over two decades will address the long-term significance of current findings. As such, it may shed some light on the MDMA controversy, while assessing a potential public health problem that remains almost totally unknown among users and prescribers of fenfluramine.

Unfortunately, the paucity of epidemiological research continues to frustrate an adequate assessment of overall extent, patterns, and changes in MDMA use over time. Almost all we have to rely on are the extremely rough estimates of use provided at the administrative law hearings in 1985. These reports cited almost exponential increases in MDMA use up to that time. What has happened since then?

A wide divergence of opinion currently exists as to whether the overall use of MDMA has increased, decreased, or stayed the same since its scheduling in 1985. With but one exception, there have been no additional published estimates of use. This exception was Peroutka's 1987 survey of 369 undergraduates, which found that an astonishing 39% had reportedly tried MDMA [70–71]. The survey provided valuable data regarding student use patterns and perceived effects. Unfortunately, the fact that it was a convenience sample seriously challenges its validity in accurately estimating the extent of MDMA use.

With the exception of a small number of deaths associated with MDMA use (examined elsewhere in this volume), the usual problem indicators remain silent, with only infrequent reports of hospitalizations, arrests, or treatment admissions. Lacking any valid estimates of use, this quiescence can be interpreted in many ways. One could alternately attribute this low reportage of MDMA-related problems to minimal levels of use and/or abuse; its illicit status, which inhibits people from seeking treatment; responsible and informed user groups; and/or low toxicity of the drug.

The current lack of valid epidemiological data seriously undermines an accurate public health appraisal of MDMA use. Even if MDMA was found to cause a particular physiological or psychological problem, the overall significance and societal implications of such a finding would largely depend on a number of epidemiological factors of which we currently know little or nothing. For example, let us imagine that research establishes that MDMA does indeed cause some form of significantly harmful neurotoxicity but only in cases involving high dosage-binge use, as opposed to the more common pattern of infrequent ingestion of low to moderate doses. Good epidemiological data would prove invaluable in providing some idea as to the general prevalence of binge use and the user groups commonly associated with it. As a result, public health warnings or interventions could be quickly and appropriately designed in a cost-effective fashion for target populations at risk.

With these considerations in mind, it is essential to obtain better epidemiological data on MDMA use. Our current lack of knowledge underscores the need for additional exploratory research. Because MDMA is such a new drug on the street, user subcultures or "social worlds" are just beginning to develop. Out of these social worlds, a body of "user folklore" evolves that informs and conditions individuals to accept certain norms as to appropriate and inappropriate use, overall expectations, and perceived benefits and harms. In essence, the user subculture becomes remarkably effective in defining and/or influencing the attitudes, use patterns, and overall drug experience of the individual user.

The power of user expectations in shaping the drug experience has significant public health implications. Utilizing marijuana and LSD as examples, Becker noted that as both substances became popularized, there appeared to be a growing consensus among users regarding appropriate set and settings, expectations, and perceived benefits and risks. He then went on to demonstrate how the development of a drug-using subculture tended to minimize adverse reactions and redefine the drug experience as something positive (rather than "going crazy") [90]. These ideas were later given credence by Bunce, who found a sharp decrease in LSD-related emergency room admissions in the late 1960s and early 1970s despite continued increases and eventual stabilization of LSD use during that time [91].

MDMA presents a particularly interesting research challenge since it happens to be a new and unique substance that may still be gaining in popularity, yet remains unknown throughout much of the country. As such, it allows us the rare opportunity to examine how the process of gradually evolving user subcultures actually unfold.

The significance of current research findings greatly depends on knowing the actual extent and various patterns of MDMA use. A larger, more representative survey of MDMA use would allow us to test many of the theories or hypotheses emerging from our exploratory research on broader populations.

For a variety of reasons, implementing such a survey would not be an easy endeavor. MDMA's illegality and small user population provide formidable obstacles to obtaining a representative sample of users. An obvious solution would be the inclusion of MDMA questions in one or both of the well-respected high school senior or household surveys [92, 93]. Owing to its recent emergence and small user population, MDMA has not been covered in these national surveys. Unfortunately, it is also unlikely to be included in the near future, considering the significant limits placed by time factors on the allowable number of separate drug categories. In fact, the tendency of most surveys has been to lump all psychedelics (often including PCP) together as a result of their comparatively small user populations. Therefore, a survey of MDMA use should also gauge levels of other psychedelic use as well.

Failing inclusion in these larger studies, the remaining strategy calls for conducting an anonymous and representative survey of a much smaller sample. Certain considerations dictate the best choice of target population in attempting to obtain a representative sample of MDMA users. Although MDMA use appears to be increasing, the number of users remains small compared to the total population of the nation. The two major user groups appear to be college students and young professionals. Attempting to obtain a representative sample of the latter group large enough to include MDMA users faces seemingly insurmountable obstacles from cost considerations alone. Although high schools provide fairly representative samples for particular age groups, the number of high school users appears to be too small at this time to justify the time and expense involved in a survey.

This leaves colleges, where we have identified what appears to be the largest proportion of MDMA users. The challenge lies in gaining the needed cooperation of selected schools across the country in constructing a survey that could be implemented anonymously to a representative sample of students. Obviously, such a survey would only tell us about users at particular colleges. Nevertheless, it would provide a needed perspective on perceived benefits, harms, and extent and patterns of use among a significant population of MDMA users. It could also prove extremely useful in examining student use of other drugs as well.

Media accounts and various professionals in the drug field have often dismissed MDMA as a short-term fad. Such an observation seems safe, considering the significant reduction in the use of most illicit drugs in the 1980s, particularly among those groups commonly associated with MDMA use. Nevertheless, MDMA may prove to be an exception to this trend. The surprising percentage of users found in Peroutka's study, combined with our preliminary findings from both dealer and user interviews, strongly suggests that MDMA remains extremely popular among current user groups, while slowly spreading to new populations.

In a manner reminiscent of media portrayals of LSD two decades before, a recent *New York Times* article proclaimed that MDMA "has soared in popularity this year, occupying center stage in a wider social drama combining fashion, music, and youthful restlessness" [94]. Not mentioned in the article is the fact that this new scene originated in London nightclubs, where it is commonly referred to as "Acid House." Possessing "its own music, dress code, and language," a London periodical reports:

> Record and fashion industries have been rushing to catch up with the fad, and even commercial radio disc jockeys have drawn on the ecstatic commentary devised by their counterparts in nightclubs. What many appear to ignore is that the drug may not be so much part of the cult, as the point of it [95].

As of this writing, the only sizeable "Acid House" following in the United States appears to be in Manhattan. Whether the popularity of MDMA grows as a result of this phenomenon remains to be seen. However, one should not underestimate the potent combination of marketing savvy and media sensationalism in contributing to increased curiosity throughout the country.

Regardless of any particular "fad" appeal, there remain a number of more enduring reasons why MDMA has become a "drug of choice" for many Americans. Whether taking MDMA for primarily therapeutic or recreational purposes, most users praise the remarkable ease with which the high itself is usually experienced. The therapeutic and euphoric qualities associated with MDMA, combined with this relative ease of experience, are likely to attract new users in spite of the current anti-drug climate. As the author of a recent article titled "Drugless in L.A." described it, "For veterans of the 60s, it is interesting to note that the major new drug of the 80s, Ecstasy, has been hyped as a drug that is not really a drug" [96].

As this chapter has sought to demonstrate, MDMA is indeed a powerful drug with potential benefits and harms that are likely to have profound public health implications. Given how little we really know about MDMA and its users, the obvious recommendation is for more research exploring all facets of this fascinating yet controversial substance.

ACKNOWLEDGEMENTS

The author gratefully acknowledges the efforts of Drs. Marsha Rosenbaum and Patricia Morgan, as well as those of other researchers involved with this study: Joel Brown, Jennifer Ham, Deborah Harlow, Lynne Jackson, Doug McDonnell, and Sheigla Mörphy.

REFERENCES

1. Shulgin, A., 1985. What is MDMA? PharmChem. Newsletter 14(3):3–5, 10–11.
2. Greer, G., 1985. Written Testimony Submitted on Behalf of Drs. Grinspoon and Greer,

Professors Bakalar and Roberts, United States Department of Justice, Drug Enforcement Administration Hearings, Docket No. 84-48.
3. Grinspoon, L., 1985. Written Testimony Submitted on Behalf of Drs. Grinspoon and Greer, Professors Bakalar and Roberts, United States Department of Justice, Drug Enforcement Administration Hearings, Docket No. 84-48.
4. Lynch, R.D., 1985. Written Testimony Submitted on Behalf of Drs. Grinspoon and Greer, Professors Bakalar and Roberts, United States Department of Justice, Drug Enforcement Administration Hearings, Docket No. 84-48.
5. Strassman, R.J. 1985. Written Testimony Submitted on Behalf of Drs. Grinspoon and Greer, Professors Bakalar and Roberts, United States Department of Justice, Drug Enforcement Administration Hearings, Docket No. 84-48.
6. Wolfson, P.E. 1985. Written Testimony Submitted on Behalf of Drs. Grinspoon and Greer, Professors Bakalar and Roberts, United States Department of Justice, Drug Enforcement Administration Hearings, Docket No. 84-48.
7. United States Department of Justice, Drug Enforcement Administration, 1985. Fact Sheet.
8. Sapienza, F., 1985. Written Testimony Submitted on Behalf of Drug Enforcement Administration. United States Department of Justice, Drug Enforcement Administration Hearings, Docket No. 84-48.
9. Holsten, D.W. and Scheister, D.W., 1985. Controls over the manufacture of MDMA. California Society for the Treatment of Alcohol and Other Drug Dependencies News 12: 14–15.
10. Beck, J., 1986. MDMA. The popularization and resultant implications of a recently controlled psychoactive substance. Contemp. Drug Problems 13(1):23–63.
11. Nichols, D.E., 1986. Differences between the mechanism of action of MDMA, MBDB, and the classic hallucinogens. Identification of a new therapeutic class: entactogens. J. Psychoactive Drugs (18):305–313.
12. Seymour, R.B., 1986. MDMA. San Francisco: Haight Ashbury Publications.
13. Dye, C., 1982. XTC: The chemical pursuit of pleasure. Drug Survival News 10(5):8–9.
14. Mullen, F.M., 1984. Schedules of controlled substances placement of 3,4-methylenedioxymethamphetamine into Schedule I. Fed. Regis. 49(146):30210–30211.
15. Getting high on 'ecstasy.' Newsweek, April 15, 1985, p. 96.
16. Kueny, S., 1980. Report on a study to examine the feasibility of using 3,4-methylenedioxymethamphetamine (MDMA) to facilitate psychotherapy. Unpublished paper.
17. Greer, G., 1983. MDMA: A new psychotropic compound and its effects in humans. Self-published.
18. Beck, J. and Rosenbaum, M. 1988. The scheduling of MDMA ("Ecstasy"). In *Handbook of Drug Control in the United States* (Inciardi, J.A., ed.). Westport, CT: Greenwood Press.
19. Sapienza, F. 1986. MDMA and the Controlled Substances Act. Paper presented at MDMA: A Multidisciplinary Conference, San Francisco.
20. Lawn, J.C., 1985. Schedules of controlled substances. Temporary placement of 3,4-methylenedioxymethamphetamine (MDMA) into Schedule I. Fed. Regis. 50(105):23118–23120.
21. Beck, J. and Morgan, P.A., 1986. Designer drug confusion: A focus on MDMA. J. Drug Educ. 16(3):287–302.
22. Ricaurte, G., Bryan, G., Strauss, L., Seiden, L., and Schuster, L., 1985. Hallucinogenic amphetamine selectively destroys brain serotonin nerve terminals. Science 229:986–988.
23. The trouble with Ecstasy, 1985. Life, September, pp. 88–94.
24. Cohen, S., 1960. Lysergic acid diethylamide: Side effects and complications. J. Nerv, Ment. Dis. 130:30–40.
25. McGlothlin, W.H. and Arnold, D.O., 1971. LSD revisited. A ten-year follow-up of non-medical LSD use. Arch. Gen. Psychiat. 24:35–49.
26. Brecher, E.M. and the editors of Consumer Reports, 1972. *Licit and Illicit Drugs*. Mt. Vernon, NY: Consumers Union.
27. Grinspoon, L. and Bakalar, J.B., 1979. *Psychedelic Drugs Reconsidered*. New York: Basic Books.
28. Kleinman, J.E., 1985. Rebuttal Testimony Submitted on Behalf of Drug Enforcement Administration, United States Department of Justice, Drug Enforcement Administration Hearings, Docket No. 84-48.

29. Sheahan, J.M., 1985. Rebuttal Testimony Submitted on Behalf of Drug Enforcement Administration, United States Department of Justice, Drug Enforcement Administration Hearings, Docket No. 84-48.
30. Snyder, L., 1985. Rebuttal Testimony Submitted on Behalf of Drug Enforcement Administration, United States Department of Justice, Drug Enforcement Administration Hearings, Docket No. 84-48.
31. Tocus, E.C., 1985. Written Testimony Submitted on Behalf of Drug Enforcement Administration, United States Department of Justice, Drug Enforcement Administration Hearings, Docket No. 84-48.
32. Shulgin, A.T., 1988. Personal communication.
33. Docherty, J.P., 1985. Written Testimony Submitted on Behalf of Drug Enforcement Administration, United States Department of Justice, Drug Enforcement Administration Hearings, Docket No. 84-48.
34. Shannon, H.E., 1985. Rebuttal Testimony Submitted on Behalf of Drug Enforcement Administration, United States Department of Justice, Drug Enforcement Administration Hearings, Docket No. 84-48.
35. Downing, J., 1986. The psychological and physiological effects of MDMA on normal volunteers. J. Psychoactive Drugs 18:335–340.
36. Docherty, J.P., 1985. Oral Testimony Given at Drug Enforcement Administration Hearings — Washington, D.C., October 11.
37. Young, F.L., 1986. Opinion and Recommended Ruling, Findings of Fact, Conclusions of Law and Decision of Administrative Law Judge, Submitted in the Matter of MDMA Scheduling. Docket 84-48, May 22.
38. Stone, S.E. and Johnson, C.A., 1986. Government's Exceptions to the Opinion and Recommended Ruling, Findings of Fact, Conclusions of Law and Decision of the Administrative Law Judge, Submitted in the Matter of MDMA Scheduling. Docket No. 84-48, June 13.
39. Lawn, J.C., 1986. Schedules of controlled substances. Scheduling of 3,4-methylenedioxymethamphetamine (MDMA) into Schedule I of the Controlled Substances Act Fed. Regis. 51(198):36552–36560.
40. Harbin, H., 1988. MDMA. Narcotics, Forfeiture and Money-Laundering Update. United States Department of Justice 2(1):14–19.
41. United States vs. Caudle, 1987. 828 F. 2d 1111 (5th Circuit Court).
42. United States vs. Spain, 1987. 825 F. 2d 1426 (10th Circuit Court).
43. United States vs. William Waldo Emerson, 1988. 88 F. 2d 3106 (9th Circuit Court).
44. Grinspoon, M.D. vs. Drug Enforcement Administration, 1987. 828 F. 2d 881 (1st Circuit Court).
45. Lawn, J.C., 1988. Schedules of controlled substances: Scheduling of 3,4-methylenedioxymethamphetamine (MDMA) into Schedule I of the Controlled Substances Act; remand. Fed. Regis. 53(54):5158.
46. Controlled Substances Analogue Enforcement Act, 1986. 21 U.S.C. 802(32) and 813.
47. Doblin, R., 1988. A Proposal for Orphan Pharmaceuticals, Inc. Self-published.
48. U.S. Department of Health and Human Services, 1987. *Drug Abuse and Drug Abuse Research.* The Second Triennial Report to Congress from the Secretary of Health and Human Services. Rockville, MD: NIDA.
49. Seymoure, R.B., Wesson, D.R., and Smith D.E., 1986. Editor's introduction, MDMA: Proceedings of the conference. J. Psychoactive Drugs 18(4).
50. Cohen, S., 1965. LSD and the anguish of dying. Harper's 231:69–78.
51. Nichols, D.E., 1987. Discovery of novel psychoactive drugs: Has it ended? J. Psychoactive Drugs 19(1).
52. Smith, D.E. and Seymour, R.B., 1985. How the Federal Government classifies drugs: From medical use to abuse potential. High Times, August 30.
53. Barnes, D.M., 1988. New data intensify the agony over Ecstasy. Science 239:864–866.
54. Ricaurte, G.A., Forno, L.S., Wilson, M.A., DeLanney, L.E., Irwin, I., Molliver, M.E., and Langston, J.W., 1988. 3,4-methylenedioxymethamphetamine selectively damages central serotonergic neurons in nonhuman primates. JAMA 260(1):51–55.
55. Ricaurte, G., DeLanney, L., Irwin, I., and Langston, W., 1988. Toxic effects of MDMA on central serotonergic neurons in the primate: Importance of route and frequency of drug administration. Brain Res. 446:165–168.

56. Schmidt, C.J., 1987. Neurotoxicity of the psychedelic amphetamine, methylenedioxymethamphetamine. J. Pharmacol. Exp. Therap. 240:1.
57. Doblin, R., 1988. MDMA: Risk assessment and the FDA. Unpublished paper.
58. Researchers say "ecstasy" is dangerous, 1986. Associated Press, Jan. 16.
59. Cotton, R., 1988. Letter dated 6/5/88 to Secretary Margaret Heckler, Department of Health and Human Services, Washington, D.C.
60. Schuster, C.R., Lewis, M., and Seiden, L.S., 1986. Fenfluramine: neurotoxicity. Psychopharmacol. Bull. 22(1):148–151.
61. Johanson, C.E., 1985. Report from the University of Chicago Drug Abuse Research Center. Problems of Drug Dependence, 1984. Proceedings of the 46th Annual Scientific Meeting. The Committee on Problems of Drug Dependence, Inc. (Harris, L., ed.), NIDA Research Monograph 55. Rockville, MD: NIDA, pp. 78–81.
62. *1988 Physician's Desk Reference*, 1988. Oradell, NJ: Medical Economics Company, Inc., pp. 1694–1695.
63. Renfroe, C.L., 1986. MDMA on the street: Analysis Anonymous. J. Psychoactive Drugs 18:363–369.
64. A crackdown on Ecstasy, 1985. Time, June 10, p. 64.
65. Ecstasy: the lure and the peril, 1985. Washington Post, June 1, pp. 1,4.
66. Siegel, R.K. 1985. Direct Testimony Submitted on Behalf of the Drug Enforcement Administration, United States Department of Justice, Drug Enforcement Administration Hearings, Docket No. 84-48.
67. Doblin, R., 1988. Personal communication.
68. Beck, J., 1987. MDMA. Drug Abuse Information and Monitoring Project, California Department of Alcohol and Drug Programs.
69. Siegel, R.K., 1986. MDMA: Nonmedical use and intoxication. J. Psychoactive Drugs 18:349–354.
70. Peroutka, S.J., Newman, H., and Harris, H., 1988. Recreational use of 3,4-methylenedioxymethamphetamine (MDMA, Ecstasy). *Neuropsychopharmacol.* 1:273–277.
71. Peroutka, S.J., 1987. Incidence of recreational use of 3,4-methylenedioxymethamphetamine (MDMA, "Ecstasy") on an undergraduate campus. N. Engl. J. Med. 317:1542.
72. Rosenbaum, M. and Morgan, P.A., 1988. Ecstasy use among professionals. Paper presented at the American Society of Criminology, Chicago, November.
73. Smith, D.E. and Seymour, R.B., 1985. Clarification of "designer" drugs. U.S. J. of Drug and Alc. Abuse, November.
74. Seymour, 1986. Personal communication.
75. Griffiths, R.R., Lamb, B., and Brady, J.V., 1985. A preliminary report on the reinforcing effects of racemic 3,4-methylenedioxymethamphetamine in the baboon. Document Submitted on Behalf of the Drug Enforcement Administration, United States Department of Justice, Drug Enforcement Administration Hearings, Docket No. 84:48.
76. Harris, L.S., 1985. Preliminary report on the dependence liability and abuse potential of methylenedioxymethamphetamine (MDMA). Document Submitted on Behalf of the Drug Enforcement Administration, United States Department of Justice, Drug Enforcement Administration Hearings, Docket No. 84:48.
77. Newmeyer, J.A., 1986. Some considerations on the prevalence of MDMA use. J. Psychoactive Drugs 18(4):361–362.
78. Smith, D.E., Wesson, D.R., and Buffum, J., 1985. MDMA: "Ecstasy" as an adjunct to psychotherapy and a street drug of abuse. California Society for the Treatment of Alcoholism and Other Drug Dependences News 12(2):1–3.
79. Young, F.L., 1985. Opinion and recommended decision on preliminary issue. Submitted in the Matter of MDMA Scheduling, United States Department of Justice, Drug Enforcement Administration Hearings, Docket No. 84-48. June 1.
80. Reidlinger, J.E., The scheduling of MDMA: A pharmacist's perspective. J. Psychoactive Drugs 17(3): 167–171.
81. Baum, R.M., 1985. New variety of street drugs poses growing problem. Chem. Eng. News, September 9, pp. 7–16.
82. Shulgin, A.T., 1988. The Controlled Substances Act: A Resource Manual of the Current Status of the Federal Drug Laws. Self-published.
83. Woody, G.E., O'Brien, C.P., and Greenstein, R., 1975. Misuse and abuse of diazepam: An

increasingly common medical problem. Int. J. Addict. 10:843–848.
84. Budd, R.D., Walkin, E., Jain, N.C., and Sneath, T.C., 1979. Frequency of use of diazepam in individuals on probation and in methadone maintenance programs. Am. J. Drug Alc. Abuse 6:511–514.
85. Peroutka, S.J., Pascoe, N., and Faull, K.F., 1987. Monoamine metabolites in the cerebrospinal fluid of recreational users of 3,4-methylenedioxymethamphetamine (MDMA "Ecstasy"). Res. Commun. Substance Abuse 8:125–138.
86. Naranjo, C., 1973. The Healing Journey — New Approaches to Consciousness. New York: Random House.
87. Naranjo, C., Shulgin, A.T., and Sargent, T., 1967. Evaluation of 3,4-methylenedioxyamphetamine (MDA) as an adjunct to psychotherapy. Med. Pharmacol. Exper. 17:359–364.
88. Turek, I.S., Soskin, R.A., and Kurland, A.A., 1974. Methylenedioxyamphetamine (MDA) subjective effects. J. Psychedelic Drugs 6(1):7–13.
89. Yensen, R., DiLeo, F.B., Rhead, J.C., Richards, W.A., Soskin, R.A., Turek, B., and Kurland, A.A., 1976. MDA-assisted psychotherapy with neurotic outpatients: A pilot study J. Nerv. Ment. Dis. 163(4):233–245.
90. Becker, H.S., 1967. History, culture and subjective experience: An exploration of the social bases of drug-induced experiences. J. Health Soc. Behav. 8(9):163–176.
91. Bunce, R., 1979. The social and political sources of drug effects: The case of bad trips on psychedelics. J. Drug Issues, Spring:213–233.
92. Johnston, L.D., Bachman, J.G., and O'Malley, P.M., 1987. National Trends in Drug Use and Related Factors Among American High School Students and Young Adults. Rockville, MD: NIDA.
93. Miller, J.D., et al., 1983. *National Survey on Drug Abuse: Main Findings 1982*. Rockville, MD: NIDA.
94. Foderaro, L.W., 1988. A drug called "Ecstasy" emerges in nightclubs. The New York Times, December 11, p. 26.
95. The hyping of Ecstasy, 1988. The Illustrated London News, October, pp. 29–30, 32.
96. Kaye, E., 1986. Drugless in L.A.: The new trend is sobriety. New Age Magazine, May.

7. STRUCTURE-ACTIVITY RELATIONSHIPS OF MDMA AND RELATED COMPOUNDS: A NEW CLASS OF PSYCHOACTIVE AGENTS?

DAVID E. NICHOLS AND ROBERT OBERLENDER

1. INTRODUCTION

It has been hypothesized that MDMA and substances that possess a psychopharmacological effect similar to MDMA are members of a novel pharmacological class named entactogens [1–3]. In this chapter evidence will be presented to support this, through a discussion of the data acquired in efforts directed toward testing this hypothesis. Although these studies are far from complete, the results gathered thus far, together with those from other laboratories, support the view that the pharmacology of entactogens is clearly different from other known classes of compounds.

1.1. Entactogens, hallucinogens, and stimulants

By definition, the identification of a new drug class results from pharmacological studies that clearly show that its members cannot be included within other known categories. The two drug classes most often mentioned as similar to MDMA are the hallucinogens and central stimulants. As medicinal chemists, our approach to this work has focused on the molecular features (i.e., structure-activity relationships) of the latter in comparison to MDMA-like compounds. The approach employed utilizes the synthesis of a series of structurally related congeners of MDMA and the measurement of their biological activity in terms of MDMA-like, hallucinogen-like, and stimulant-like effects in animal models.

If the hypothesis of novel activity is correct, specific structural modifications of prototypic molecules should have differing consequences for the three types of activities. These differences would be envisioned to arise from the lack of overlap between the mechanisms by which members of these psychopharmacological classes produce their distinct effects. If the hypothesis is incorrect, molecular modifications in the entactogen class would produce parallel functional changes in one of the other drug classes, indicating that MDMA-like compounds are actually included in one of the known categories.

The chemical structures of representative compounds are illustrated in Figure 1. These are all derivatives of β-phenethylamine. The stimulant amphetamine (1) is simply α-methyl phenethylamine while DOM (2) illustrates the type of aromatic substitution typical for the most potent hallucinogenic phenethylamine derivatives. MDA (3) and MDMA (4) possess a 1,3-dioxole ring fused to the aromatic nucleus.

1.2. Defining activity — therapeutic versus pharmacologic

With respect to drugs that have centrally-mediated psychopharmacological effects in humans, the biological action can be defined in two ways. Definitions are commonly derived from the therapeutic goals toward which an activity is applied in the treatment of medical disease states or symptoms. Thus some examples of familiar drug classes include antipsychotics, antidepressants, anxiolytics, and analgesics.

Alternatively, activity may relate to the primary pharmacological effect produced by a drug. Amphetamine and cocaine, for example, are categorized together as stimulants, although the former has been prescribed, among other things, as an appetite suppressant, whereas the latter is useful as a local anesthetic.

For a substance like MDMA, with no currently accepted medical use, primary pharmacological effects must be used to define activity. Since the effects of psychopharmacological agents may be manifested, for example, as changes in mood, perception, and thought, the key question is whether or not humans can distinguish MDMA from other classes of drugs, based on the way it makes them "feel." In other words, does the psychopharmacology of MDMA-like drugs differ in a significant way from that of other drug classes?

Such a determination would ideally be approached by conducting a series of double-blind clinical trails, comparing MDMA with a variety of known substances, such as LSD, mescaline, amphetamine, and cocaine. A range of doses for each agent would be employed. In order to minimize the effects of individual variation, a large number of subjects would be studied. It is probably safe to say that such a detailed comparison will never be made. Nevertheless, there are numerous anecdotal reports that suggest that MDMA is, in fact, different from other known classes of drugs. These are, unfortunately, the best data available at the present time.

The qualitative activity of hallucinogens is generally recognized to vary from subtle at low doses to profound with progressively higher doses. The

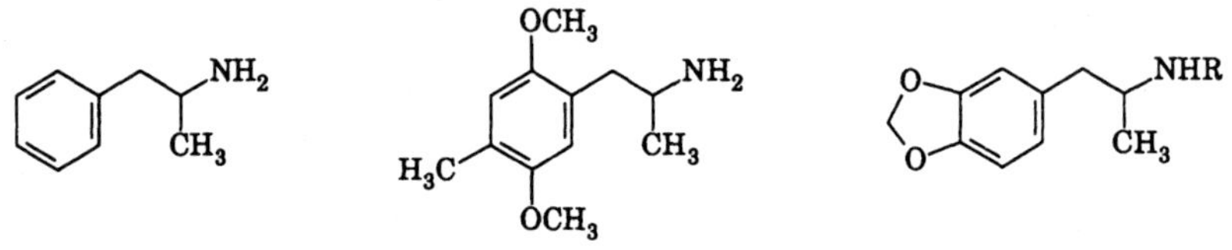

AMPHETAMINE (1) DOM (2) MDA (3) R = H

MDMA (4) $R = CH_3$

Figure 1.

distinguishing feature of these drugs, according to at least one widely used pharmacology text [4], "is their capacity to realiably induce or compel states of altered perception, thought, and feeling that are not (or cannot be) experienced otherwise except in dreams or at times of religious exaltation." Stimulant effects are characterized [5] by wakefulness, alertness, elevation of mood (often with elation and euphoria), a decreased sense of fatigue, increased initiative, self-confidence, ability to concentrate, and increased motor and speech activity.

In contrast to hallucinogens, the effects of MDMA seem to change in intensity yet remain qualitatively similar over the typical dose range of 75–200 mg. An increase in duration and side effects are noted at larger doses. Sensory disruption and loss of contact with reality have not been commonly reported with MDMA. Rather, the primary effect seems to involve enhanced closeness and communication with others, accompanied by positive changes in feelings and attitudes [6]. Like stimulants, MDMA also seems to produce increased talkativeness and mood elevation, but this is apparently not accompanied by increases in initiative, motor activity, or ability to concentrate.

1.3. Determining activity — biochemical and behavioral pharmacology

The type of information gained from anecdotal reports may be valuable in comparing subjective drug effects, yet it lacks the rigor and precision needed for reliable scientific conclusions. For that reason, behavioral and biochemical pharmacological activities have been assayed in a variety of animal models. The results of these investigations have also generally supported the hypothesis that MDMA-like activity is distinct from other known drug classes.

The following discussion of behavioral pharmacology will primarily involve results from drug discrimination (DD) experiments in rats. In this paradigm, drug "states" produced by relatively low doses can be studied in a qualitative and quantitative manner in terms of discriminative stimulus properties. The characteristic subjective effects of a specific training drug at a specific dose, time, and route of administration serve as an interoceptive cue. In the typical procedure, the presence or absence of this cue allows the subject to choose one of two possible operant responses, in this case pressing a lever in order to obtain a reward. Animals thus learn to discriminate a drug versus non-drug state through training with differential reinforcement; responses on the correct lever are rewarded, whereas responses on the incorrect lever are not.

After the animals acquire the discrimination, substitution tests with new substances are performed. Testing several animals (larger numbers give more reliable results) at several doses, a test drug can be evaluated for the degree of substitution for the training drug, based either on the percentage of tested animals selecting the drug-appropriate lever or the percentage of total responses on the drug lever. Complete substitution (80% or greater drug-appropriate responding) reflects a similarity of activity; lack of substitution (less than 60%) reflects a lack of similarity, and partial substitution (60%–79%) may reflect some degree of overlap.

Using substitution tests in the drug discrimination paradigm, an objective evaluation of the subjective effects of drugs is possible. If two compounds produce essentially similar effects (i.e., are members of the same drug class), one will fully substitute for the other at doses that produce relatively little behavioral disruption. If the profiles of action only partially overlap, complete substitution may still be observed, but relatively large doses might be required to provide the shared effects with sufficient intensity [7]. This means of classification [8] is extremely powerful in studies of centrally active drugs.

It should be kept in mind, however, that a number of variables can influence the results, such as the reinforcement schedule, the numbers of animals and doses tested, and the type of reinforcement used. Therefore, as Overton points out [9], the results of substitution tests have no fixed meaning and must be interpreted with reference to the training paradigm employed.

One of the limitations of DD worth noting is that substitution tests can only indicate whether or not a test drug is *similar* to a training drug [9]. In addition, the DD paradigm is not, in strict terms, a completely valid animal model for human activity, since false positives have been observed. However, it does represent an excellent "first guess" approach to behavioral activity, especially in those cases where the two drugs being compared have been "cross-tested" under similar conditions. If the psychopharmacology of MDMA-like compounds is novel, a series of DD experiments using a variety of training drugs can help to elucidate the nature of its pharmacological characteristics relative to known drug categories.

We have carried out extensive drug discrimination studies using rats trained to discriminate saline from LSD, saline from (+)-amphetamine, saline from (±)-MDMA, and saline from (+)-MBDB (see section 4.1). Table 1 summarizes the results of those experiments. In general, the data demonstrate the similarity between the MDMA and (+)-MBDB training cues. Individual experiments will be considered and we will refer back to the data in Table 1 numerous times in the course of the discussion in this chapter.

Table 1. Drug discrimination results of substitution testing in rats: degree of substitution is expressed as complete (ED_{50} value in μmol/kg), partial (PS), none (NS) or not tested (NT) in this laboratory. All injections were given intraperitoneally, 30 minutes prior to the session.

Test drug	Training drug			
	LSD[a]	(+)-AMP[b]	MDMA[c]	(+)-MBDB[d]
LSD	0.025	NS	PS	PS
DOM	0.61	NS	NS	NS
Mescaline	33.0	NT	NT	NS
(+)-AMP	NS	1.68	4.22	NS
Cocaine	NT	NT	13.9	PS
MDA	4.52	NS	4.06	2.09
(+)-MDA	NS	NS	1.63	1.43
(−)-MDA	2.94	NS	2.27	3.09
MDMA	NS	NS	3.40	3.35
(+)-MDMA	NS	NS	1.92	1.67
(−)-MDMA	NS	NS	5.03	3.09
MBDB	NS	NS	4.19	2.92
(+)-MBDB	NS	NS	3.67	3.28
(−)-MBDB	NS	NS	6.71	6.51

[a] LSD tartrate training dose = 0.08 mg/kg (0.186 μmol/kg).
[b] (+)-amphetamine sulfate training dose = 1.0 mg/kg (5.43 μmol/kg).
[c] MDMA hydrochloride training dose = 1.75 mg/kg (7.63 μmol/kg).
[d] (+)-MBDB hydrochloride training dose = 1.75 mg/kg (7.19 μmol/kg).

2. STRUCTURE-ACTIVITY RELATIONSHIPS — GENERAL CONSIDERATIONS

One can envision at least three areas for structural modification of MDMA. These are illustrated in Figure 2. First, the nature of the amine substituents can be varied; other N-alkyls can be studied, or the nitrogen could be incorporated into a ring system. A second point for structural modification might be the side chain. the alpha-methyl can be extended, or replaced by α,α-dialkylation or the side chain can be incorporated into a ring system fused to the aromatic nucleus to produce rigid analogues. In the following discussion, examples of these types of structural modification and their effects on activity will be presented.

3. STRUCTURE-ACTIVITY RELATIONSHIPS FOR AMINE SUBSTITUENTS

The "parent" compound MDA (3), although classified as an hallucinogenic amphetamine and available on the illicit market for about 20 years, had gained a reputation as the "love drug" [10]. It had been recognized for many years both by recreational drug users and by clinicians [11] that MDA had unique psychoactive properties that were different from hallucinogens, such as LSD or mescaline.

In animal studies, Shannon [12] concluded, based on the work accumulated at that time (1980), that although MDA had effects like both LSD and amphetamine, its mode of action might be like neither of these, and it therefore might represent a unique class of drugs. Subsequent DD studies have shown that

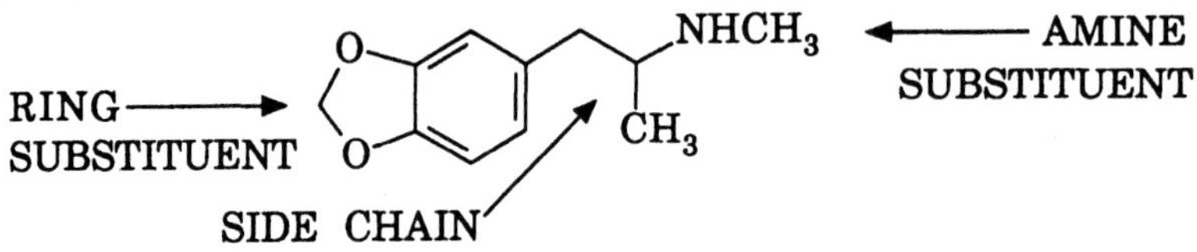

Figure 2.

MDA has stimulus properties similar to hallucinogens [1,13–16], whereas equivocal results have been obtained when the drug was compared with stimulants. Clearly, however, MDA is a psychoactive agent that is not easily categorized.

The results of DD studies in our laboraotry (Table 1) and Shannon's data [12] indicate that MDA does not substitute for (+)-amphetamine. However, complete substitution for (+)-amphetamine was reported by Glennon et al. [16]. One possible explanation for this is that MDA and amphetamine have partially overlapping pharmacological profiles (e.g., on dopaminergic systems). Depending on the paradigm used, shared effects may either be masked by the non-overlapping activity (e.g., on serotonergic systems) or may become apparent at higher doses [7]. The dose of MDA (2.75 mg/kg) required for complete substitution for (+)-amphetamine was, in fact, high and resulted in disruption of two out of the four animals tested [16]. A similar situation was observed for the complete substitution of (+)-amphetamine in MDA-trained rats [16]. This topic is discussed further in section 4.3.

While MDA, especially in high doses, appears to be hallucinogenic or psychotomimetic [17], it seems not to have been used for this action but rather for its effects on mood: production of a sense of decreased anxiety and enhanced self-awareness. Even early reports described the desire of MDA users to be with and to talk to other people [18]. MDA is also the only substituted amphetamine that received serious clinical study as an adjunct to psychotherapy [19, 20].

An important structural feature of MDMA that distinguishes it from hallucinogenic amphetamines is the fact that it is a secondary amine. That is, the basic nitrogen is substituted with an N-methyl, while hallucinogenic amphetamines are most potent as primary amines. In either 3,4,5-substituted or 2,4,5-substituted phenethylamine derivatives, N-methylation decreases hallucinogenic potency by up to an order of magnitude [17].

When MDA is ingested, the hallucinogenic effects last typically 10 to 12 hours, similar to the duration of LSD or mescaline [17]. By contrast, MDMA has a much shorter action, with perhaps a three to five hour duration of effects [21]. There is no evidence that typical doses of MDMA lead to hallucinogenic effects in a significant proportion of users, although in high dosages hallucinogenic effects have been reported [22]. Thus the simple addition of the N-methyl group limits the temporal course of the action to less than half that of MDA and attenuates or abolishes the hallucinogenic effects that occur with MDA itself.

Results from drug discrimination experiments indicate that the cues produced by MDA and MDMA are very similar, since the former completely substitutes for the latter at relatively low doses [3]. Similar results were reported for the substitution of MDMA in MDA-trained rats (1.5 mg/kg) [23] and in fenfluramine-trained rats (2 mg/kg) [24] (see section 5.3). By contrast, relatively high doses, accompanied by significant disruptive effects, were necessary for the complete substitution of (+)-amphetamine in MDMA-trained rats [3].

The drug discrimination results provide evidence for the attenuated hallucinogenic activity of MDMA relative to the primary amine, MDA. Racemic MDMA does not substitute for DOM [14] or LSD [1]. In MDMA-trained rats, DOM does not substitute, whereas LSD at relatively high doses produces partial substitution accompanied by significant behavioral disruption [3].

As with MDA, results from substitution experiments with MDMA in (+)-amphetamine-trained animals are once again equivocal, since complete substitution occurs in some paradigms [23, 25, 26] but not in others [3]. In rhesus monkeys, MDMA was found to be more like amphetamine than MDA, but unlike the training drug, (+)-amphetamine, drug-appropriate responding after both MDMA and MDA was accompanied by large decreases in response rates [26]. These data are discussed in terms of the mechanism of action of MDMA in section 4.3.

A number of investigators have examined the N-ethyl congener of MDMA, MDE (or "MDEA," 5), which seems also to have gained some popularity on the illicit market. In drug discrimination studies, MDE has pharmacological effects that are similar to those of MDMA [27]. Braun et al. [28] have reported that of the N-substituted MDA derivatives that were studied for analgesic action and human psychopharmacology, only the N-methyl (4), N-ethyl (5), and N-hydroxy (6) compounds (Figure 3) were active. The N-hydroxy (6) compound may serve merely as a prodrug for MDA, being metabolically reduced to the primary amine, as has been observed for N-hydroxy-para-chloroamphetamine [29]. It has recently been reported, however, that in drug discrimination experiments, N-OH-MDA, like MDE, failed to substitute for DOM or (+)-amphetamine, while MDA substituted for both in a similar paradigm [30]. The potential metabolic conversion of N-OH-MDA to MDA will have to be studied, especially with respect to time course, before this situation is clarified.

In a recent report by Noggle *et al.* [31], the toxicity of a series of N–alkyl-substituted MDA derivatives was reported and none exceeded the toxicity of MDA itself. Interestingly, these authors point out that the effect of N-methylation on relative toxicity serves as additional evidence that MDA-derivatives

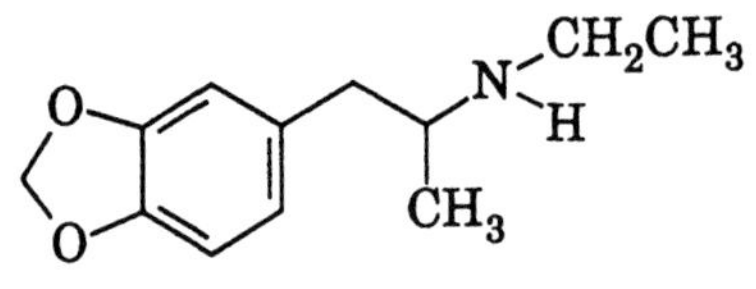

MDE (5)

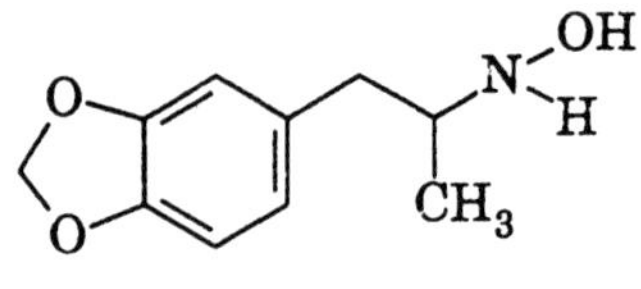

N-OH-MDA (6)

Figure 3.

alter neurotransmission through a mechanism different than amphetamine. That is, N-methylation of amphetamine to yield metamphetamine results in an increase in toxicity, whereas a decrease in toxicity occurs upon N-methylation of MDA to MDMA [32].

Since the range of modification of N-substitution seems so limited, it appears unlikely that studies of N-substituted MDA analogues will offer additional insight into mechanism of action. However, different N-alkyl groups may affect regional brain distribution and pharmacokinetic properties. For example, the N-ethyl analogue (MDE, 5) has a much shorter biological half-life in rats [27] and humans than does MDMA [17]. N,N-Dialkylation of the nitrogen has been previously reported to abolish the activity of MDA [33].

4. STRUCTURE-ACTIVITY RELATIONSHIPS FOR SIDE CHAIN MODIFICATIONS

4.1. Extension of the α-methyl to an α-ethyl

The most important support for the hypothesis that entactogens represent a unique drug class came from the discovery that the alpha-ethyl homologue of MDMA, MBDB (7), possessed MDMA-like properties in man and in the drug discrimination paradigm in rats [1, 3]. It was known that homologation of the alpha-methyl of the hallucinogenic amphetamines completely abolished hallucinogenic activity [34]. For example, the more active optical isomer of the alpha-ethyl homologue of DOM, BL-3912 (8), was evaluated by a major pharmaceutical firm and found to lack hallucinogenic activity at doses more than 100-fold higher than those effective for DOM [35]. This additional feature of the entactogens, that the alpha-ethyl homologues retained activity, was a most powerful argument that MDMA and certainly MBDB could not fit within the well-established structure-activity relationships of the hallucinogenic amphetamines. The structures of MBDB and BL-3912 are illustrated in Figure 4.

Several studies have characterized MDMA as an amphetamine or cocaine-like agent, based on its stimulus properties or its self-administration in primates [23, 25, 26, 36, 37]. It is well-known that both amphetamine and cocaine have powerful effects on dopamine pathways in the brain, and it seems likely that

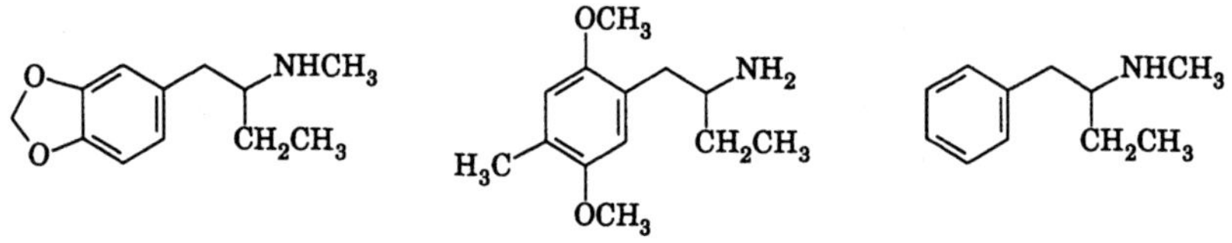

MBDB (7) BL3912 (8) (9)

Figure 4.

drugs that release dopamine or stimulate dopamine receptors have reinforcing properties that lead to self-administration and dependence liability [38].

One could not anticipate that the extension of the alpha-methyl of MDMA to an alpha-ethyl would also attenuate the effects of the compound on dopaminergic pathways in the brain. In contrast to MDMA, MBDB has no significant effect either on inhibition of uptake of dopamine into striatal synaptosomes [39] or on release of dopamine from caudate slices [40]. Drug discrimination data support this idea, since amphetamine substitutes for MDMA but not for MBDB. Furthermore, while cocaine fully substitutes in MDMA-trained rats, only partial substitution occurs in (+)-MBDB-trained rats (Table 1). This is further evidence of the decreased effect of MBDB on catecholaminergic systems. These data could be interpreted to suggest that MBDB would not be self-administered in animal models of dependence behavior and, hence, might have low abuse potential. Howver, (+)-MBDB also produces serotonin neurotoxicity in rats, although MBDB is somewhat less toxic than MDMA [Johnson, M.P. and Nichols D.E., unpublished findings].

Based on these data, it seemed likely that an alpha-ethyl moiety might attenuate the ability of other phenethylamines to interact with dopaminergic systems. To test this hypothesis, the alpha-ethyl homolog of methamphetamine (9, Figure 4) was synthesized. This compound (9) was also tested in the drug discrimination paradigm in (+)-amphetamine trained rats and compared with (+)-methamphetamine. The racemic alpha-ethyl homologue was found to possess approximatley one-tenth the potency of (+)-methamphetamine. This supported the speculation that the alpha-ethyl group was generally effective in reducing the impact of phenethylamines on dopaminergic pathways.

Thus for structure-activity studies of MDMA-like substances, emphasis has been placed on the use of (+)-MBDB as the training drug, since it seems to possess a primary psychopharmacology similar to that of MDMA but lacks the "psychostimulant" component of MDMA. That is, MBDB is pharmacologically less complex. Symmetrical transfer of the MDMA and MBDB stimuli indicates that their primary discriminative stimulus effects are very similar. Furthermore, the complete substitution of MDA for both MDMA and (+)-MBDB suggests that the parent compound also possesses similar stimulus properties.

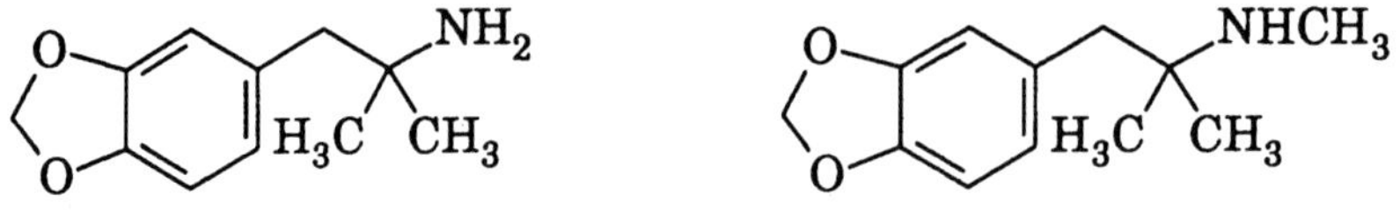

(10a) (10b)

Figure 5.

4.2. α,α-Dialkylation

Several side chain modified analogues of MDMA and MBDB have now been examined. The earliest studies were of the α,α-dimethyl analogue, 3,4-methylenedioxyphentermine (10a), and its N-methyl derivative (10b), shown in Figure 5. This latter compound proved to lack MDMA-like activity [Shulgin, A.T., unpublished findings]. Interestingly, this compound also lacked the ability to stimulate the release of [^{3}H]-serotonin from prelabeled rat brain synaptosomes [41].

4.3. Stereochemistry and mechanism of action

An important difference between MDMA and the hallucinogenic amphetamines is an observed reversal of stereoselectivity. For every substituted hallucinogenic amphetamine that has been studied, it is the isomer with the *R* absolute configuration in the side chain that is more potent in animal models, in a variety of *in vitro* assays, and in humans. The two isomers differ in potency by a factor of three to ten, depending on the assay system [42]. By contrast, it is the *S* isomer of MDMA that is more potent. This was first reported in experiments with rabbits and in clinical studies [43], and it has recently been confirmed in other animal models [3, 44].

These findings led to the proposal [43], based on the stereoselectivity for the *S*-enantiomer of MDMA, that rather than having a direct effect at serotonin receptors, perhaps MDMA was a neurotransmitter releasing agent acting in a manner similar to amphetamine, where the *S* enantiometer is also more active than the *R*. A subsequent study indicated that the *S* isomers of MDA and MDMA were indeed potent releasers of [^{3}H]-serotonin from prelabeled rat brain synaptosomes [41]. Recently, we reported that MDA and MDMA were also potent releasers of serotonin from superfused hippocampal slices prelabeled with [^{3}H]-serotonin [40]. In all studies to date, whether of release of monoamines from synaptosomes or brain slices or in inhibiting monoamine reuptake into synaptosomes [39], the *S* enantiomer of MDMA is either equipotent to the *R* isomer or is more potent.

It is difficult to trivialize the significance of this argument, since the stereoselectivity of biological receptors is accepted as a basic tenet of pharmacology. There is no rationale or experimental precedent for believing that the 3,4-methylenedioxy substitution should do anything that would cause the receptor(s) involved in hallucinogen action to accomodate a side chain stereochemistry reversed from that for phenylisopropylamines with other aromatic substituents.

It is clear from the drug discrimination results that the common activity of MDMA and MBDB is not like that of the typical hallucinogenic drugs, LSD, DOM, and mescaline. While LSD seems to be more similar to MDMA than to MBDB, transfer of the training stimulus does not occur to MDMA or MBDB in animals trained to discriminate LSD from saline [1]. Similarly, when DOM was used as a training drug, no substitution was observed for MDMA [14]. In entactogen-trained rats, DOM did not substitute for either MDMA [3] or (+)-MBDB (Table 1).

These results can be interpreted as reflecting the different mechanisms by which entactogens and hallucinogens produce their stimulus effects. Although additional experiments will be necessary to understand the neurochemical events mediating the entactogen cue, they seem to involve presynaptic actions at serotonergic neurons. The evidence for this will be discussed in detail later. By contrast, a variety of experimental data suggests that the stimulus properties of DOM are related to post-synaptic activity at 5-HT_2 receptors [45].

Various experiments have provided evidence for this conclusion. First, selective 5-HT_1 agonists do not substitute for DOM, which, along with several related derivatives, lacks high affinity for 5-HT_1 binding sites. Second, these same compounds have high affinities for 5-HT_2 sites, which are significantly correlated with both ED_{50} values for DOM-stimulus generalization and estimates of human potencies. Third, 5-HT_2 antagonists block the stimulus effects of DOM, as well as DOM-stimulus generalization to mescaline, 5-MeO-DMT, LSD, and quipazine [45]. Similar experiments also implicate a role for 5-HT_2 receptors in the stimulus properties of LSD in rats [46].

However, affinity of MDMA for 5-HT_2 receptors probably cannot be considered significant to its mechanism of action, since the more active *S* isomer has a higher K_I than does the *R* isomer [47]. In pilot experiments, the MDMA or (+)-MBDB cues were not blocked with the selective 5-HT_2 antagonist ketanserin. A similar 5-HT_2 antagonist, pirenpirone, was reportedly ineffective in reversing the disruption of operant responding in mice caused by either isomer of MDMA [48].

Although Callahan and Appel [49] reported that (+)-MDMA and (−)-MDMA substituted for mescaline, (+)-MBDB-trained rats did not recognize the mescaline cue as similar to the training drug (Table 1). Trulson et al. [50] have discussed the involvement of dopaminergic mechanisms in the behavioral effects of mescaline, and it seems possible that its combined activities at serotonin and dopamine neurons may account for a similarity to MDMA and lack of similarity to MBDB.

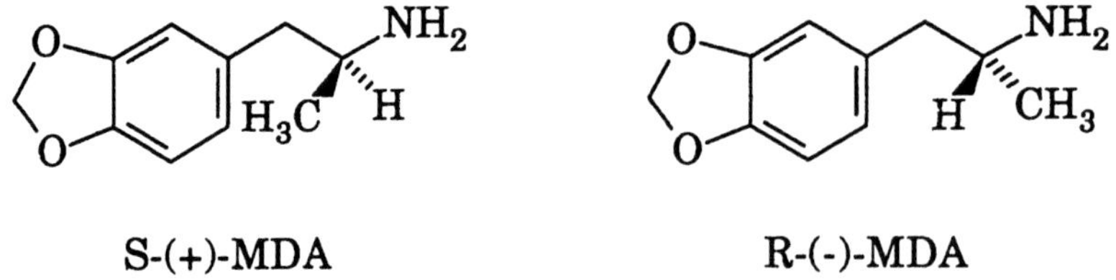

Figure 6.

The results of experiments with the isomers of the parent compound MDA (Figure 6) provide an important perspective for a discussion of stereoselectivity. Several studies have now clearly shown that it is the *R* enantiomer of MDA that has the hallucinogenic effects of the racemate, while it is the *S* enantiomer that possesses more potential MDMA-like properties in animal models (1, 3, 14, 17, 41, 43]. Although *S*-(+)-MDA sometimes appears similar to stimulants in the drug discrimination assay in rats [23, 51], it is not generally realized that the effects of (+)-MDA in humans qualitatively resemble those of MDMA rather than amphetamine [Shulgin, A.T., personal communication].

One can view this as a rather unique situation. Both enantiomers of MDA are active but differ in qualitative effect. Thus, if the psychopharmacology of (+)-MDA is like that of MDMA, then N-methylation has little effect on the entactogenic properties of this enantiomer but serves primarily to attenuate the hallucinogenic activity of *R*-(−)-MDA. Drug discrimination data provide evidence for this since (−)-MDA substitutes for hallucinogenic training drugs [1, 13, 14], whereas (−)-MDMA does not [1, 14]. However, (−)-MDA also substitutes for MDMA, and one could envision that the psychopharmacology of racemic MDA might be viewed as comprised of the hallucinogenic and entactogenic properties of the (−)-isomer and the entactogenic and psychostimulant properties of the (+)-isomer. This is a perfect example of why detailed studies of the mechanism of action of psychoactive compounds should be done with the pure optical isomers!

The net effect of (±)-MDA can really be viewed as the result of the simultaneous actions and interactions of two different drugs that happen to be enantiomers. Varying the dose of racemate can alter the psychopharmacological properties in a manner that depends on the potency of each isomer in producing its distinct activity.

Some confusion may currently exist as to which isomer of MDA is more potent, as a consequence of the differing qualitative effects of the enantiomers. In contrast to initial reports [17], there is now evidence that the activity of (+)-MDA is actually greater than that of (−)-MDA. For example, (+)-MDA was found to be the most potent compound tested in substituting for both MDMA and (+)-MBDB, although the (−)-isomer also has entactogen activity (Table

1). Similarly, when racemic MDA (1.5 mg/kg) was used as a training drug [23], (+)-MDA was found to be more potent than (−)-MDA. Thus, entactogens, as studied in rats trained to discriminate MDA [23], MDMA, or (+)-MBDB from saline, consistently demonstrate *stereoselective* action (*S* > *R*).

Similar stereoselectivity is observed for the stimulant activity of amphetamine and related compounds, such as cathinone [52]. However, the *R* isomers of MDA, MDMA, and MBDB do not substitute for (+)-amphetamine, whereas *S*-(+)-MDA and *S*-(+)-MDMA substitute for (+)-amphetamine in some tests [23, 30] but not in others [3, 51, Table 1]. Thus, the "amphetamine-like" activity of both MDA and MDMA, in cases where it has been observed, is stereo*specific* [53], rather than stereo*selective*.

Ariëns [54] has noted that if a compound has multiple pharmacological actions and if the eudismic ratios (activity of the more active stereoisomer ÷ activity of the less active stereoisomer) differ for the different effects, this indicates that these effects are based on different mechanisms involving different receptors. It seems unlikely, therefore, that MDMA-like and (+)-amphetamine-like activities are identical.

Furthermore, although the N-ethyl derivative of MDA shares stimulus properties with MDMA [27], and N-OH-MDA is reportedly similar in its actions to MDA [28], neither substitutes for (+)-amphetamine under conditions identical to those used when complete substitution of MDA and MDMA for (+)-amphetamine was reported [30]. By contrast, the N-ethyl and N-OH derivatives of amphetamine were observed to completely substitute for (+)-amphetamine [30]. Thus, either an ethyl or a hydroxy substitutent on the nitrogen abolishes the "amphetamine-like" effects of MDA but not amphetamine itself. This difference in structure-activity relationships lends further support to the concept of different mechanisms for entactogen and stimulant activities.

5. AROMATIC SUBSTITUENT-ACTIVITY RELATIONSHIPS

At the present time, little is known about requirements for particular aromatic ring substituents for a compound to possess MDMA-like activity. The largest group of substituted amphetamines with significant hallucinogenic potency possess either 3,4,5- or 2,4,5-trisubstitution patterns [42], whereas stimulant activity is generally attenuated by any aromatic substitution [53].

5.1. The 3,4-disubstitution

One of the structural features of MDMA that is somewhat unusual is the fact that it is 3,4-disubstituted. MDA, MDMA, and MBDB all possess the 3,4-methylenedioxy function, and there apparently are no other *active* compounds presently known that fall within the substituted amphetamine class that have substituents only in the 3 and 4 positions.

The isomer of MDA with a 2,3-methylenedioxy substituent has been found to completely substitute for 3,4-MDA, with a potency estimated at 20% of the

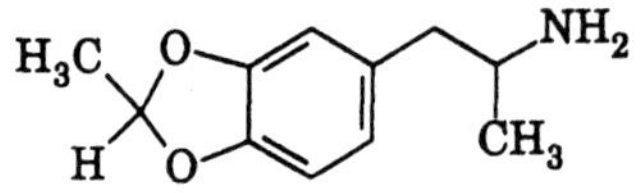

(11)

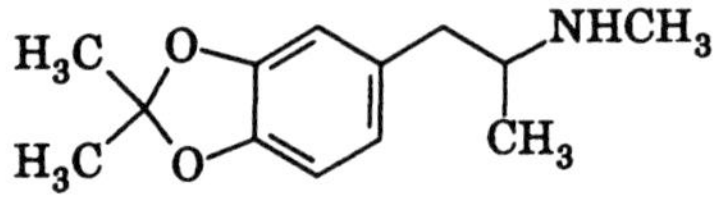

(12)

Figure 7.

latter, but it does not substitute for either DOM or (+)-amphetamine [55]. This example once again illustrates that compounds with entactogen-like activity may differ from both hallucinogens and stimulants.

5.2. Additional methoxy groups

Shulgin [33] has previously summarized the effects on hallucinogenic activity when either one or two methoxy groups are added to 3,4-MDA or 2,3-MDA. Many of these derivatives are potent hallucinogens, but they have not as yet been evaluated for MDMA-like activity. Interestingly, while the addition of an ortho-methoxy to amphetamine results in a compound that retains stimulant activity [53], the same modification to MDMA abolishes its ability to substitute for (+)-amphetamine [56].

5.3. Alkylation of the dioxole ring

Substitution on the methylene group between the two oxygens of MDMA's benzodioxole ring has been examined. The 3,4-ethylidenedioxy and 3,4-isopropylidenedioxy compounds (11 and 12, Figure 7) were tested for ability to substitute in LSD-trained or MDMA-trained rats in the drug discrimination paradigm. Both compounds gave full substitution in rats trained to either drug. Those results and comparison data for MDA are given in Table 2. Addition of steric bulk to the dioxole ring reduces but does not abolish CNS activity, whether defined as LSD-like or MDMA-like. At the training doses employed, both 11 and 12 were more potent in mimicking MDMA than LSD, based on their ED_{50} values.

Table 2. Results of drug discrimination substitution testing of dioxole-ring-alkylated MDA analogues in rats: ED_{50} values in μmol/kg (95% confidence interval).

Test drug	Training drug[a]	
	LSD	MDMA
MDA (3)	4.52 (3.11–6.57)	4.06 (2.59–6.38)
EDA (11)	13.39 (7.13–25.12)	8.09 (4.28–15.31)
IDA (12)	29.25 (14.75–57.99)	21.41 (12.51–36.66)

[a] Doses, presession time interval, and route of administration same as in Table 1.

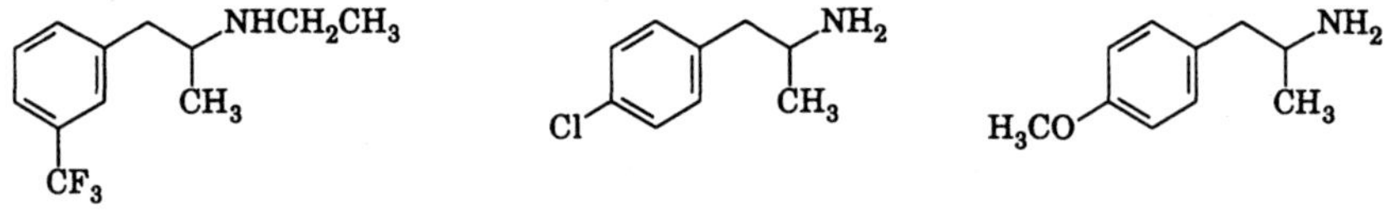

FENFLURAMINE (13) PCA (14) PMA (15)

Figure 8.

5.3. Fenfluramine — The meta-trifluoromethyl substituent

Fenfluramine (13) was the product of a search for antiobesity agents with decreased stimulant, toxic, and hypertensive effects relative to amphetamine [57]. Clinical studies with this compound have demonstrated its effectiveness in weight-loss programs and its lack of mood elevating properties [58]. In fact, fenfluramine seems to produce overall effects characterized as unpleasant, sedative, and qualitatively different from those of amphetamine [59, 60] and, apparently, from those of entactogens. Figure 8 shows the structures of fenfluramine and two other pharmacologically similar compounds with para-substituents, para-chloroamphetamine (PCA) and para-methoxyamphetamine (PMA).

In contrast to these differences in human psychopharmacology, the discriminative stimulus properties of racemic fenfluramine have been well studied and, in many respects, appear similar to those of (+)-MBDB. Fenfluramine substitutes for (+)-MBDB at relatively low doses (Table 1), and Schechter [24] has reported that MDMA substituted for fenfluramine. In studies where stimulants have been tested in fenfluramine-trained rats, no substitution occurred for (+)-amphetamine [61–63]. Furthermore, the dopamine antagonist haloperidol, which effectively blocks the discriminative stimulus properties of both (+)-amphetamine and cocaine [64], does not affect the fenfluramine cue [61]. Similarly, haloperidol was found to have no effect when given prior to the training dose of (+)-MBDB (see Table 3).

In exploratory studies, pretreatment of MDMA-trained rats with haloperidol failed to block the MDMA discriminative stimulus. These results may reflect the non-critical nature of the dopamine component in the discriminative stimulus properties of MDMA. It seems likely that the serotonergic component may be more significant to the behavioral effects. Support for this speculation is found in Schechter's work [24], where MDMA was observed to substitute for the serotonergic agent fenfluramine at lower doses and with less disruption than was observed for the substitution of MDMA for the stimulant l-cathinone.

The interpretation of experimental results involving racemic fenfluramine is complicated by the pharmacological differences between the enantiomers [65], the activity of the N-dealkylated metabolite, norfenfluramine [62, 63, 66],

Table 3. Results of attempts to block the cue produced by the training dose of (+)-MBDB (1.75 mg/kg) in drug discrimination testing.

Drug	Action	Dose (mg/kg)	Time[a]	N	Result[b]
Fluoxetine	Serotonin	2.5	90	4	100%
	Uptake	5	90	6	83%
	Inhibitor	10	90	4	100%
Metergoline	Serotonin	0.125	90	5	80%
	Antagonist	0.25	90	6	50%
		0.5	90	7	29%
Haloperidol	Dopamine	0.05	45	5	100%
	Antagonist	0.1	45	5	100%
		0.25	45	6	83%
		0.5	45	5	D[c]

[a] Presession time interval in minutes; (+)-MBDB given at the usual time of 30 minutes.
[b] Results are expressed as the percentage of rats selecting the drug lever.
[c] D = 4/5 rats were disrupted or failed to finish 50 presses on one lever in five minutes. One rat responded on the (+)-MBDB appropriate lever.

and the temporal variation in effects [67]. It does seem likely, though, that the similarity between fenfluramine and entactogens may relate to common neuronal actions. The more potent *S*-(+)-isomer of fenfluramine seems to produce its effects through a release of serotonin [68].

Since fenfluramine has been extensively studied with an increasing use of individual enantiomers, much can be learned about entactogens by comparing their effects with those of fenfluramine. An interesting example of this can be found in studies with cocaine, which, compared with (+)-amphetamine, may produce stimulus effects that are more MDMA-like. Broadbent et al. [51] reported that neither isomer of MDA substituted for (+)-amphetamine, but (+)-MDA and, to a lesser extent, (−)-MDA substituted for cocaine. As a test drug, cocaine completely substitutes for MDA [15] and MDMA but only partially substitutes for (+)-MBDB (Table 1) and fenfluramine [61].

The serotonergic properties of cocaine may account for some of these differences with (+)-amphetamine. White and Appel [61] found that fenfluramine-appropriate responding after cocaine was reduced by haloperidol and cyproheptadine. The data were interpreted to mean that cocaine partially mimicked fenfluramine through a serotonergic mechanism secondary to dopamine stimulation. Cocaine's ability to increase synaptic serotonin levels may lead to a greater similarity between its effects and those of MDA, MDMA, MBDB [39, 40], and fenfluramine [66–68]. Thus cocaine may be able to completely substitute for drugs such as MDA and MDMA, which share its serotonergic actions *and* some of its dopaminergic activity. Partial substitutions may result with compounds such as (+)-MBDB and fenfluramine, which share with cocaine the former but not the latter.

Given the apparent similarities between the in vivo and in vitro effects of fenfluramine and entactogens in rats, this would seem to imply that the dioxole ring is not essential. However, as previously mentioned, the psychopharmacology of fenfluramine is quite different from that of MDMA. While MDMA produces CNS stimulation and euphoria, fenfluramine, in common doses, is sedative and dysphoric. A detailed comparison of the pharmacology of fenfluramine and MDMA may be necessary before we understand exactly how MDMA works.

Other similarities between (+)-MBDB and fenfluramine include the partial substitution of the selective serotonin uptake inhibitor fluoxetine [61]. In addition, the serotonin antagonists cyproheptadine and methiothepin decreased fenfluramine-appropriate responding from 95% to 30% [61]. Pretreatment in (+)-MBDB-trained rats with the serotonin antagonist metergoline demonstrated a similar effect. More detailed studies with other serotonin antagonists will be necessary before conclusions may be reached. Table 3 summarizes the results of attempts to block the training cue in (+)-MBDB-trained rats.

Based on the modest ability of the (+)-isomers of MDMA and MBDB to inhibit the reuptake of norepinephrine into hypothalamic synaptosomes [39], it seemed possible that noradrenergic pathways might be involved in the cue. In another series of drug discrimination experiments designed to test this hypothesis, the specific norepinephrine uptake inhibitor (−)-tomoxetine [69] was tested for stimulus transfer in doses up to 10 mg/kg in MDMA-trained rats. At 5 mg/kg, 67% of the animals responded on the drug lever, indicating that some similarity in pharmacology may exist. However, pretreatment with tomoxetine of six rats trained to discriminate MDMA from saline had no effect on discrimination of a subsequent dose of MDMA. One may anticipate that, eventually, appropriate pharmacological manipulations will be found that will give useful information about the mechanism of action for entactogens.

5.4. PCA — the 4-Cl substituent

One might also speculate that para-chloroamphetamine (PCA, 14) would have an effect similar to MDMA. Indeed, the early clinical data for PCA suggest that it possessed antidepressant activity [70]. This would suggest that the human psychopharmacology of PCA may well be closer to that of MDMA than is fenfluramine, but it is unlikely that clinical experiments can be carried out to study this. In drug discrimination studies, PCA substitutes for both a low dose [61] and a high dose [63] of fenfluramine but has not yet been tested in entactogen-trained animals.

5.5. PMA — the 4-methoxy substituent

In another study underway in our laboratory, we have begun to examine the effect of para-methoxyamphetamine (PMA, 15) in MDMA-trained rats. Complete substitution of PMA for MDMA would be consistent with a mechanism of action for MDMA where serotonin release is important, since PMA is a potent releasing agent of serotonin both in vivo [71] and in vitro

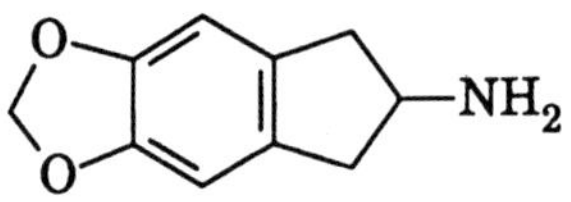

(16)

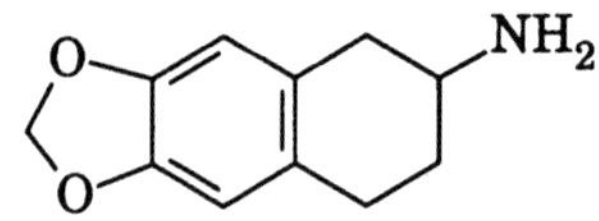

(17)

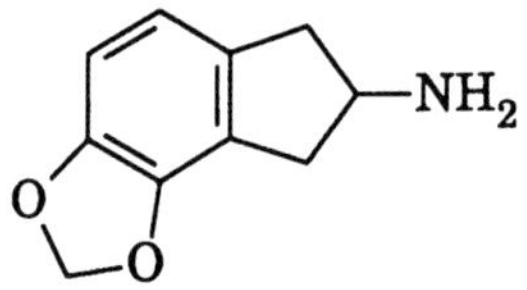

(18)

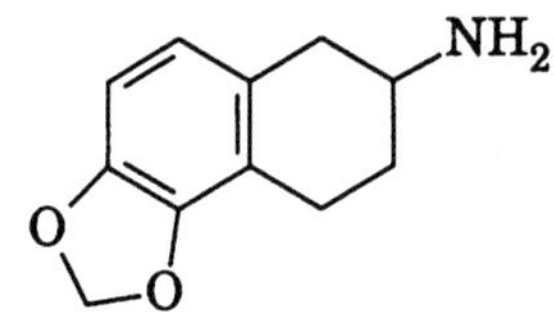

(19)

Figure 9.

[41, 72]. PMA also is a potent releaser of norepinephrine in peripheral tissues [73], but the blockade of its behavioral effects by chlorimipramine [71] suggests that serotonin release may be important in the mechanism of action. PMA did make a brief appearance on the illicit market in the early 1970s but was responsible for several deaths [74], and its use subsequently declined.

5.6. Rigid analogues

Most recently, the tetralin and indan analogues of MDA (16–19, Figure 9) have been examined. It has been previously shown that when hallucinogenic amphetamine derivatives are incorporated into similar structures, the hallucinogen-like activity in animal models is lost [75]. Thus, one might anticipate that a similar strategy with MDMA could lead to congeners that would lack MDA-like hallucinogenic effects. Furthermore, by examination of the two methylenedioxy positional isomers, one could infer the binding conformation of MDMA itself at the target site. As shown in Table 4, one positional isomer is clearly preferred for MDMA-like activity.

Furthermore, the indan derivative 16 substitutes for MDMA with a potency at least comparable to that of the training drug and was more potent than MDMA in substituting for (+)-MBDB. When administered to (+)-amphetamine-trained rats, compound 16 failed to elicit a drug lever choice from a single rat out of 29 tested at four doses. Together with a lack of substitution for LSD, the data indicate that this compound provides a good example of a drug that can potently mimic both MDMA and MBDB, while at the same time showing significant differences from hallucinogenic and stimulant training drugs.

Table 4. Drug discrimination results of substitution tests of rigid analogues of MDA in rats trained to discriminate MDMA (1.75 mg/kg) from saline.

Structure no.	ED_{50} (μmol/kg) and 95% confidence interval
16	2.75 (1.61–4.69)
17	5.68 (3.31–9.73)
18	Partial substitution
19	Partial substitution

Thus, with this series, definition has begun for some of the conformational preferences of the receptor or target sites with which MDMA interacts, at least in producing its discriminative cue. The results of these studies have also been useful for contrasting stimulant and entactogen activities. When Glennon et al. [52] tested the unsubstituted analogues of 16 and 17 in (+)-amphetamine-trained rats, 2-aminotetralin reportedly had about twice the potency of 2-aminoindan, and it was concluded that the former compound best mimics the conformation of amphetamine for producing amphetamine-like stimulus effects. The results described above for 16–19 strongly suggest that MDMA-like drugs probably adopt a different active conformation at their target site than does amphetamine.

6. DIFFERENCES BETWEEN ENTACTOGENS AND HALLUCINOGENS

6.1. EEG studies

Recently, collaborative studies with Dr. W. Dimpfel have been employed using quantitative radioelectroencephalography in the rat to characterize the EEG "fingerprint" of hallucinogenic amphetamines and of MDMA and MBDB. With this technique, four bipolar stainless steel electrodes are chronically implanted in each of four brain regions in rats: frontal cortex, hippocampus, striatum, and reticular formation [76]. The rats are freely-moving, and transmission of field potentials is accomplished using a telemetric device. The EEG is analyzed by Fourier analysis. Power density spectra are computed for periods of four seconds, are segmented into six frequency bands, and are averaged on each channel over time blocks of 15 minutes.

Using this method, a variety of hallucinogenic and non-hallucinogenic compounds were examined. As previously reported [77], hallucinogens produce a marked increase of power in the α_1 frequency (7.0–9.50 Hz) in the striatum. The ability to increase power in this region of the EEG has been observed for other classes of serotonergic drugs, including the 5-HT_{1A} agonists ipsapirone, gepirone, and buspirone, and with serotonin uptake inhibitors [78]. However, with 5-HT_{1A} agonists, an increase in α_1 power is recorded only from the frontal cortex and hippocampus.

Doses of DOM, DOB, or DOI of 0.2, 0.1, and 0.1 mg/kg, respectively, produced a pronounced and long-lasting increase in α_1 power recorded from the striatum. By contrast, doses of (+)-MDMA and (+)-MBDB up to 1.6

mg/kg did not elicit this characteristic feature in the EEG. Thus in this sensitive, quantitative EEG procedure, neither MDMA nor MBDB elicited an EEG "fingerprint" (four electrodes x six frequency bands per electrode) that resembled the fingerprint produced by the hallucinogenic amphetamines DOM, DOB, or DOI, or by LSD. These data are consistent with the results obtained in other models and further support the hypothesis that MDMA and MBDB cannot be classified as hallucinogenic phenethylamines.

6.2. Summary of structure-activity relationship differences

There are four structural features that currently contrast the structure-activity relationships of entactogens with those of hallucinogenic amphetamines.

1. Ring substitution at only the 3,4- positions does not give active hallucinogens, except for MDA. However, this substitution is active for entactogenic agents.
2. N-methylation greatly attenuates hallucinogenic activity but has no significant effect on potency of entactogens. N-ethylation also seems to allow compounds to retain entactogenic activity.
3. The more active stereochemistry of the entactogens is *S*, while that of the hallucinogenic amphetamines is *R*.
4. Extension of the alpha-methyl to an alpha-ethyl abolishes hallucinogenic activity but only has a minor effect on entactogens.

At the present time these contrasts seem sufficient to distinguish between the two drug classes. The stereochemical argument and the effects of alpha-ethylation are extremely powerful.

7. DIFFERENCES BETWEEN ENTACTOGENS AND STIMULANTS

There are several lines of evidence that distinguish entactogens from stimulants. The drug discrimination data are consistent with the view that although the dopaminergic activity of MDA and MDMA may account for similarities observed with stimulants, depending on the experimental conditions, their primary pharmacological activity is similar to that of MBDB and is probably related to effects in serotonin pathways.

Although amphetamine substitutes for MDMA in our studies, this only occurs at doses that disrupt a significant number of animals, indicative of a partially overlapping profile of action [7]. Furthermore, the large ED_{50} for amphetamine substitution in MDMA-trained rats is certainly not consistent with the known potency of amphetamine in measures of its stimulant activity. That is, in humans or in animal assays of its activity as a CNS stimulant, amphetamine is perhaps ten times more potent than MDA or MDMA. Thus its large ED_{50} in MDA-trained or MDMA-trained rats, relative to that of the enantiomers of MDA or MDMA, seems to suggest strongly that the primary discriminative cue of MDMA cannot simply be "amphetamine-like."

In animals trained to discriminate (+)-amphetamine from saline, some investigators have reported stimulus transfer with MDMA. In our paradigm, no substitution occurred. Differences in experimental design or in numbers of animals and doses tested may account for this discrepancy. Thus, in our experiments, symmetrical transfer did not occur between MDMA and amphetamine. The MDMA cue certainly seems complex and may have some similarity to amphetamine. However, suggestions that the pharmacology of MDMA is essentially the same as that of amphetamine are clearly not warranted by the data.

In summary, the following points highlight the differences between entactogens and stimulants.

1. The primary pharmacological effects of entactogens are serotonergically, not dopaminergically, mediated. This is clearly indicated by the lack of stimulant-related activity for MBDB, which can be attributed to the ability of an alpha-ethyl substituent on phenethylamines to attenuate potency at presynaptic dopamine release sites. Further evidence is provided by the ability of metergoline to block (+)-MBDB, the inability of haloperidol to block the cues produced by either MDMA or (+)-MBDB, and by the similarity between (+)-MBDB and the serotonergic drug fenfluramine.
2. A stimulant component is not necessary for entactogen activity. Drugs such as fenfluramine, MDE, 2,3-MDA, and (16) substitute for entactogens but do not substitute for (+)-amphetamine.
3. The active conformation of the side chains of MDMA and amphetamine are probably different, as indicated by the relative potencies of their respective aminoindan and aminotetralin analogues.
4. The stereo*selective* activity of entactogens in entactogen-trained animals contrasts with the stereo*specific* activity when generalization occurs in (+)-amphetamine substitution tests. Although entactogens and stimulants display similar stereoselectivity, when tested within their respective classes, the *R* isomers of MDA, MDMA, and MBDB substitute for MDMA and (+)-MBDB but not for (+)-amphetamine. In fact, contrary to other estimates of potency, the *R* isomer of MDA is more potent than the *S* isomer of amphetamine in substituting for MDMA.
5. The effects on stimulant and entactogen activities differ for N-ethyl and N-OH substitutions.
6. N-methylation leads to increased toxicity for amphetamine but decreased toxicity for MDA.

8. OUTLOOK FOR THE FUTURE

If entactogens are a distinct pharmacological class, the next question must concern the therapeutic utility of such novel agents. The term entactogen was chosen [1] after a consideration of the potential therapeutic applications of the drug class it described. The name is meant to apply to agents with MDMA-like pharmacology, but would generally apply to any substance that can produce (gen) an inner (en) "touching" (tact).

Just as the word tact, with the same Latin root tactus, is meant to imply both skill and considerateness in dealing with others and the ability to do or say the appropriate thing, entactogens should ideally produce an inner state where the patient does not feel threatened or defensive. Yet, the memory cannot be dulled as it is with benzodiazepines. Indeed, memory retrieval should be facilitated, so that the ability to recall emotionally painful, repressed memories is not impaired.

Since the neurochemistry of anxiety, depression, and other basic emotional states has not yet been elucidated, it may be quite some time before the pharmacology of entactogens is fully understood. Nevertheless, when a drug like MDMA produces a unique psychoactive effect, it should be possible, given enough time, to gain an understanding of the neuronal substrates that mediate its effect(s).

The serotonin neurotoxicity of MDA and MDMA has resulted in the need to develop new compounds that may retain clinical utility but be devoid of potentially harmful side effects. Is it possible that non-neurotoxic entactogens can be developed? As with most technologies, this is a two-edged sword. A major concern might be that a non-neurotoxic entactogen could become popular as a recreational drug. Although the possibility of neurotoxicity with MDMA should be a deterrent to potential users of this drug, it is not clear that this knowledge has had any effect on MDMA abuse. Even if a nontoxic entactogen were abused to the same extent as MDMA, at least concerns over neurological damage would be lessened. On the other hand, if clinical utility exists for MDMA-like substances, it cannot be explored until the issue of neurotoxicity is resolved. Hence, a non-neurotoxic MDMA congener would perhaps allow clinical testing of the potential of these compounds as adjuncts to psychotherapy. Should such drugs be proven efficacious for this use, the significance of this advance for psychiatry would far outweigh any concerns about abuse of entactogens.

Non-neurotoxic entactogens can and will be discovered. Sufficient evidence already exists to support this hypothesis. For example, Schechter [24] has shown that the discriminative stimulus properties of MDMA are largely dissipated by four hours following drug administration. On the other hand, Schmidt [79] found that MDMA has a biphasic depleting effect on cortical serotonin, with the later phase (>6 hours) associated with the long term toxicity and blocked by fluoxetine.

Schmidt and Taylor [80] administered the serotonin uptake inhibitor fluoxetine to rats three hours after treatment with MDMA and were able to prevent neurotoxicity. These workers suggested that the unique neurochemical effects of MDMA are independent of the long-term neurotoxicity. In our studies, it was shown that fluoxetine does not antagonize the MDMA discriminative cue. Battaglia et al. [81] reported that acute MDMA treatment decreased brain serotonin and 5-HIAA levels, but that multiple MDMA treatments were required to decrease the number of 5-HT uptake sites, the latter response presumably a reflection of neuron terminal degeneration. All these studies indicate that the acute pharmacology of MDMA can be dissociated from the long-term neurotoxic effects.

Further, it is also known from work with the neurotoxin para-chloroamphetamine that some structural congeners have an acute 5-HT depleting effect on brain 5-HT but lack the long-term neurotoxicity that is characteristic of PCA [82]. Since the psychopharmacological effects of MDMA have a relatively rapid onset and in rodents are largely dissipated at a time when a serotonin uptake inhibitor can still block neurotoxicity, it seems quite clear that molecules can be developed that will probably possess human psychopharmacology similar to MDMA but will lack serotonin neurotoxicity. When this is accomplished, we can look forward to a clear definition of the primary pharmacology of entactogens, and one would hope that at that time clinical studies with such a compound would be possible to determine, finally, whether entactogens represent a new technology for psychiatry.

ACKNOWLEDGEMENT

This research was supported in part by USPHS grants DA-02189 and DA-04758 from the National Institute on Drug Abuse and Biomedical Research Support Grant 2-507-RR05586-18.

REFERENCES

1. Nichols, D.E., Hoffman, A.J., Oberlender, R.A., Jacob III, P., and Shulgin, A.T., 1986. Derivatives of 1-(1,3-benzodioxol-5-yl)-2-butanamine: Representatives of a novel therapeutic class. J. Med. Chem. 29:2009–2015.
2. Nichols, D.E., 1986. Differences between the mechanisms of action of MDMA, MBDB, and the classical hallucinogens. Identification of a new therapeutic class: Entactogens. J. Psychoactive Drugs 18:305–313.
3. Oberlender, R. and Nichols, D.E., 1988. Drug discrimination studies with MDMA and amphetamine. Psychopharmacology 95:71–76.
4. Jaffe, J.H., 1985. Drug addiction and drug abuse. In *The Pharmacological Basis of Therapeutics*, Seventh Edition, (Gilman, A.G., Goodman, L.S., Rall, T.W., Murad, F. eds). New York: Macmillan Publishing Company, pp. 523–581.
5. Weiner, N., 1980. Norepinephrine, epinephrine, and the sypathomimetic amines. In *The Pharmacological Basis of Therapeutics*, Sixth Edition, (Gilman, A.G., Goodman, L.S. and Gilman, A., eds.) New York: Macmillan, pp. 138–175.
6. Greer, G. and Tolbert, R., 1986. Subjective reports of the effects of MDMA in a clinical setting. J. Psychoactive Drugs 18:319–327.
7. Stolerman, I.P. and D'Mello G.D., 1981. Role of training conditions in discrimination of central nervous system stimulants by rats. Psychopharmacology 73:295–303.
8. Barry, H., 1974. Classification of drugs according to their discriminable effects in rats. Fed. Proc. 33:1814–1824.
9. Overton, D.A., (1984). State dependent learning and drug discriminations. In *Handbook of Pyschopharmacology*, Vol. 18 (Iversen, L.L., Iversen, S.D. and Snyder, S.H., eds.) New York: Plenum Press, pp. 59–126.
10. Weil, A., 1976. The love drug. J. Psychoactive Drugs 8:335–337.
11. Turek, I.S., Soskin, R.A., and Kurland, A.A., 1974. Methylenedioxyamphetamine (MDA). Subjective effects. J. Psychedelic Drugs 6:7–14.

12. Shannon, H.E., 1980. MDA and DOM: Substituted amphetamines that do not produce amphetamine-like discriminative stimuli in the rat. Psychopharmacology 67:311–312.
13. Glennon, R.A., Rosecrans, J.A., and Young, R., 1981. Behavioral properties of psychoactive phenylisopropylamines in rats. Eur. J. Pharmacol. 76:353–360.
14. Glennon, R.A., Young, R., Rosecrans, J.A., and Anderson, G.M., 1982. Discriminative stimulus properties of MDA analogues. Biol. Psychiat. 17:807–814.
15. Glennon, R.A. and Young, R., 1984. MDA: An agent that produces stimulus effects similar to those of 3,4-DMA, LSD and cocaine. Eur. J. Pharmacol. 99:249–250.
16. Glennon, R.A. and Young, R., 1984. MDA: A psychoactive agent with dual stimulus effects. Life Sci. 34:379–383.
17. Shulgin, A.T., 1978. Psychotomimetic drugs: Structure-activity relationships. In *Handbook of Psychopharmacology*. Vol. 11 (Iversen, L.L., Iversen, S.D. and Snyder, S.H., eds.). New York: Plenum, pp. 243–333.
18. Jackson, B. and Reed Jr, A. 1970. Another abusable amphetamine. JAMA 211:830.
19. Yensen, R., DiLeo, F.B., Rhead, J.C., Richards, W.A., Soskin, R.A., Turek, B., and Kurland, A.A., 1974. MDA-Assisted psychotherapy with neurotic outpatients: A pilot study. J. Nerv. Ment. Dis. 163:233–245.
20. Naranjo, C., Shulgin, A.T., and Sargent, T., 1967. Evaluation of 3,4-methylenedioxyamphetamine (MDA) as an adjunct to psychotherapy. Med Pharmacol. Exp. 17:359–364.
21. Shulgin, A.T. and Nichols, D.E., 1978. Characterization of three new psychotomimetics. In *The Psychopharmacology of Hallucinogens* (Stillman, R.C. and Willette, R.E., eds). New York: Pergamon, pp. 74–83.
22. Siegel, R.K., 1986. MDMA: Nonmedical use and intoxication. J. Psychoactive Drugs 18:349–354.
23. Glennon, R.A. and Young, R., 1984. Further investigation of the discriminative stimulus properties of MDA. Pharmacol. Biochem. Behav. 20:501–505.
24. Schechter, M.D., 1986. Discriminative profile of MDMA. Pharmacol. Biochem. Behav. 24:1533–1537.
25. Evans, S.M. and Johanson, C.E., 1986. Discriminative stimulus properties of (±)-3,4-methylenedioxymethamphetamine, and (±)-3,4-methylenedioxyamphetamine in pigeons. Drug Alc. Depend. 18:159–164.
26. Kamien, J.B., Johanson, C.E., Schuster, C.R., and Woolverton, W.L., 1986. The effects of (±)-methylenedioxymethamphetamine and (±)-methylenedioxyamphetamine, in monkeys trained to discriminate (+)-amphetamine from saline. Drug Alc. Depend. 18:139–147.
27. Boja, J.W. and Schechter, M.D., 1987. Behavioral effects of N-ethyl-3,4-methylenedioxyamphetamine (MDE; "Eve"). Pharmacol. Biochem. Behav. 28:153–156.
28. Braun, U., Shulgin, A.T., and Braun, G., 1980. Centrally active N-substituted analogs of 3,4-methylenedioxyphenylisopropylamine (3,4-methylenedioxyamphetamine). J. Pharm. Sci. 69:192–195.
29. Fuller, R.W., Perry, K.W., Baker, J.C., Parli, C.J., Lee, N., Day, W.A., and Molloy, B.B., 1974. Comparison of the oxime and the hydroxylamine derivatives of 4-chloroamphetamine as depletors of brain 5-hydroxyindoles. Biochem. Pharmacol. 23:3267–3272.
30. Glennon, R.A., Yousif, M., and Patrick, G., 1988. Stimulus properties of 1-(3,4-methylenedioxyphenyl)-2-aminopropane (MDA) analogs. Pharmacol. Biochem. Behav. 29:443–449.
31. Noggle Jr, F.T., DeRuitter, J., Coker, S.T., and Clark, C.R., 1987. Synthesis, identification, and acute toxicity of some N-alkyl derivatives of 3,4-methylenedioxyamphetamine. J. Assoc. Off. Anal. Chem. 70:981–986.
32. Hardman, H.F., Haavik, C.O., and Seevers, M.H., 1973. Relationship of the structure of mescaline and seven analogs to toxicity and behavior in five species of laboratory animals. Toxicol. App. Pharmacol. 25:299–309.
33. Shulgin, A.T., 1980. Hallucinogens. In *Burger's Medicinal Chemistry*, Fourth Edition (Wolff, M.E., ed.). New York: Wiley-Interscience, pp. 1109–1137.
34. Standridge, R.T., Howell, H.G., Gylys, J.A., Partyka, R.A., and Shulgin, A.T., 1976. Phenylalkylamines with potential psychotherapeutic utility. 1. 2-Amino-1-(2,5-dimethoxy-4-methylphanyl) butane. J. Med. Chem. 19:1400–1404.
35. Winter, J.C., 1980. Effects of the Phenethylamine Derivatives, BL-3912, Fenfluramine, and Sch-12679, in rats trained with LSD as a Discriminative Stimulus. Psychopharmacology 68:159–162.

36. Beardsley, P.M., Balster, R.L., and Harris, L.L., 1986. Self-administration of methylenedioxymethamphetamine (MDMA) by rhesus monkeys. Drug Alc. Depend. 18:149–157.
37. Lamb, R.J. and Griffiths, R.R., 1987. Self-injection of d,1–3,4-methylenedioxymethamphetamine in the baboon. Psychopharmacology 91:268–272.
38. Wise, R.A. and Bozarth, M.A., 1987. A psychomotor stimulant theory of addiction. Psychol. Rev. 94:469–492.
39. Steele, T.D., Nichols, D.E., and Yim, G.K.W., 1987. Stereochemical effects of 3,4-methylenedioxymethamphetamine (MDMA) and related amphetamine derivatives on inhibition of uptake of [^{3}H]-monoamines into synaptosomes from different regions of rat brain. Biochem. Pharmacol. 36:2297–2303.
40. Johnson, M.P., Hoffman, A.J., and Nichols, D.E., 1986. Effects of the enantiomers of MDA, MDMA and related analogs on [^{3}H]serotonin and [^{3}H]dopamine release from superfused rat brain slices. Eur. J. Pharmacol. 132:269–276.
41. Nichols, D.E., Lloyd, D.H., Hoffman, A.J., Nichols, M.B., and Yim, G.K.W., 1982. Effects of certain hallucinogenic amphetamine analogues on the release of [^{3}H]serotonin from rat brain synaptosomes. J. Med. Chem. 25:530–535.
42. Nichols, D.E. and Glennon, R.A., 1984. Medicinal chemistry and structure-activity relationships of hallucinogens. In *Hallucinogens: Neurochemical, Behavioral, and Clinical Perspectives* (Jacobs, B. ed.). New York: Raven Press, pp. 95–142.
43. Anderson III, G.M., Braun, G., Braun, U., Nichols, D.E., and Shulgin, A.T., 1978. Absolute configuration and psychotomimetic activity. In *QuaSAR Quantitative Structure Activity Relationships of Analgesics, Narcotic Antagonists, and Hallucinogens* (Barnett, G., Trsic, M. and Willette, R.E., eds). National Institute on Drug Abuse Research Monograph 22. DHEW Pub. No. (ADM) 78–729. Washington, DC: Supt of Documents, U.S. Government Printing Office, pp. 27–32.
44. Schechter, M.D., 1987. MDMA as a discriminative stimulus: Isomeric comparisons. Pharmacol. Biochem. Behav. 27:41–44.
45. Glennon, R.A., Titeler, M., and Young, R., 1986. Structure-activity relationships and mechanism of action of hallucinogenic agents based on drug discrimination and radioligand binding studies. Psychopharmacol. Bull. 22:953–958.
46. Appel., J.B. and Cunningham, K.A., 1986. The use of drug discrimination procedures to characterize hallucinogenic drug actions. Psychopharmacol. Bull. 22:959–969.
47. Lyon, R.A., Glennon, R.A., and Titeler, M., 1986. 3,4-Methylenedioxymethamphetamine (MDMA): Stereoselective interactions at brain 5-HT_1 and 5-HT_2 receptors. Psychopharmacology 88:525–526.
48. Rosecrans, J.A. and Glennon, R.A., 1987. The effect of MDA and MDMA ("ecstasy") isomers in combination with pirenpirone on operant responding in mice. Pharmacol. Biochem. Behav. 28:39–42.
49. Callahan, P.M. and Appel, J.B., 1987. Differences in the stimulus properties of 3,4-methylenedioxyamphetamine (MDA) and N-Methyl-1-(3,4-methylenedioxyamphetamine) (MDMA) in animals trained to discriminate hallucinogens from saline. Soc. Neurosci. Abst., p. 1720 (476.2).
50. Trulson, M.E., Crisp, T., and Henderson, L.J., 1983. Mescaline elicits behavioral effects in cats by action at both serotonin and dopamine receptors. Eur. J. Pharmacol. 96:151–154.
51. Broadbent, J., Michael, E.K., Ricker, J.H., and Appel, J.B., 1987. A comparison of the discriminative stimuli of (+) and (−)-3,4-methylenedioxyamphetamine (MDA) with those of hallucinogenic and stimulant drugs. Soc. Neurosci. Abstract, p. 1720 (476.1).
52. Glennon, R.A., Young, R., Hauck, A.E., and McKenney, J.D., 1984. Structure-activity studies on amphetamine analogues using drug discrimination methodology. Pharmacol. Biochem. Behav. 21:895–901.
53. Young, R. and Glennon, R.A., 1986. Discriminative stimulus properties of amphetamine and structurally related phenalkylamines. Med. Res. Rev. 6:99–130.
54. Ariëns, E.J., 1987. Stereochemistry in the analysis of drug action. Part II. Med. Res. Rev. 7:367–387.
55. Glennon, R.A., Young, R., and Soine, W., 1984. 1-(2,3-Methylenedioxyphenyl)-2-aminopropane (2,3-MDA): A preliminary investigation. Gen. Pharmacol. 15:361–362.
56. Glennon, R.A., Yousif, M., Naiman, N., and Kalix, P., 1987. Methcathinone: A new and potent amphetamine-like agent. Pharmacol. Biochem. Behav. 26:547–551.

57. Beregi, L.G., Hugon, P., LeDouarec, J.C., Laubie, M., and Duhault, J., 1970. Structure-activity relationships in CF_3 substituted phenethylamines. in *Amphetamines and Related Compounds* (Costa, E. and Garattini, S. eds.). New York: Raven Press, pp. 21–61.
58. Woodward Jr, E., 1970. Clinical experience with fenfluramine in the United States. In *Amphetamines and Related Compounds* (Costa, E. and Garattini, S., eds). New York: Raven Press, pp. 21–61.
59. Griffith, J.D., Nutt, J.G., and Jasinski, D.R., 1975. A comparison of fenfluramine and amphetamine in man. Clin. Parmacol. Ther. 18:563–570.
60. Chait, L.D., Uhlenhuth, E.H., and Johanson, C.E., 1986. The discriminative stimulus and subjective effects of d-amphetamine, phenmetrazine and fenfluramine in humans. Psychopharmacology 89:301–306.
61. White, F.J. and Appel, J.B., 1981. A Neuropharmacological analysis of the discriminative stimulus properties of fenfluramine. Psychopharmacology 73:110–115.
62. Goudie, A.J., 1977. Discriminative stimulus properties of fenfluramine in an operant task: An analysis of its cue function. Psychopharmacology 53:97–102.
63. McElroy, J.F. and Feldman, R.S., 1984. Discriminative stimulus properties of fenfluramine: Evidence for serotonergic involvement. Psychopharmacology 83:172–178.
64. Colpaert, F.C., Niemergeers, C.J.E., and Janssen, P.A.J., 1978. Discriminative stimulus properties of cocaine and d-amphetamine, and antagonism by haloperidol: A comparative study. Neuropharmacolology 17:937–942.
65. Invernizzi, R., Berettera, C., Garattini, S., and Samanin, R., 1986. D- and L-isomers of fenfluramine differ markedly in their interaction with brain serotonin and catecholamines in the rat. Eur. J. Pharmacol. 120:9–15.
66. Borroni, E., Ceci, A., Garattini, S., and Mennini, T., 1983. Differences between d-fenfluramine and d-norfenfluramine in serotonin presynaptic mechanisms. J. Neurochem. 40: 891–893.
67. McElroy, J.F., DuPont, A.F., and Feldman, R.S., 1982. The effects of fenfluramine and fluoxetine on the acquisition of a conditioned avoidance response in rats. Psychopharmacology 77:356–359.
68. Fuller, R.W., Snoddy, H.D., and Robertson, D.W., 1988. Mechanisms of effects of d-fenfluramine on brain serotonin metabolism in rats: Uptake inhibition versus release. Pharmacol. Biochem. Behav. 30:715–721.
69. Wong, D.T., Threlkeld, P.G., Best, K.L., and Bymaster, F.P., 1982. A new inhibitor of norepinephrine uptake devoid of affinity for receptors in rat brain. J. Pharmacol. Exp. Ther. 222:61–65.
70. Verster, J. and Van Prag, H.M., 1970. A comparative investigation of methylamphetamine and 4-chloro-N-methylamphetamine in healthy test subjects. Pharmako-Psychiatrie Neuropsychopharmacologie. 3:239–248.
71. Tseng, L–F, Harris, R.A., and Loh, H.H., 1978. Blockade of para-Methoxyamphetamine-induced serotonergic effects by chlorimipramine. J. Pharmacol. Exp. Ther. 204:27–38.
72. Tseng, L–F, Menon, M.K., and Loh, H.H., 1976. Comparative actions of monomethoxy-amphetamines on the release and uptake of biogenic amines in brain tissue. J. Pharmacol. Exp. Ther. 197:263–271.
73. Cheng, H.C., Long, J.P., Nichols, D.E., and Barfknecht, C.F. 1974. Effects of para-methoxyamphetamine (PMA) on the cardiovascular system of the dog. Arch. Int. Pharmacodyn. Ther. 212:83–88.
74. Cimbura, G. 1974. PMA Deaths in Ontario. Can. Med. Ass. J. 110:1263–1265.
75. Nichols, D.E., Barfknecht, C.F., Long, J.P., Standridge, R.T., Howell, H.G., Partyka, R.A., and Dyer, D.C., 1974. Potential psychotomimetics 2: Rigid analogs of 2,5-Dimethoxy-4-methylphenylisopropylamine (DOM, STP). J. Med. Chem. 17:161–166.
76. Dimpfel, W., Spüler, M., Nickel, B., and Tibes, U., 1986. "Fingerprints" of Central Stimulatory Drug Effects by Means of Quantitative Radioelectroencephalography in the Rat (Tele-Stereo-EEG). Neuropsychobiology 15:101–108.
77. Spüler, M. and Nichols, D.E., 1988. Effects of the hallucinogenic drugs LSD, DOM and Scopolamine on the frequency content of field potentials from the rat brain (Tele-Stereo-EEG). Deutsch Gesellschaft für Pharmakologie and Toxikologie., Abstract 451.
78. Dimpfel, W., Spüler, M., Traber, J., and Nichols, D.E., 1988. Tele-Stereo-EEG in the rat after injection of drugs interacting with serotonergic transmission. Intl. Pharmacol. EEG

Group Symposium, Abstracts, Kobe, Japan.
79. Schmidt, C.J., 1987. Neurotoxicity of the psychedelic amphetamine, methylenedioxymethamphetamine. J. Pharmacol. Exp. Ther. 240:1–7.
80. Schmidt, C.J. and Taylor, V.L., 1987. Acute Effects of methylenedioxymethamphetamine (MDMA) on 5-HT synthesis in the rat brain. Pharmacologist 29:(224).
81. Battaglia, G., Yeh, S.Y., and DeSouza, E.G., 1988. MDMA-Induced neurotoxicity: Parameters of degeneration and recovery of brain serotonin neurons. Pharmacol. Biochem. Behav. 29:269–274.
82. Fuller, R.W., Wong, D.T., Snoddy, H.D., and Bymaster, F.P., 1977. Comparison of the effects of 6-chloro-2-aminotetralin and of Org 6582, a related chloroamphetamine analog, on brain serotonin metabolism in rats. Biochem. Pharmacol. 26:1333–1337.

8. NEUROCHEMICAL EFFECTS OF MDMA

JAMES W. GIBB, DONNA STONE, MICHEL JOHNSON, AND GLEN R. HANSON

1. INTRODUCTION

Since 1971 we have extensively investigated the neurochemical effects of amphetamine and related congeners. Early in those studies, we observed that methamphetamine, given in large repeated doses (10–15 mg/kg, s.c., every six hours for five doses), caused a dose-related decrease in tyrosine hydroxylase (TH) activity in the neostriatum [1, 2] and substantia nigra [3]. A parallel decline in concentrations of dopamine (DA) and its metabolites, dihydroxyphenylacetic acid (DOPAC) and homovanillic acid (HVA) [4], accompanied the decrease in enzyme activity.

We first suspected that dopamine may be involved in the response to methamphetamine when dopamine antagonists prevented the methamphetamine effects [5, 6]. More convincing evidence for the role of dopamine was obtained when we observed that inhibition of dopamine synthesis with α-methyl-p-tyrosine (MT), administered concurrently, prevented the methamphetamine-induced decline in tyrosine hydroxylase activity and dopamine content. When the inhibited step in the biosynthesis of dopamine was circumvented by administering L-DOPA and a peripheral decarboxylase inhibitor, the methamphetamine-induced decrease in tyrosine hydroxylase and dopamine content recurred [7].

Additional evidence for the possible role of dopamine in the methamphetamine-induced response was obtained by employing the dopamine uptake inhibitor, amfonelic acid. When amfonelic acid was administered concurrently with methamphetamine, neither tyrosine hydroxylase activity nor dopamine content was compromised [4].

These experiments provided evidence that dopamine is involved in the methamphetamine-induced response. Whether dopamine or a reactive metabolite(s) is responsible for the apparent neurotoxicity associated with methamphetamine administration needs further study.

2. RESULTS AND DISCUSSION

2.1. Methamphetamine effects on the serotonergic and other neurotransmitter systems

We considered whether methamphetamine in large doses caused a generalized effect on all transmitter systems or whether specific systems were selectively affected. The activity of enzymes served as a marker to assess the effect of methamphetamine on these neurotransmitter systems. Repeated large doses of methamphetamine did not alter neostriatal acetylcholinesterase nor glutamic acid decarboxylase activity, which suggested that neither the cholinergic nor the GABAergic systems, respectively, were adversely affected by methamphetamine [8].

In many of the brain regions examined, tryptophan hydroxylase (TPH) activity and concentrations of 5-hydroxytryptamine (5-HT) and 5-hydroxyindoleacetic acid (5-HIAA) were decreased rapidly after a single dose of methamphetamine [9,10]. When only one dose was administered, the serotongeric parameters returned to normal within two weeks. However, when repeated doses of methamphetamine were administered, the effects of methamphetamine were more pronounced and persisted for as long as 110 days after the fifth and final dose of the drug [11]. The rate and extent of recovery varied for each brain region.

We characterized further the effects of methamphetamine on the serotonergic system by determining whether agents that prevent or attenuate the response in the dopaminergic systems had a similar protective effect in the serotongic system. As in the dopaminergic system, haloperidol blocked the effect of repeated doses of methamphetamine on the serotonergic system [9]. Surprisingly, when dopamine synthesis was inhibited with MT, methamphetamine had no effect on neostriatal or hippocampal TPH activity [9]. When synthesis of dopamine was reinstated by administering L-DOPA and a peripheral decarboxylase inhibitor, the decline in TPH activity and content of 5-HT and 5-HIAA returned to control levels [12]. These data suggested that dopamine is essential for the methamphetamine-induced neurotoxicity in the serotongic system.

Involvement of dopamine in the serotonergic response was further defined by disrupting innervation to a specific brain area and determining whether that dopamine-depleted area, but not regions with intact dopaminergic input, was selectively protected from the methamphetamine [13]. 6-Hydroxydopamine (6-0H-DA) was injected bilaterally into the substantia nigra and methamphetamine was administered 10–14 days later. The decline in neostriatal TPH activity observed in the nonlesioned rat after methamphetamine, was prevented in the 6OHDA-lesioned neostriatum. In the frontal cortex, however, the decrease in enzyme activity persisted, while in the hippocampus there was an attenuation of the methamphetamine effect. We concluded that dopaminergic innervation is essential for the neurotoxicity in the serotonergic system.

The methamphetamine-induced decrease in TPH activity was prevented by a dopamine uptake inhibitor, amfonelic acid [4]. Moreover an inhibitor of 5-HT uptake, fluoxetine [9, 14], was also effective in blocking the methamphetamine-induced decrease in TPH activity.

In summary, methamphetamine compromises both the dopaminergic and serotonergic systems. Because the effects of multiple doses of methamphetamine persist long after the drug administration is discontinued, we suggest that methamphetamine given in large doses is neurotoxic to both the dopaminergic and serotonergic systems. Histological changes [15, 16], as well as impairment of DA uptake [17] by these drugs, provide additional evidence of a neurotoxic response. It appears that dopamine and/or its reactive metabolite(s), may be responsible for the neurotoxic response in both systems.

2.2. Studies with designer drugs

Seiden and his coworkers [18] reported that 3,4-methylenedioxyamphetamine (MDA) caused a long-lasting decrease in rat neostriatal 5-HT uptake and in 5HT and 5-HIAA content; alterations of Fink–Heimer staining suggested that nerve terminal degeneration occurred.

We investigated the effects of 3,4-methylenedioxymethamphetamine (MDMA) or MDA on dopaminergic and serotonergic systems, by employing techniques previously used in our laboratory to determine alterations of the two transmitter systems after methamphetamine. Three hours after a single injection of either drug (10 mg/kg), neostriatal TPH activity was markedly decreased (Figure 1, Ref. 19). MDMA also reduced enzyme activity in the hippocampus (52% of control) and cerebral cortex (30% of control). Similar responses were observed for MDA. In contrast to methamphetamine, tyrosine hydroxylase activity was unaltered by either drug. MDMA or MDA also caused decreases in 5-HT and 5-HIAA concentrations in the three brain areas (Figure 2). When repeated doses of either agent (10 mg/kg) were administered and rats were sacrificed 18 hours after the last dose, the decrease in TPH activity was further depressed as were the concentrations of 5-HT and 5-HIAA in the three brain regions (data not shown).

In rats that received a single dose of MDMA or MDA, neostriatal dopamine concentrations were elevated to approximately 140% of control; MDMA increased homovanillac acid and MDA decreased dihydroxyphenylacetic acid concentrations (data not shown). It is interesting that although tyrosine hydroxylase activity was not altered, there were transient changes of dopamine and its metabolites. These observations provide evidence that although MDMA and MDA may not cause neurotoxicity in the dopamine system as

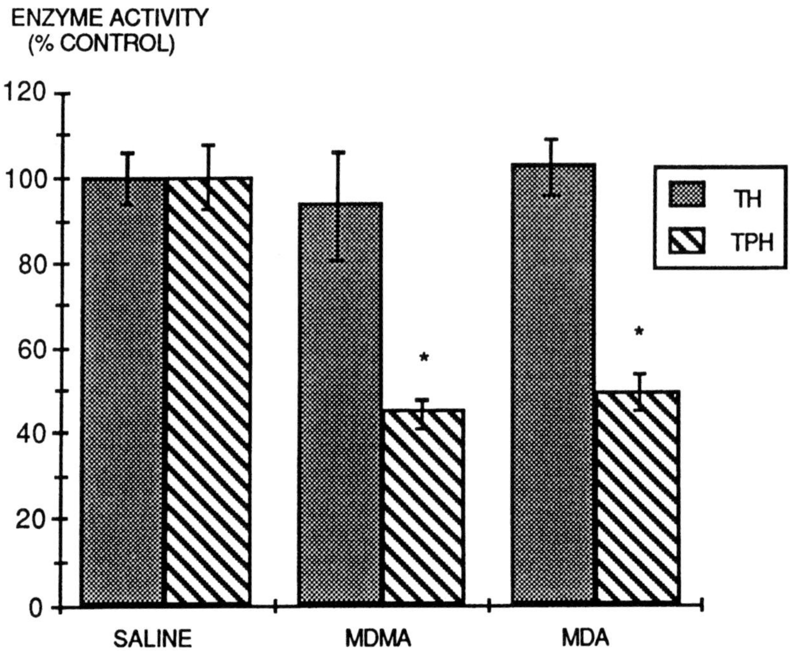

Figure 1. Effect of acute drug treatments on neostriatal tyrosine hydroxylase (TH) and trytophan hydroxylase (TPH) activities. Rats were killed 3 hours after a single 10 mg/kg injection of MDA or MDMA. Results are presented as the means ± SEM (n=6) and expressed as percent control. Control values for TH and TPH activities were 2178.8 and 38.2 nmol/g tissue per hour, respectively. *P < 0.001 versus control, by the two-tailed Student's t-test. (After Stone et al. [19]. Courtesy Eur. J. Pharmacol.)

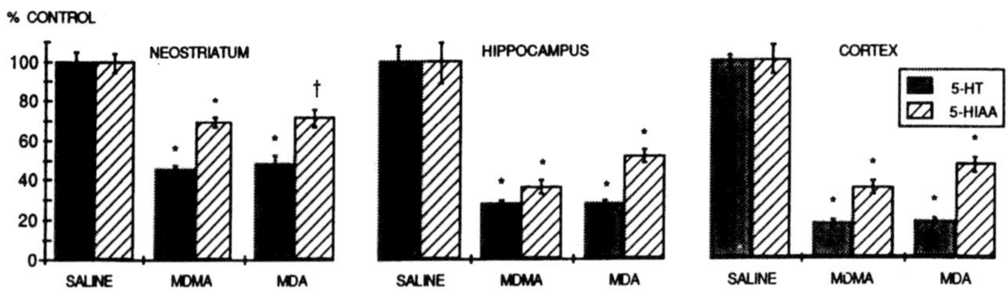

Figure 2. Effect of acute drug treatments on 5-hydroxytryptamine (5-HT) and 5-hydroxyindoleacetic acid (5-HIAA) concentrations. Experimental conditions are described in Figure 1. The means ± SEM from 6 animals are presented as percent of control. Control values (in μg/g tissue), with 5-HT concentration listed first, were: neostriatum, 0.421 and 0.517; hippocampus, 0.277 and 0.375; cortex, 0.487 and 0.267. *P < 0.001, †P < 0.005 versus corresponding control, by the two-tailed Student's t-test (After Stone et al. [19]. Courtesy Eur. J. Pharmacol.)

defined by decreases in tyrosine hyroxylase activity, there is, however, a significant effect on dopamine metabolism, as evidenced by the alteration of the content of dopamine and its metabolites.

These responses to MDMA and MDA are strikingly similar to those observed after another amphetamine analogue, p-chloroamphetamine [20–24]. After a single dose of p-chloroamphetamine, it was discovered by Sanders-Bush et al. [22] that decreases in 5-HT and 5-HIAA concentrations occurred, which persisted for at least four months; TPH activity was also decreased after p-choloamphetamine.

2.3. Isomers of MDMA and MDA

We [25] compared the effects of the two isomers of MDMA and MDA on brain serotonergic parameters. Three doses (3.5, 5.0, or 10.0 mg/kg, every six hours for five administrations) of either isomer were administered; TPH activity and 5-HT and 5-HIAA content were determined in the neostriatum, hippocampus, and frontal cortex. Both isomers of each drug caused qualitatively similar but quantitatively different responses. The d-isomer was more potent than the l-isomer of MDMA, at both the 5 and 10 mg/kg dose, in decreasing TPH activity in all three brain areas (Figure 3). There was no quantitative difference between the effects of the d- and l-isomers of MDA on TPH activity.

There were parallel quantitative differences in the effects of the two isomers of MDMA on 5-HT and 5-HIAA concentrations. The d-isomer was more potent than the l-isomer of MDMA at the two higher doses in all three brain regions (data not shown). In those areas where there was a quantitative difference in the decrease of 5-HT and 5-HIAA concentrations caused by 3.5 mg/kg of MDA, the d-isomer was more potent than the l-isomer. At higher doses, the isomers of MDA were equipotent.

In summary, when quantitative differences do occur, the d-isomer of MDMA is more potent than the l-isomer.

2.4. N-ethyl-3,4-methylenedioxyamphetamine

We were interested as to how the response of other congeners of MDMA might differ from that observed for the parent compound. Like MDMA and MDA, the N-ethylated derivative of MDA, N-ethyl-3,4-methylenedioxyamphetamine (MDE) decreased TPH activity and lowered concentrations of 5-HT and 5-HIAA in the various brain areas; moreover, the N-ethylated analog did not alter tyrosine hydroxylase activity [26, 27]. Interestingly, MDE was much less potent than MDMA or MDA. Three hours after a single dose of MDE, neostriatal TPH activity was decreased to approximately 70% of control (Figure 4); neostriatal enzyme activity three hours after MDMA was normally depressed to approximately 45% of control (data not shown).

The rate of recovery of TPH activity in concentrations of 5-HT and 5-HIAA, after multiple doses of MDE, was more rapid than after MDMA or

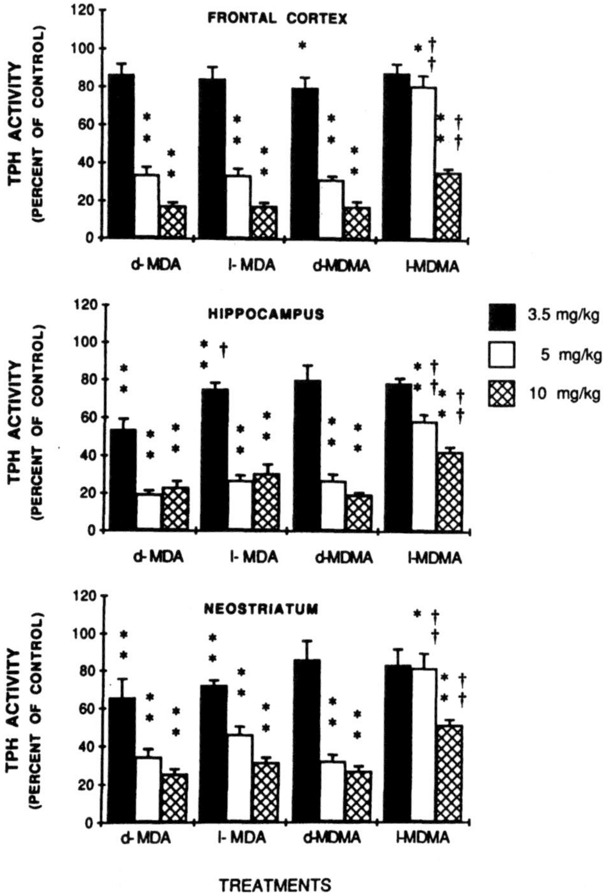

Figure 3. Effects of MDA and MDMA isomers on TPH activities within frontal cortex, hippocampus, and neostriatum. Isomers of MDMA or MDA (3.5, 5, or 10 mg/kg s.c.) were administered for 5 doses at 6-hour intervals, and animals were killed 18 hours later. Results are expressed in percentage of control values (saline treatment) and represent means ± SEM of 6–18 rats/group. Actual control values for the 10 mg/kg treatment follow: frontal cortex, 66.5 ± 1.9 nmol tryptophan oxidized/hr/g tissue; hippocampus, 54.5 ± 2.9 nmol tryptophan oxidized/hr/g tissue; neostriatum, 49.5 ± 2.4 nmol tryptophan oxidized/hr/g tissue. $^{*}P < .05$, $^{**}P < .01$ versus respective control; $^{\dagger}P < .05$, $^{\dagger\dagger}P < .01$ versus corresponding d isomer group. (After Johnson et al. [25]. Courtesy J. Pharmacol. Exp. Ther.)

MDA [26, 27]. With MDE there was significant recovery within 18 hours after the last of five doses of the drug (Figure 5); however, there was no evidence of recovery 18 hours after multiple doses of MDMA or MDA (data not shown). MDE is less potent and the effects are more short-lived than for MDMA and MDA. It is interesting that MDE is less potent than MDMA or MDA in releasing dopamine (28).

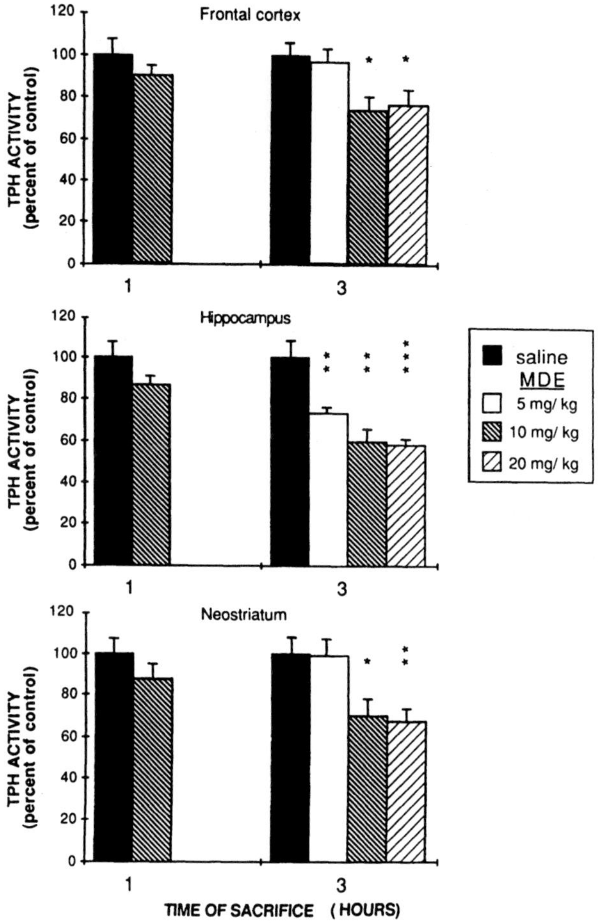

Figure 4. Effects of MDE (10 mg/kg) 1 hour after a single injection and of MDE (5, 10, or 20 mg/kg) 3 hours after the injection on the frontal cortex, hippocampal, and neostriatal TPH activity. The enzymatic activities are expressed as percent ± SEM of the control group (injected with saline) for the respective time of sacrifice. Enzymatic activities of the control groups (3 hours), expressed in nmol of hydroxylated tryptophan/hr/g of tissue, were: 68.0 ± 4.3 in the frontal cortex; 52.0 ± 4.5 in the hippocampus; and 49.5 ± 4.0 in the neostriatum. Statistical analyses of the enzymatic activity between means of the MDE groups and control were performed with Student's t-test. *$P < 0.05$, **$P < 0.01$, and ***$P < 0.001$ ($N=6$). (After Johnson et al. [27]. Courtesy Biochem. Pharmacol.)

2.5. Immediate and long-term effects of MDMA

The immediate and long-term responses of serotonergic parameters in four different brain regions at varying times after a single dose of MDMA are illustrated in Figure 6 [29]. TPH activity declined significantly in the frontal cortex within 15 minutes and in the neostriatum, hippocampus, and hypo-

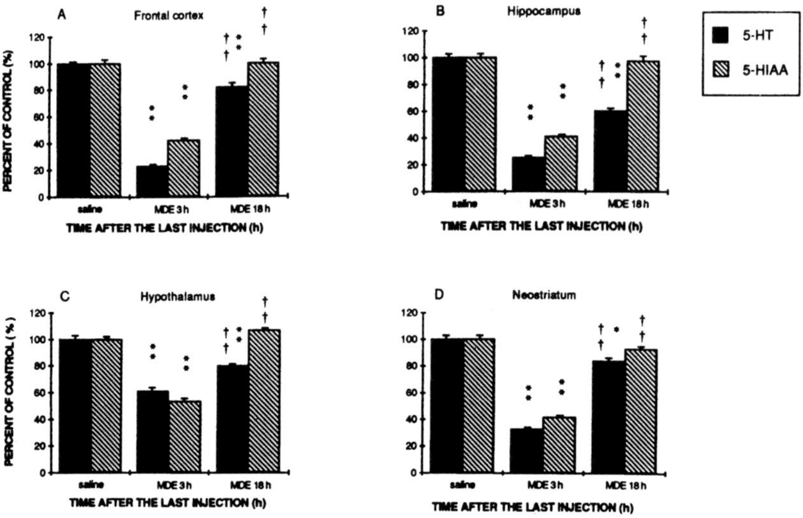

Figure 5. Effects of multiple administrations of MDE (10 mg/kg) on 5-HT and 5-HIAA concentrations 3 and 18 hours after treatment. The means of the 5-HT and 5-HIAA concentrations in the frontal cortex (A), hippocampus (B), hypothalamus (C), and neostriatum (D) are expressed in a percentage of control (saline) ± SEM. As there was no significant difference between the two control groups at each time of sacrifice, these determinations were combined and expressed as a single control group in order to simplify the figure. Means of 5-HT concentrations of the control (saline) group, expressed as μg/g tissue, were: 0.62 ± 0.01 in the frontal cortex, 0.40 ± 0.02 in the hippocampus, 0.86 ± 0.04 in the hypothalamus, and 0.43 ± 0.02 in the neostriatum. Concentrations of control 5-HIAA were: 0.19 ± 0.01 in the frontal cortex, 0.32 ± 0.02 in the hippocampus, 0.45 ± 0.02 in the hypothalamus, and 0.41 ± 0.02 in neostriatum. Statistical analyses were performed with a one-way ANOVA test, while a Student–Newman–Keuls test was used for the multiple comparisons analysis. Key: *P < 0.05, and **P < 0.01 versus respective control, and ††P < 0.01 versus corresponding 3-hour group (N = 16–18 for control, N = 6 for the 3-hour group, N = 13–14 for the 18-hour group). (After Johnson et al. [27]. Courtesy of Biochem. Pharmacol.)

thalamus within 60 minutes; 5-HT content followed a similar pattern, while 5-HIAA content declined at a slower rate than the other two serotonergic parameters. Within three to six hours after a single dose of MDMA, all parameters had reached their nadir. There was a rebound towards normal in 5-HT and 5-HIAA concentrations between six and 24 hours in all areas. In the hippocampus and frontal cortex, there was a secondary decline in 5-HT concentrations. The recovery of TPH activity in all three brain regions was slower than for the other two serotonergic parameters.

After a single dose of MDMA (10 mg/kg), dopamine was elevated within one hour and persisted at elevated levels for at least three hours; homovanillac acid concentrations were significantly elevated by three hours. In contrast, dihydroxyphenylacetic acid concentrations were immediately decreased, but returned to normal by 24 hours (data not shown).

When assessing the potential neurotoxicity of MDMA, the long-term

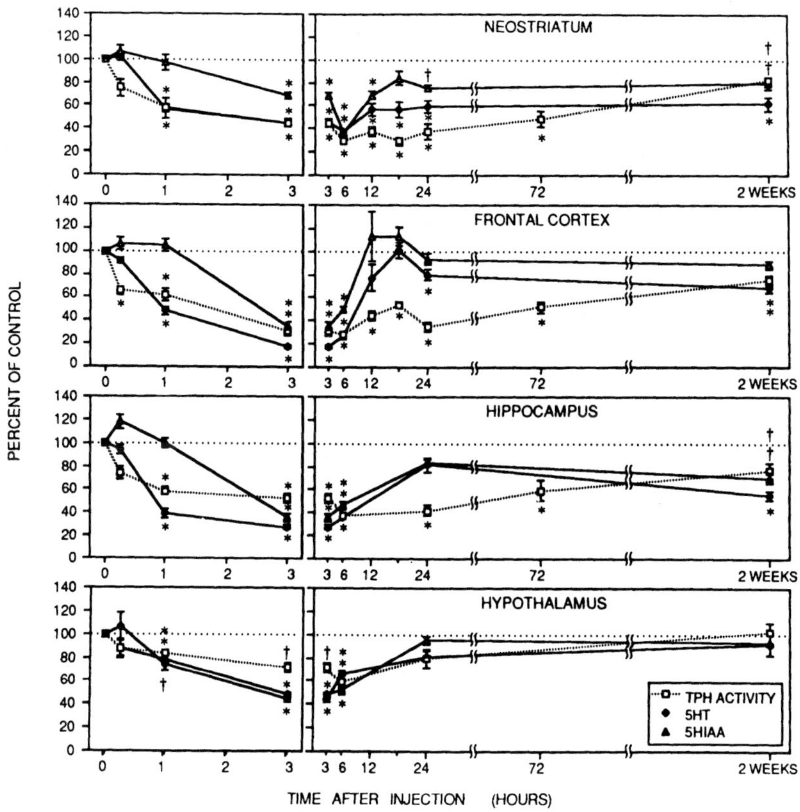

Figure 6. Time course of the regional serotonergic effects of acute administration of MDMA. A single dose of MDMA (10 mg/kg) or saline (control) was injected subcutaneously; rats were killed at specified times thereafter. Each point represents the mean ± SEM from 4–6 rats, expressed as a percentage of the corresponding control. Immediate effects (up to 3 hours after injection) are represented in the left panel; the right panel diagrams longer-term regional responses (from 3 hour-2 weeks) after injection). One hour control values ± SEM for neostriata (n), frontal cortex (fc), hippocampus (h), and hypothalamus (ht) were as follows: activity of tryptophan hydroxylase (TPH) (in nmol/g tissue/hr); n = 39.9 ± 3.6, fc = 80.1 ± 4.9, h = 63.0 ± 2.6, ht = 249.3 ± 7.0; concentrations of 5-HT and 5-HIAA, respectively (in μg/g tissue): n = 0.533 ± 0.019 and 0.557 ± 0.023, fc = 0.518 ± 0.018 and 0.218 ± 0.013, h = 0.359 ± 0.040 and 0.338 ± 0.009, ht = 0.905 ± 0.064 and 0.396 ± 0.018. Control values at other times did not vary significantly from those listed above. †$P < 0.05$, *$P < 0.005$ versus corresponding control by the two-tailed Student's t-test. (After Stone et al. [29]. Courtesy of Neuropharmacology.)

persisting alterations of serotonergic parameters may be more pertinent than are the transient changes. As can be seen in Figure 6, the effects of a single large dose of MDMA subsided considerably by two weeks. However, when five doses of MDMA (10 mg/kg, given at six-hour intervals) were administered, the decrease in TPH activity persisted in the neostriatum for 110 days (Figure 7). Similar responses were observed in the hippocampus and frontal

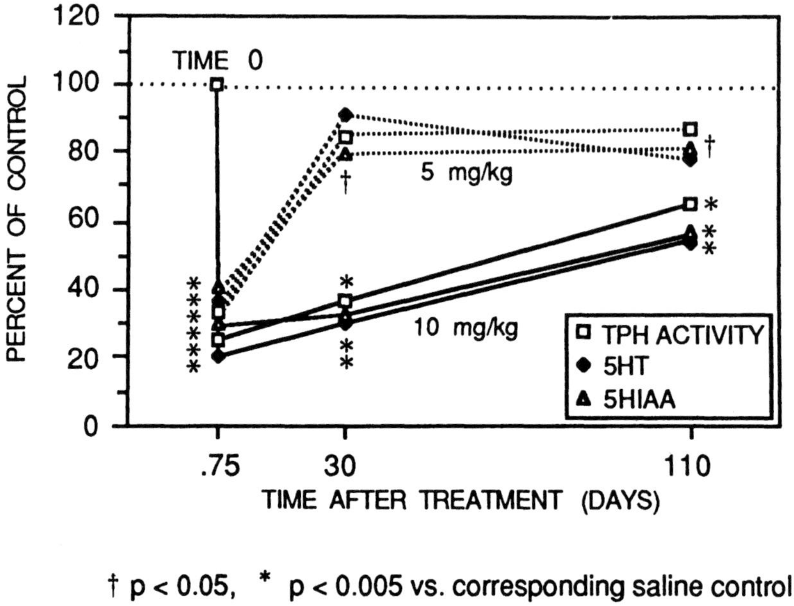

Figure 7. Long-term recovery of serotonergic parameters in the neostriatum after multiple doses of MDMA. Rats were given 5 doses of saline or MDMA (1 dose every 6 hours) and killed at either 18 hours, 30 days, or 110 days after cessation of treatment. Each point represents the mean value (error bars have been omitted for the sake of clarity) from 6–10 rats, expressed as a percentage of the corresponding control. Doses of 5 mg/kg (dotted lines) and 10 mg/kg (solid lines) are represented. Control TPH activity (in nmol/g tissue/hour) at 18 hour, 30 days and 110 days, respectively, was: 45.1 ± 3.5, 42.8 ± 2.9, and 45.1 ± 3.6. Control concentrations of 5-HT and 5-HIAA, respectively, (in μg/g tissue) were: 0.469 ± 0.024 and 0.362 ± 0.023 at 18 hour, 0.365 ± 0.061 and 0.381 ± 0.034 at 30 days; and 0.527 ± 0.047 and 0.453 ± 0.032 at 110 days. †$P < 0.05$ (5-HIAA only), *$P < 0.005$ versus corresponding control, by the two-tailed Student's t-test. (After Stone et al. [29]. Courtesy of Neuropharmacology.)

cortex [29]. These persisting serotonergic deficits suggest that MDMA, administered to rats in repeated large doses, is neurotoxic to 5-HT-containing neurons.

2.6. MDMA response in the mouse

Since little is known about the neurochemical effects of MDMA in species other than the rat, we [30] characterized the effects of MDMA in mice, a species that displays a different metabolic profile than that of the rat [31]. The remarkable differences in the time required for the MDMA-induced depressed tryptophan hydroxylase activity to return to normal after a single dose of MDMA in the mouse and in the rat are depicted in Figure 8. As previously reported, the effect in the rat persisted for at least two weeks. In contrast to the rat, however, mouse TPH activity was not significantly altered

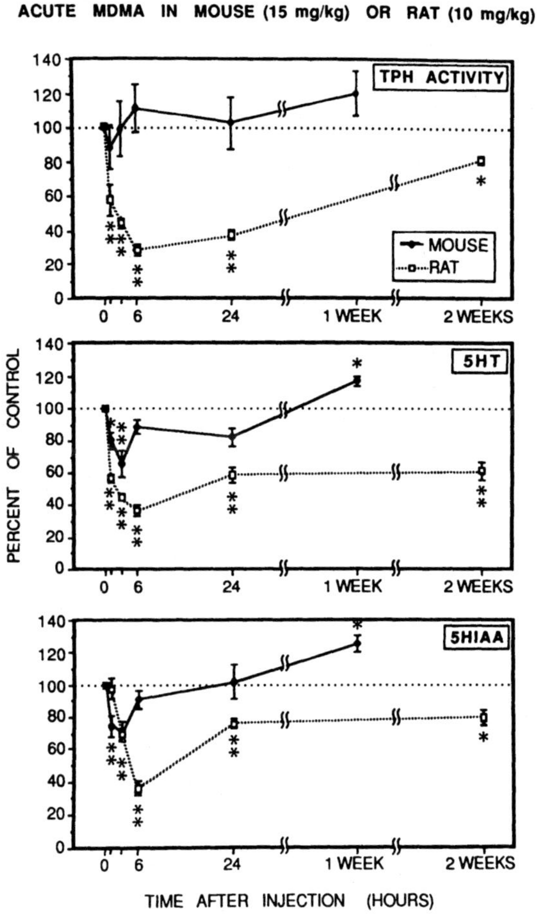

Figure 8. Time course of the neostriatal serotonergic effects of a single dose of MDMA in mouse and rat. MDMA was dissolved in saline and administered as a single subcutaneous injection to mice (15 mg/kg) or rats (10 mg/kg); animals were killed at specified time points thereafter. Points represent means ± SEM for n = 6 − 10 animals, and are expressed as a percent of corresponding time-matched control (vehicle-injected) animals. Representative control values (24-hour time point) for mice and rats, respectively, were: tryptophan hydroxylase (TPH) activity (in nmol/g tissue/hour): 27.0 ± 4.0 and 40.0 ± 2.5; 5-hydroxytryptamine (5-HT) concentration (in μg/g tissue): 0.379 ± 0.021 and 0.503 ± 0.042; 5-hydroxyindoleacetic acid (5-HIAA) concentration (in μg/g tissue); 0.228 ± 0.021 and 0.430 ± 0.028. Data were statistically analyzed by a two-way analysis of variance followed by the Student–Newman–Keuls multiple comparisons test. *$P<0.05$, **$P<0.01$ versus corresponding control. (After Stone et al. [29]. Courtesy of Neuropharmacology.)

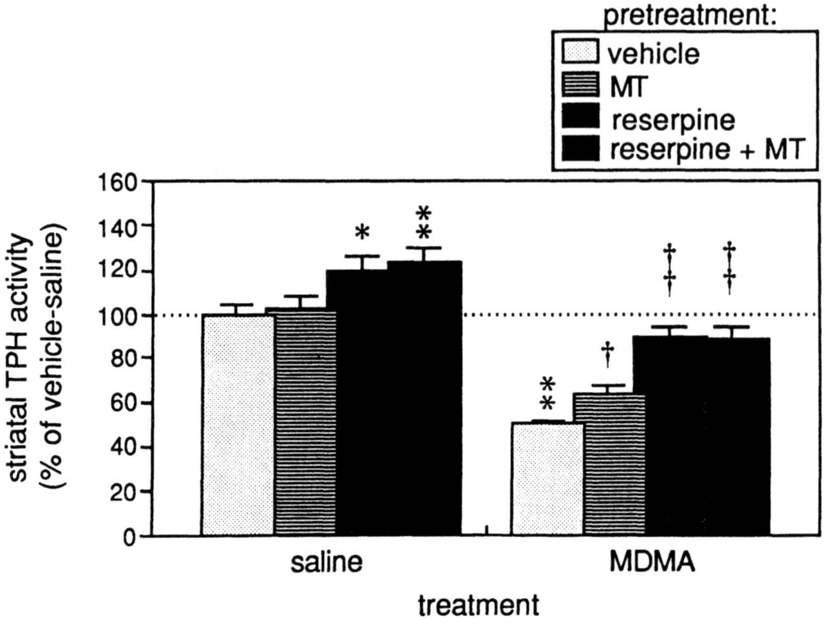

Figure 9. Effect of prior dopamine depletion on the immediate MDMA-induced loss of neostriatal TPH activity. Rats were pretreated with MT (120 mg/kg, i.p.), reserpine (5 mg/kg, i.p.), or reserpine + MT (5 mg/kg and 60 mg/kg, respectively, i.p.) 90 min, 12 hours, or 12 hours + 90 min, respectively, prior to acute MDMA (5 mg/kg, s.c.) or saline (control); animals were killed 3 hours later. Results presented are the means ± SEM (n = 6 − 11), expressed as a percent of control (vehicle-saline). Control value for TPH activity was 49.2 ± 1.9 nmol/g tissue/hour. $^{*}P < .05$, $^{**}P < .01$ versus vehicle-saline, $^{\dagger}P < .05$, $^{\dagger\dagger}P < .01$ versus vehicle-MDMA. Because reserpine pretreatments alone significantly elevated TPH activity, values from MDMA-treated rats were expressed as a percentage ± SEM of their respective (same pretreatment) saline-treated control mean: TPH activity for the reserpine-MDMA and reserpine + MT-MDMA groups, respectively, were 74.6 ± 3.4% and 71.7 ± 4.2% versus 50.3 ± 1.3% for vehicle-MDMA; pretreatment versus vehicle ($p < 0.01$). (After Stone et al. [33]. Courtesy of J. Pharmacol. Exper. Ther.)

after a single dose of MDMA. Mouse 5-HT and 5-HIAA concentrations were significantly decreased at three hours but had returned to normal within six hours after the MDMA was given.

When multiple larger doses of MDMA were administered to mice at more frequent intervals (six doses, 15 mg/kg, at four-hour intervals), TPH activity was significantly decreased in the neostriatum (60% of control) and hippocampus (35% of control) three hours after the last dose and remained significantly depressed in the hippocampus one week after treatment. Similar responses were observed for 5-HT and 5-HIAA (data not shown).

In summary, at comparable doses, MDMA is less toxic in the mouse than in the rat. These studies provide evidence for the importance of comparing responses to potential neurotoxins in a variety of species in order to assess their risks in humans.

2.7. Role of dopamine in MDMA effects

Because of our previous observations that dopamine is necessary for methamphetamine to cause serotonergic deficits, it was important to determine the possible role of dopamine in the MDMA-induced alterations of the 5-HT system. MDMA is known to elevate dopamine [19, 32], which suggests that dopamine release is enhanced in vivo by MDMA. We previously reported that inhibition of dopamine synthesis by MT prevents methamphetamine-induced alterations of both the dopaminergic [7] and the serotonergic system [9, 12]. When dopamine synthesis was reinstated by administering L-DOPA and a peripheral decarboxylase inhibitor, the methamphetamine-induced deficit in the 5-HT system recurred [12].

A similar experiment was conducted with MDMA to determine whether dopamine is essential for the MDMA-induced neurotoxic effects. In these experiments, dopamine concentrations were depleted by MT, reserpine, or with reserpine + MT prior to acute MDMA or saline; animals were sacrified three hours later (Figure 9). MDMA caused the decrease in neostriatal TPH activity [33]. Prior administration of α-methyl-p-tyrosine (MT) partially prevented the MDMA-induced TPH decline, while reserpine, administered with or without MT, effectively prevented the decline of neostriatal enzyme activity in the MDMA-treated rats.

When multiple doses of MT and MDMA were given concurrently, MT attenuated the longer-term (18 hours) effect of MDMA (5 mg/kg) on TPH and 5-HT content, but did not alter the responses to the higher dose of MDMA (10 mg/kg). In a separate experiment we administered MT 90 minutes before a single larger dose of MDMA (20 mg/kg), and rats were sacrificed three days later. MT attenuated the effects of MDMA on TPH activity and 5-HT content but did not alter the effects of MDMA on 5-HIAA content. Using a paradigm in which reserpine (5 mg/kg) was substituted for MT, reserpine completely prevented the long-term (three-day) response to MDMA, given in high doses (Figure 10) [33]. It is interesting that when the attenuation by MT or reserpine of the response to MDMA and methamphetamine is compared, MT appeared less effective than reserpine in preventing the response to MDMA. This could possibly be attributed to the greater degree of dopamine depletion by reserpine as compared to MT. It is also possible that depletion of 5-HT,in addition to dopamine, by reserpine may also be important in the protection. We are currently conducting experiments to explore these possibilities.

Two additional experiments were conducted to explore the role of dopamine in the serotonergic response to MDMA. We have previously demonstrated that a selective lesion of the dopaminergic input to the neostriatum from the substantia nigra prevented the effects of a single dose of methamphetamine on TPH activity [13]. A similar experiment was conducted with MDMA. Bilateral injections of 6OHDA into the substantia nigra were performed seven to ten days before MDMA. Three hours after MDMA was

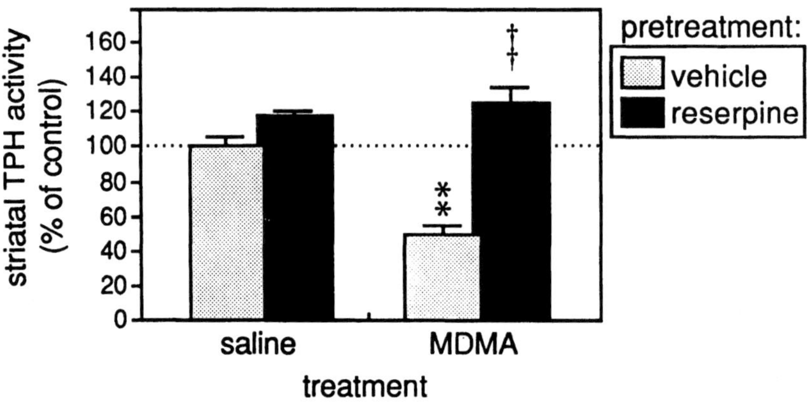

Figure 10. Effect of reserpine pretreatment on the persistent MDMA-induced reduction in striatal TPH activity. Reserpine (5 mg/kg) or vehicle was administered i.p. 12 hours prior to a single dose of MDMA (20 mg/kg, s.c.) or saline; rats were killed 3 days later. Results are represented as the means ± SEM (n = 6 − 7), expressed as a percent of control (vehicle-saline). Control TPH activity was 43.4 ± 2.6 nmol/g tissue/hour. $^{**}P < .01$ versus vehicle-saline; $^{\dagger\dagger}P < .01$ versus vehicle-MDMA. (After Stone et al. [33]. Courtesy J. Pharmacol. Exp. Ther.)

administered, the rats were killed and TPH activity was compared in the neostriatum, frontal cortex, and hippocampus (Figure 11). In the neostriatum, where there was no longer dopaminergic input, the MDMA-induced decrease in TPH activity was essentially prevented; however, in the frontal cortex and hippocampus, where dopaminergic innervation was intact, MDMA still caused a significant decrease in TPH activity. At the lower dose of MDMA there was some protection in the hippocampus, which is thought to have some dopaminergic innervation from the substantia nigra [34].

We [4] previously reported that amfonelic acid, a dopamine uptake blocker, attenuated the effects of methamphetamine on the 5-HT system. In a similar fashion, we (Figure 12) [33] investigated the effects of another specific dopamine uptake inhibitor, GBR 12909, on the MDMA response. GBR 12909 (20 mg/kg) was administered 15 minutes prior to a single dose of MDMA (20 mg/kg); rats were killed three days later. The dopamine uptake inhibitor effectively attenuated the MDMA-induced decrease in TPH activity and in 5-HT and 5-HIAA content.

The above experiments provide convincing evidence that dopamine plays a role in the MDMA-induced changes in the serotonergic system. When dopamine was depleted with MT or reserpine, the MDMA effects were attenuated. Moreover, when dopaminergic input was disrupted by lesioning the nigrostriatal pathway with 6OHDA, the response to MDMA was attenuated. Finally, when dopamine uptake was blocked with GBR 12909, MDMA again was less effective in eliciting long-term serotonergic deficits.

If dopamine is involved with the MDMA and methamphetamine-induced

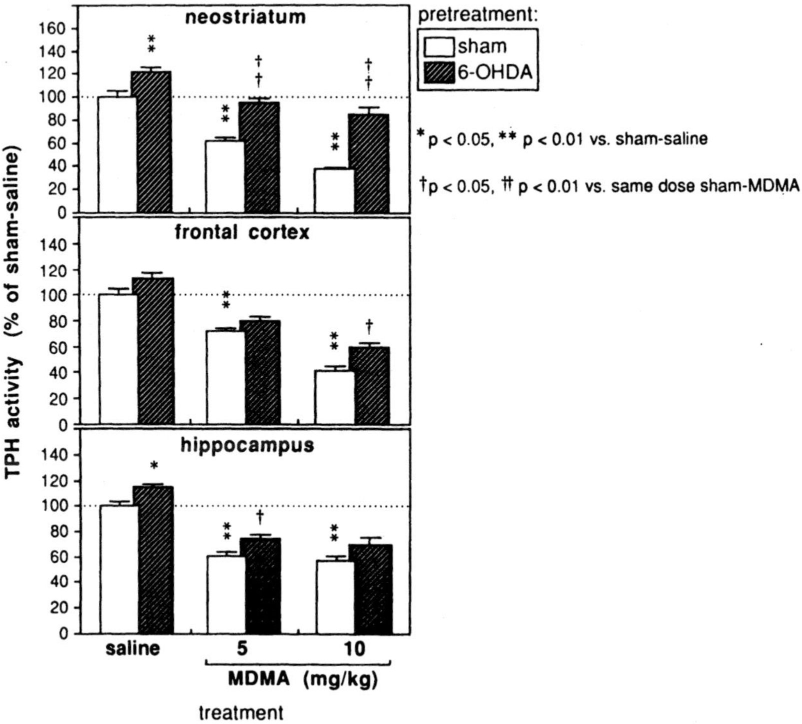

Figure 11. Effect of prior substantia nigral lesions on the immediate MDMA-induced decreases in regional TPH activity. Lesions were induced bilaterally by local injection of 4 μg 6OHDA/8 μl 0.1% ascorbate saline/side. Control rats received sham lesions of ascorbate vehicle alone. Following a 7–10 day recovery period, acute MDMA (5 or 10 mg/kg) was administered s.c. and rats were killed 3 hours later. Results are the means ± SEM, expressed as a percent of sham-saline (n = 22 for sham-saline group, n = 14 for 6OHDA-saline group, n = 6 – 12 for MDMA-treated groups). Control TPH activities (in nmol/g tissue/hour) were: striatum, 42.2 ± 2.3; frontal cortex, 77.3 ± 3.7; hippocampus, 52.2 ± 1.8. *P < .05, **P < .01 versus sham-saline, †P < .05, ††P < .01 versus corresponding sham-MDMA. By 2-way ANOVA and Newman Keuls multiple comparisons test. Because 6OHDA itself significantly elevated TPH activity, values from MDMA-treated rats were expressed as percentage ± SEM of their respective saline-treated control mean: in the neostriatum, TPH activity for the 6OHDA-MDMA group was 67.6 ± 5.1% versus 37.5 ± 2.3% for sham-MDMA, P < 0.01 by Students' t-test. When similarly expressed, no significant differences were found between sham-MDMA and 6OHDA-MDMA groups in the hippocampus or frontal cortex. (After Stone et al. [33]. Courtesy of J. Pharmacol. Exp. Ther.)

neurochemical effects as we suggest, the exact mechanism remains elusive. It is known that these drugs release dopamine and that dopamine can be readily oxidized to reactive metabolites that could possibly destroy nerve terminals [35, 36]. Moreover, inhibition of MAO by these drugs [37] could enhance this response. The possibility that 6OHDA is formed resulting in destruction of nerve terminals, as suggested by Seiden et al. [38], also is an important consideration.

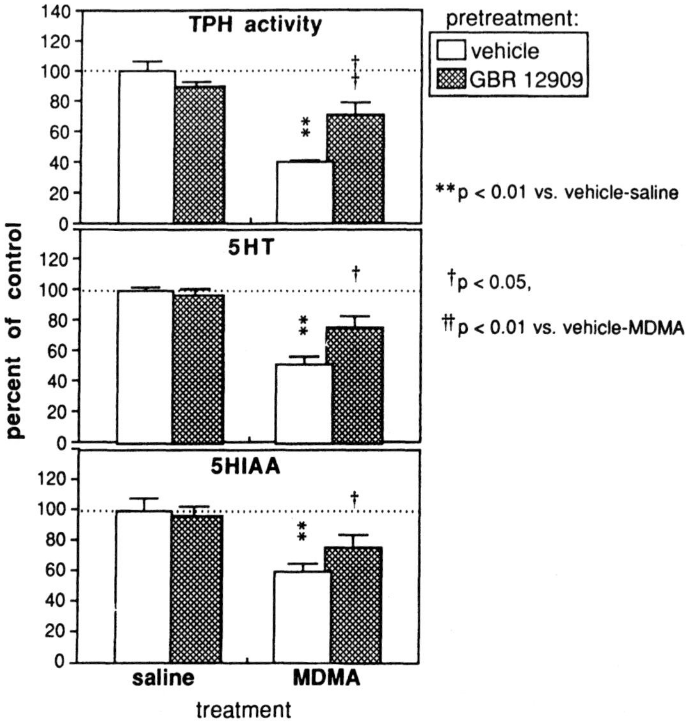

Figure 12. Effect of dopamine-uptake inhibition on the toxic serotonergic deficits induced by acute MDMA. GBR 12909 (20 mg/kg, i.p.) or vehicle was administered 15 minutes prior to a single dose of MDMA (20 mg/kg, s.c.); rats were killed 3 days later. Results depicted are the means ± SEM (n = 5 − 6), expressed as a percent of control (vehicle-saline). Control values were: TPH activity (in nmol/g tissue/hour), 53.7 ± 3.1; 5-HT and 5-HIAA (in μg/g tissue), 0.533 ± 0.013 and 0.502 ± 0.040, respectively. **P < .01 versus vehicle-saline; †P < .05, ††P < .01 versus vehicle-MDMA. (After Stone et al. [33]. Couresty of Pharmacol. Exp. Ther.)

3. CONCLUSIONS

We have observed that the dopaminergic and serotonergic systems are dramatically altered by methamphetamine. The response of these transmitter systems to the methylenedioxy-derivatives of methamphetamine have been compared. Although MDMA perturbs both the dopaminergic and serotonergic systems, the serotonergic, but not the dopaminergic, system is persistently altered. We have provided evidence that dopamine plays a role in the changes in the serotonergic system induced by both methamphetamine and MDMA.

ACKNOWLEDGEMENTS

Supported by USPHS grants DA 00869 and DA 04221. The authors also thank the National Institute on Drug Abuse for the methamphetamine HCl,

MDMA, MDA and MDE. Appreciation is extended to NOVO Industrials for the GBR 12909.

REFERENCES

1. Koda, L.Y. and Gibb, J.W., 1971. The effect of repeated large doses of methamphetamine on adrenal tyrosine hydroxylase. Pharmacologist 13:253.
2. Koda, L.Y. and Gibb, J.W., 1973. Adrenal and striatal tyrosine hydroxylase activity after methamphetamine. J. Pharmacol. Exp. Ther. 185:42–48.
3. Kogan, F.J., Nichols, W.K., and Gibb, J.W., 1976. Influence of methamphetamine on nigral and striatal tyrosine hydroxylase activity and on striatal dopamine levels. Eur. J. Pharmacol. 36:363–371.
4. Schmidt, C.J. and Gibb, J.W., 1985. Role of the dopamine uptake carrier in the neurochemical response to methamphetamine: Effects of amfonelic acid. Eur. J. Pharmacol. 109:73–80.
5. Buening, M.E. and Gibb, J.W., 1974. Influence of methamphetamine and neuroleptic drugs on tyrosine hydroxylase activity. Eur. J. Pharmacol. 26:30–34.
6. Sonsalla, P.K., Gibb, J.W., and Hanson, G.R., 1986. Roles of D_1 and D_2 dopamine receptor subtypes in mediating the methamphetamine-induced changes in monoamine systems. J. Pharmacol. Exp. Ther. 238:932–937.
7. Gibb, J.W. and Kogan, F.J., 1979. Influence of dopamine synthesis on methamphetamine-induced changes in striatal and adrenal tyrosine hydroxylase activity. N-S. Arch. Pharmacol. 310:185–187.
8. Hotchkiss, A.J., Morgan, M.E., and Gibb, J.W., 1979. The long-term effects of multiple doses of methamphetamine on neostriatal tryptophan hydroxylase, tyrosine hydroxylase, choline acetyltransferase and glutamate decarboxylase activities. Life Sci. 25:1373–1378.
9. Hotchkiss, A.J. and Gibb, J.W., 1980. Long-term effects of multiple doses of methamphetamine on tryptophan hydroxylase and tyrosine hydroxylase activity in rat brain. J. Pharmacol. Exp. Ther. 214:257–262.
10. Bakhit, C. and Gibb, J.W., 1981. Methamphetamine-induced depression of tryptophan hydroxylase: Recovery following acute treatment. Eur. J. Pharmacol. 76:229–233.
11. Bakhit, C., Morgan, M.E., Peat, M.A., and Gibb, J.W., 1981. Long-term effects of methamphetamine on the synthesis and metabolism of 5-hydroxytryptamine in various regions of the rat brain. Neuropharmacology 20:1135–1140.
12. Schmidt, C.J., Ritter, J.K., Sonsalla, P.K., Hanson, G.R., and Gibb, J.W., 1985. Role of dopamine in the neurotoxic effects of methamphetamine. J. Pharmacol. Exp. Ther. 233: 539–544.
13. Johnson, M., Stone, D.M., Hanson, G.R., and Gibb, J.W., 1987. Role of dopaminergic nigrostriatal pathway in methamphetamine-induced depression of the neostriatal serotonergic system. Eur. J. Pharmacol. 135:231–234.
14. Schmidt, C.J. and Gibb, J.W., 1985. Role of the serotonin uptake carrier in the neurochemical response to methamphetamine: Effects of citalopram and chlorimipramine. Neurochem. Res. 10:637–648.
15. Ellison, G., Eison, M.S., Haberman, H.S., and Daniel, F., 1978. Long-term changes in dopaminergic innervation of caudate nucleus after continuous amphetamine administration. Science 201:276–278.
16. Ricaurte, G.A., Guillery, R.W., Seiden, L.S., Schuster, C.R., and Moore, R.Y., 1982. Dopamine nerve terminal degeneration produced by high doses of methylamphetamine in the rat brain. Brain Res., pp. 93–103.
17. Wagner, G.C., Ricaurte, G.A., Seiden, L.S., Schuster, C.R., Miller, R.J., and Westley, J., 1980. Long-lasting depletions of striatal dopamine and loss of dopamine uptake sites following repeated administration of methamphetamine. Brain Res., pp. 151–160.
18. Ricaurte, G., Bryan, G., Strauss, L., Seiden, L., and Schuster, C., 1985. Hallucinogenic amphetamine selectively destroys brain serotonin nerve terminals. Science 229:986–988.
19. Stone, D.M., Stahl, D.C., Hanson, G.R., and Gibb, J.W., 1986. The effects of 3,4-methylenedioxymethamphetamine (MDMA) and 3,4-methylenedioxyamphetamine (MDA) on monoaminergic systems in the rat brain. Eur. J. Pharmacol. 128:41–48.
20. Harvey, J.A., McMaster, S.E., and Yunger, L.M., 1975. p-Chloroamphetamine: Selective

neurotoxic action in brain. Science 187:841–843.
21. Fuller, R.W., Hines, C.W., and Mills, J., 1965. Lowering of brain serotonin level by chloramphetamines. Biochem. Pharmacol. 14:483–488.
22. Sanders–Bush, E., Bushing, J.A., and Sulser, F., 1972. Long-term effects of p-chloroamphetamine on tryptophan hydroxylase activity and on the levels of 5-hydroxytryptamine and 5-hydroxyindole acetic acid in brain. Eur. J. Pharmacol. 20:385–388.
23. Sanders–Bush, E., Bushing, J.A., and Sulser, F., 1975. Long-term effects of p-chloroamphetamine and related drugs on central serotonergic mechanisms. J. Pharmacol. Exp. Ther. 192:33–41.
24. Peat, M.A., Warren, P.F., Bakhit, C., and Gibb, J.W., 1985. The acute effects of methamphetamine, amphetamine and p-chloroamphetamine on the cortical serotonergic system of the rat brain: Evidence for differences in the effects of methamphetamine and amphetamine. Eur. J. Pharmacol. 116:11–16.
25. Johnson, M., Letter, A.A., Merchant, K., Hanson, G.R., and Gibb, J.W., 1988. Effects of 3,4-methylenedioxyamphetamine and 3,4-methylenedioxymethamphetamine isomers on central serotonergic, dopaminergic and nigral neurotensin systems of the rat. J. Pharmacol. Exp. Ther. 244:977–982.
26. Stone, D.M., Johnson, M., Hanson, G.R., and Gibb, J.W., 1987. A comparison of the neurotoxic potential of methylenedioxyamphetamine (MDA) and its N-methylated and N-ethylated derivatives. Eur. J. Pharmacol. 134:245–248.
27. Johnson, M., Hanson, G.R., and Gibb, J.W., 1987. Effects of N-ethyl-3,4-methylenedioxyamphetamine (MDE) on central serotonergic and dopaminergic systems of the rat. Biochem. Pharmacol. 23:4085–4093.
28. Schmidt, C.J., 1987. Acute administration of methylenedioxymethamphetamine: Comparison with the neurochemical effects of its N-desmethyl and N-ethyl analogs. Eur. J. Pharmacol. 136:81–88.
29. Stone, D.M., Merchant, K.M., Hanson, G.R., and Gibb, J.W., 1987. Immediate and long-term effects of 3,4-methylenedioxymethamphetamine on serotonin pathways in brain of rat. Neuropharmacol. 12:1677–1683.
30. Stone, D.M., Hanson, G.R., and Gibb, J.W., 1987. Differences in the central serotonergic effects of methylenedioxymethamphetamine (MDMA) in mice and rats. Neuropharmacol. 26:1657–1661.
31. Caldwell, J., 1976. The metabolism of amphetamine in mammals. Drug Metab. Rev. 5: 219–280.
32. Schmidt, C.J., Wu, L., and Lovenberg, W., 1986. Methylenedioxymethamphetamine: A potentially neurotoxic amphetamine analogue. Eur. J. Pharmacol. 124:175–178.
33. Stone, D.M., Johnson, M., Hanson, G.R., and Gibb, J.W., 1988. Role of dopamine in the central serotonergic deficits induced by 3,4-methylenedioxymethamphetamine (MDMA). J. Pharmacol. Exp. Ther. 247:7987.
34. Bjorklund, A. and Lindvall, O., 1984. *Handbook of Chemical Neuroanatomy*, Vol. 2. Elsevier Science Publishers B.V.), p. 55.
35. Graham, D.G., 1978. Oxidative pathways for catecholamines in the genesis of neuromelanin and cytotoxic quinones. Mol. Pharmacol. 14:633–643.
36. Maker, H.S., Weiss, C., and Brannan, T.S., 1986. Amine-mediated toxicity: The effects of dopamine, norepinephrine, 5-hydroxytryptamine, 6-hydroxydopamine, ascorbate, glutathione and peroxide on the in vitro activities of creatine and adenylate kinases in the brain of the rat. Neuropharmacol. 25:25–32.
37. Susuki, O., Hattori, H., Oya, M., and Katsumata, Y., 1980. Inhibition of monoamine oxidase by d-methamphetamine. Biochem. Pharmacol. 29:2071–2073.
38. Seiden, L.S. and Vosmer, G., 1984. Formation of 6-hydroxydopamine in caudate nucleus of the rat brain after a single large dose of methylamphetamine. Pharmacol. Biochem. Behav. 21:29–31.

9. NEUROCHEMICAL EFFECTS OF METHYLENEDIOXYMETHAMPHETAMINE IN THE RAT: ACUTE VERSUS LONG-TERM CHANGES

CHRISTOPHER J. SCHMIDT AND VICKI L. TAYLOR

INTRODUCTION

Amphetamine-like central stimulants are one of the most well-studied classes of pharmacological agents known today. This is due to the fact that their behavioral activity is believed to be mediated primarily by the monoamines, which themselves have been scrutinized sufficiently to have earned the title "classical transmitters." In spite of the attention given this class of agents and their relatively well-described neurochemical activities, there remains a great deal about these drugs that we do not understand. The neurotoxicity associated with high doses of many of these agents is one of these unexplained actions. Although the term high dose is used, it is important to point out that these doses are often in the range of which humans are exposed. This is particularly true in the case of 3,4-methylenedioxymethamphetamine (MDMA), as has been described elsewhere in this book. This neurotoxicity of the amphetamines is selective, in that it primarily affects the neuronal systems through which the drugs mediate their behavioral effects, i.e., the monoaminergic systems. Thus amphetamine, which is believed to cause the majority of its stimulant activities through dopamine release, causes persistent damage selectively to dopaminergic processes [1]. Methamphetamine, which is also a potent releaser of 5-HT, is neurotoxic to both the dopaminergic and serotonergic systems [2]. Finally, the selective serotonergic neurotoxicity of p-chloroamphetamine (PCA) correlates with 5-HT release as the presumed basis of most, though not all, of its behavior effects [3]. This pattern and some of our own data (to be reviewed here) have suggested to us that transmitter release may also be involved, in some manner, in the long-term effects of these drugs. This hypothesis has to a large extent determined the focus of our research effort on MDMA.

Due to this empirical relationship between behavior, acute neurochemistry, and neurotoxicity, our studies of MDMA have dealt exclusively with the effects of single administration of the drug. This reflects the human abuse situation more accurately and allows us to discern the acute neurochemical changes produced by the drug, through the use of both in vitro and in vivo models. Not surprisingly, results from these studies indicate that MDMA is similar in a number of regards to previously studied amphetamines. In its overall spectrum of neurochemical effects, however, MDMA most closely resembles the serotonergic neurotoxin PCA.

2. NEUROCHEMICAL EFFECTS OF MDMA

The administration of MDMA to rats produces rapid and pronounced reductions in central concentrations of 5-HT. This depletion of transmitter concentrations is widespread and involves a majority of the terminal field of the ascending serotonergic system. In contrast, the effects of MDMA on the dopaminergic system consist of a modest increase in dopamine concentrations, which return to control in a few hours [4]. Figure 1A shows the changes for cortical, hippocampal, and striatal 5-HT three hours following a single injection of MDMA (10 mg/kg, s.c.). Interestingly, when 5-HT concentrations are measured 24 hours following MDMA administration, the only significant effects observed are in the striatum (Figure 1B), suggesting a complete recovery in the other two regions. Further examination revealed that MDMA has a complex pattern of effect on 5-HT concentration, which differed slightly between brain regions. As shown in Figure 2 for the cortex and striatum, MDMA produces two distinct phases of transmitter depletion in the rat brain. This is most obvious in the cortex, where the transient recovery of 5-HT concentrations reaches control levels before the second phase of depletion begins. Although a recovery phase does seem to be present in the striatum, 5-HT concentrations do not return completely to control levels [4].

A number of investigators [5–7] have reported similar biphasic changes in 5-HT concentrations following the administration of PCA to rats. Studies from these laboratories have revealed that the initial effects of PCA are due to reversible alterations in serotonin synthesis, uptake, and release, whereas the long-term effect of the drug on serotonergic neurons is neurotoxic, involving a selective degeneration of the nerve terminals. PCA apparently produces little or no neurochemical or histological damage at the level of the cell bodies [7–9]. Our own results indicate that these effects of PCA also occur in the case of MDMA. Examination of various parameters of serotonergic function after MDMA administrations confirmed that one of the first alterations occurs in the activity of the rate-limiting enzyme for 5-HT synthesis,

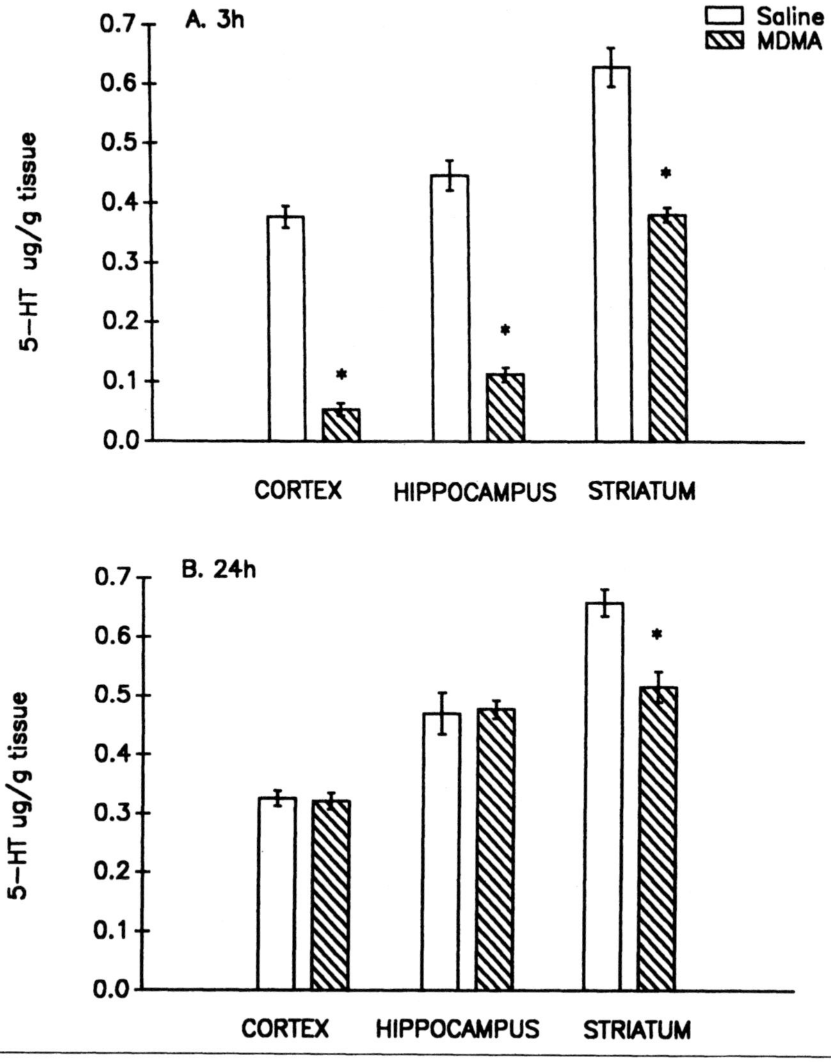

Figure 1. Effect of MDMA on regional brain concentrations of 5-HT in the rat at 3 (A) and 24 (B) hours after a single administration (10 mg/kg, s.c.). All error bars represent the SEM. $^*P < 0.05$ versus saline.

tryptophan hydroxylase (TPH). This change actually precedes the decrease in 5-HT concentrations, as shown for the striatum in Figure 3. Similar results are observed in the cortex and the hippocampus, while the brainstem appears largely resistant to the effects of MDMA [10]. The data in Figure 3 also show that the decline in tryptophan hydroxylase activity begins as quickly as 15 minutes after drug administration. In most cases this is prior to any behavioral

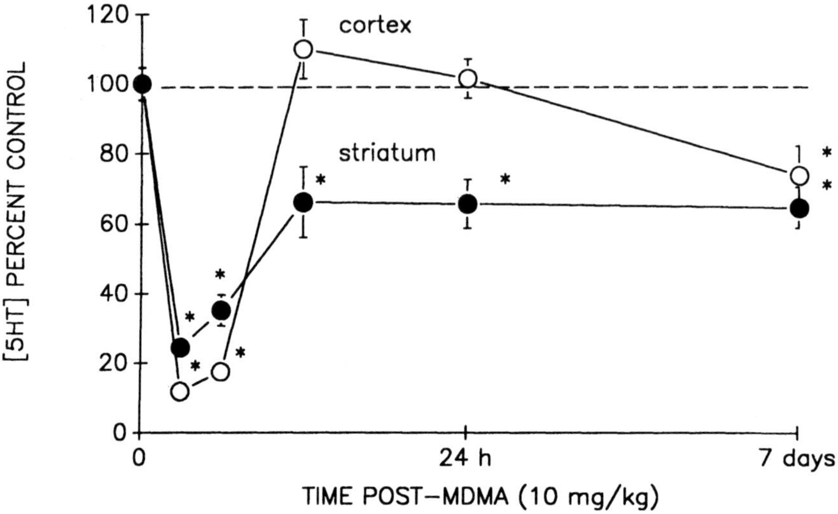

Figure 2. Time course of the changes in 5-HT concentrations in the cerebral cortex and striatum after a single 10 mg/kg dose of MDMA. *$P < 0.05$ versus control.

effect of the drug becoming apparent. A slight elevation in 5-HIAA at the earlier time points suggests that an increase in 5-HT turnover may be an initial event in these changes as well. This is consistent with results from in vitro release studies using striatal slices preloaded with [^{3}H]5-HT [11]. These experiments showed MDMA to be a potent 5-HT releasing agent, much like PCA. Collectively, these findings indicate that the first effects of MDMA on serotonergic neurons are a decrease in TPH activity and an increase in transmitter turnover. With TPH no longer acting to maintain transmitter levels, the releasing action of MDMA begins to deplete the nerve terminal of 5-HT. During this period (one to four hours) the animals exhibit most of the behavioral effects of MDMA. The recovery of 5-HT concentrations, shown in Figure 2, begins after most of the behavioral effects of MDMA have started to wane (four to six hours after drug administration), suggesting that the drug is no longer producing 5-HT release. Under these conditions, even the small 20% to 30% residum of TPH activity in the terminal can gradually restore levels of 5-HT to near control concentrations. Therefore, the acute effect of MDMA on 5-HT concentrations is due to both actions of the drug occurring more or less simultaneously at the nerve terminal. This requirement for both processes is demonstrated experimentally in the results shown in Figure 4. Administration of the 5-HT uptake inhibitor citalopram, which blocks the carrier-mediated release of transmitter [11], completely prevents the depletion of 5-HT measured at three hours after MDMA, in spite of tryptophan hydroxylase activity still being depressed 50% by MDMA. While these results

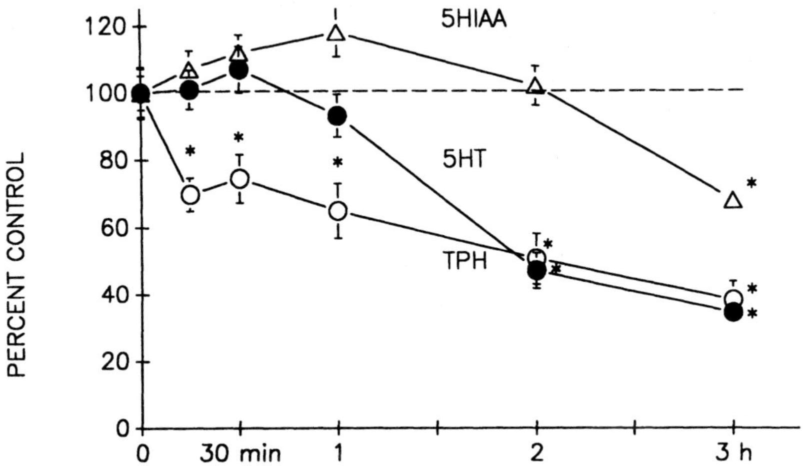

Figure 3. Acute time course of the effects of 10 mg/kg MDMA on serotonergic parameters in the striatum. $^*P < 0.05$ versus control.

identify the processes operating at the level of the nerve terminal that lead to the gross neurochemical changes, the mechanism(s) that ultimately initiates these processes remains to be determined.

As part of our initial efforts to identify this mechanism(s), we began to characterize the acute and long-term effects of MDMA. Figure 5 displays results from an experiment examining the stereochemical requirements of each effect by comparing the optical isomers of MDMA for their ability to produce 5-HT depletion at three hours or at seven days. As shown in Figure 5A, either stereoisomer reduces striatal 5-HT concentrations acutely, with the 10 mg/kg dose already giving the maximum effect. In Figure 5B the results for the 20 mg/kg dose at seven days are plotted together with measurements of whole brain synaptosomal [^{3}H]5-HT uptake. At this time point, the (−)-stereoisomer of MDMA is without any significant effect on either parameter, while (+)-MDMA produces both a depletion in striatal 5-HT concentrations and a reduction in the uptake of [^{3}H]5-HT [11, 12]. Similar results were observed when cortical indoles were measured [12]. The reduction in the uptake of [^{3}H]5-HT has been demonstrated to be due to a decrease in the V_{max} of the 5-HT transporter, without any alteration in its affinity for 5-HT as shown in Table 1 [12]. Consequently, these changes in transmitter uptake represent a loss of functional 5-HT uptake sites and provide biochemical evidence of neurotoxicity at the terminal level.

The results shown in Figure 5, therefore, indicate that the mechanisms leading to the acute and neurotoxic effects of MDMA have different stereochemical requirements or are somehow influenced differentially by the stereochemistry of the drug. An obvious explanation would be a difference in the

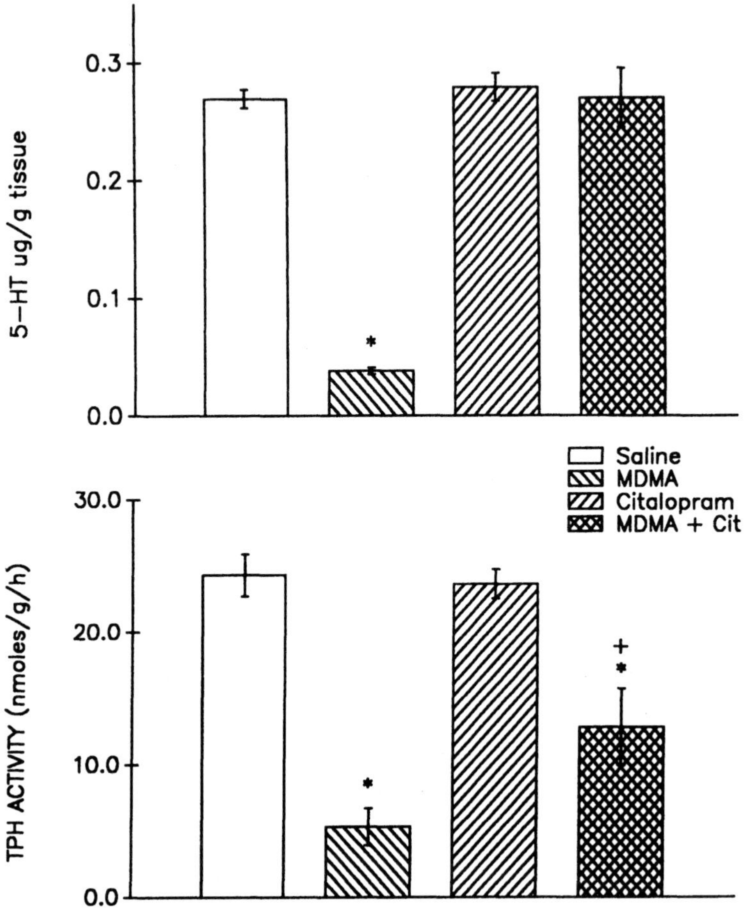

Figure 4. Effect of the simultaneous administration of the 5-HT uptake inhibitor citalopram on the acute reduction in cortical 5-HT concentrations (top) and cortical TPH activity (bottom) produced by 10 mg/kg MDMA at 3 hours. *$P < 0.05$ versus MDMA alone.

rate of metabolism of the two enantiomers, leading to a difference in their brain concentrations. However, we believe this divergence in the neurochemical effects of the two stereoisomers at one week involves more than differences in their rate of metabolism. This reasoning is based in part on comparison with PCA, which has been shown to have an identical stereochemical preference for the production of its acute, versus neurotoxic, effects in rats. When brain concentrations of (+)-PCA and (−)-PCA were measured, the half-lives of the two enantiomers were found to be identical; hence, a difference in the duration of drug action cannot account for the greater neurotoxicity of (+)-PCA compared to (−)-PCA [13]. Similar studies of the metabolism

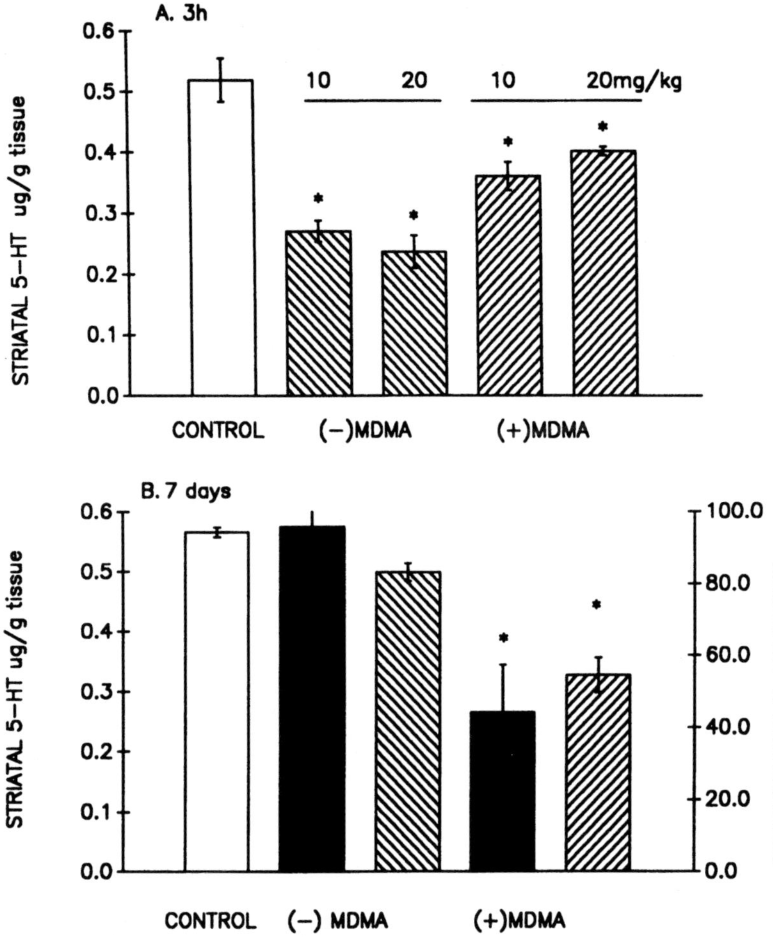

Figure 5. Comparison of the acute (A) and long-term effects (7 days) of the stereoisomers of MDMA. Data for both 10 and 20 mg/kg are shown for the 3 hour time point while the effects of the higher dose are shown in the lower figure. Synaptosomal uptake of [^{3}H]5-HT (black bars) was measured in whole brain synaptosomes. *$P < 0.05$ versus control.

of MDMA and its stereoisomers are required to determine if variations in the rate of metabolism could account for the differences in the long-term effects of the two enantiomers.

Interestingly, the effect of (+)-MDMA and (−)-MDMA on neurotransmitter release in vitro shows a correlation with the apparent difference in the stereochemical stringency observed for the acute and neurotoxic effects of MDMA in vivo. Whereas there was little difference between the stereoisomers in terms of their effect on [^{3}H]5-HT release, (+)-MDMA was a considerably more potent releasing agent for [^{3}H]-dopamine than was (−)-MDMA [11].

Table 1. Kinetic parameters for the uptake of $[^3H]$5-HT by rat P_2 synaptosomes 7 days following (+)-MDMA (20 mg/kg) or saline administration. *$P < 0.05$ versus control.

		Uptake	
	Cortical 5-HT (μg/g)	V_{max} (dpm/mg/5 min)	K_m (μM)
Control	0.39 ± 0.03	105080 ± 6568	0.037 ± 0.003
MDMA	0.21 ± 0.02*	73546 ± 10720*	0.044 ± 0.004

Similar results were observed by Johnson et al. [14] and Steele et al. [15].

The existence of several homologues of MDMA provides another strategy for examining the relationship between MDMA's acute and neurotoxic effects. Table 2 contains data comparing the acute and long-term effects of MDMA with those of its desmethyl and N-ethyl analogues, methylenedioxyamphetamine (MDA) and N-ethyl-methylenedioxyamphetamine (MDE), respectively. At the single dose of 20 mg/kg used, all three drugs produced massive depletions of cortical 5-HT at the three hour time point. However, at seven days the same dose of the compounds produced depletions only in animals administered MDA or MDMA. These drugs also produced the decrease in the uptake of $[^3H]$5-HT by whole brain synaptosomes, indicative of neurotoxicity [16]. These data are thus similar to the results observed for the stereoisomers of MDMA, in that the drugs appeared similar at the three hours time point, with significant differences only becoming apparent one week later. The lack of any residual effect of MDE, and (−)-MDMA in the previously discussed experiment, proves the effect of these drugs on tryptophan hydroxylase activity is reversible in as little as one week and is not due to a neurotoxic response at the nerve terminal. This further suggests that different mechanisms are responsible for the production of the two effects. The divergence in the neurochemical response to the three homologues after one week could be attributed to differences in their rates of metabolism, as already discussed for the stereoisomers of MDMA. Even if this is the case, the results indicate at least a pharmacokinetic difference exists in the requirements for the production of the acute effect, versus the neurotoxicity. Metabolic studies are required to determine if such a difference in the disposition of these drugs does exist.

The corresponding in vitro release experiments with the three homologues also yielded results similar to the experiments with the enantiomers of MDMA. All three drugs were found to be essentially identical in terms of their potency for producing $[^3H]$5-HT release. When their effects on $[^3H]$-dopamine release were determined, however, the three drugs were significantly different, with a rank order of potency of MDA > MDMA > MDE [16]. Thus, $[^3H]$-dopamine release follows the same rank order as does he neurotoxicity of the three drugs.

The rapid onset and rate of decrease in TPH activity following the administration of MDMA or its homologues to rats are among the most dramatic neurochemical effects produced by any class of CNS active agents. This phenomenon also occurs acutely after the administration of PCA [5, 6], methamphetamine [17, 18], or fenfluramine [19]. The reversibility of this effect indicates it is not due to damage to the integrity of the serotonergic nerve terminal. Since all these agents release large quantities of 5-HT from the terminal, it seems entirely possible that a decrease in the activity of TPH might merely be a homeostatic response on the part of the neuron. To determine if an allosteric modification of the enzyme is responsible for the loss of TPH activity following MDMA administration, we compared the kinetic characteristics of the enzyme in cortical homogenates from animals treated three hours previously with MDMA to that of saline-treated animals. Figure 6 shows there was no change in the affinity of the enzyme for either its substrate, tryptophan, or the synthetic cofactor, 6-methyl-tetrahydropterine, after MDMA administration. There was however a significant decrease in the V_{max} activity of the enzyme, in this case amounting to approximately 50% [20]. We have shown this is not due to a direct effect of MDMA on the enzyme in vitro [20], and experiments combining cortical homogenates from saline- and MDMA-treated rats failed to yield any suggestion of a metabolite that might directly affect TPH activity [Schmidt and Taylor, unpublished results]. These data and the V_{max} decrease in enzyme activity suggest a complete inactivation of tryptophan hydroxylase, as does the lack of a recovery phase during the same period as the transient recovery of 5-HT concentrations. Complete recovery of TPH activity may depend upon synthesis of a new enzyme in the cell bodies, with subsequent transport from the brainstem to the terminals. Assuming that this process requires several days, replenishment of the terminals with enzyme would eventually be blocked by the MDMA-induced degeneration of the nerve terminals. In the case of (−)-MDMA and MDE, where no long-term or neurodegenerative effects develop, complete recovery of TPH activity and, hence, transmitter concentrations could occur.

We have been unable to show a MDMA-induced loss of tryptophan hydroxylase activity using either synaptosomes or superfused slices of cerebral cortex exposed to high concentrations of the drug [10], although it is apparent that MDMA does release 5-HT from such preparations [21]. This suggests that an intact neuronal network is necessary for this effect or that in vivo metabolism of the drug is required. To address the first possibility, we made direct injections of MDMA stereotaxically into the brains of rats under metophane anesthetic. Three injection sites were selected, with each group having their own saline injected controls. All animals were allowed to survive for three hours after injection, to observe any behavioral effecs of the drug. TPH activity as well as 5-HT concentrations were determined in a number of regions for each injection site. Results from the assay of cortical TPH activity are shown in Figure 7. The injection of 300 μg of MDMA directly into the

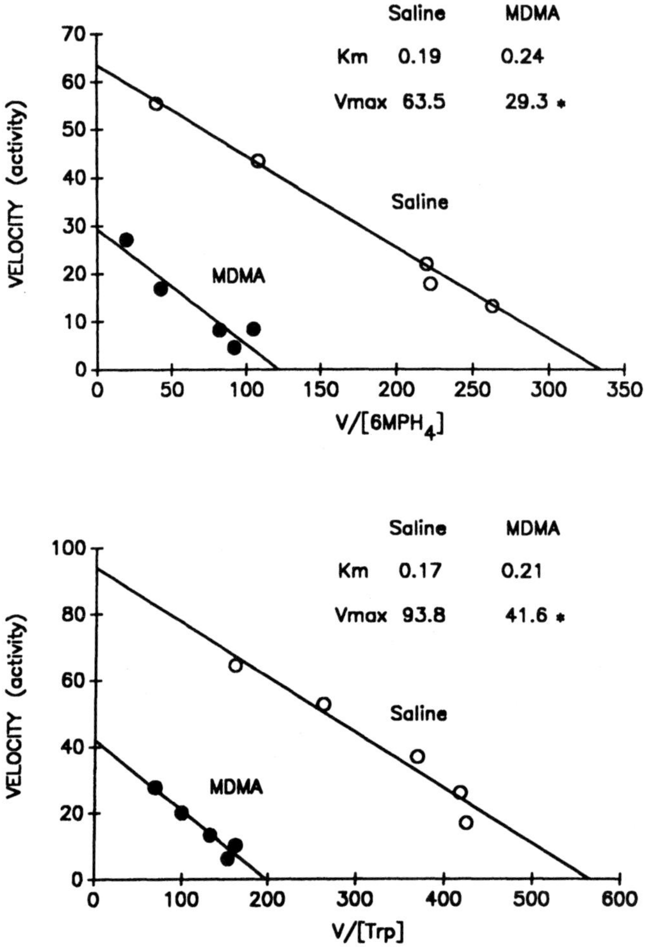

Figure 6. Effect of the acute MDMA (10 mg/kg) on the kinetics of cortical TPH at 3 hours with respect to $6MPH_4$ (top) or tryptophan (bottom). The data are shown as Eadie–Hofstee plots. Values for the K_m and V_{max} of the enzyme for control and MDMA-treated rats are provided in the figure. $^*P < 0.005$ versus control.

substantia nigra, the dorsal raphe, or the cerebral ventricles had no effect on cortical TPH activity. Similar results were observed for TPH activity and 5-HT concentrations in the striatum and hippocampus, regardless of the injection site [10]. Repeating the i.c.v. injections using pentobarbital as the anesthetic did not alter the outcome of the experiment. Hence, direct application of MDMA to either the terminal field or cell bodies of serotonergic neurons did not reproduce the acute effects of peripheral administration. In addition to the lack of neurochemical effects, these injections produced no

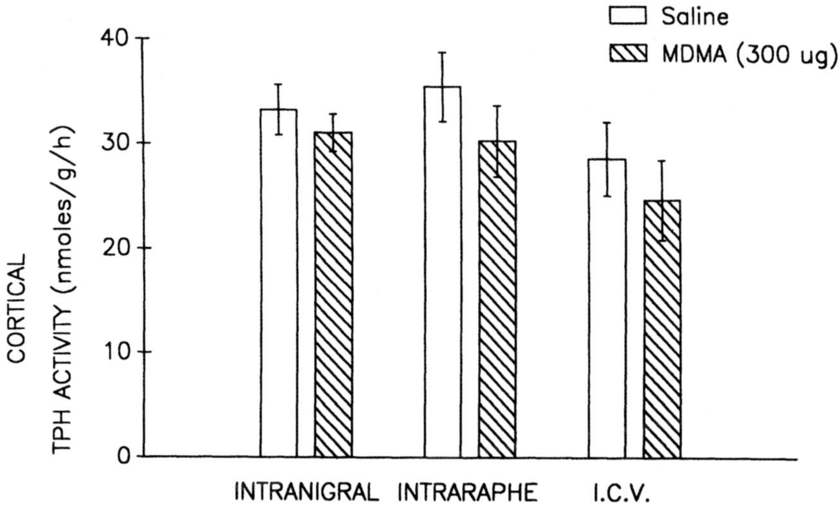

Figure 7. Lack of effect of direct injections of MDMA (300 μg) into various brain sites on the activity of TPH in the cerebral cortex. The drug was stereotaxically injected under a light metophane anesthetic and the animals were sacrificed 3 hours later. Similar results were observed for 5-HT concentrations and in all other areas examined.

obvious behavioral effects in the animals, with the exception of some contralateral turning in the nigra-injected rats. This led us to question the validity of using local injections of a small quantity of a lipophilic drug such as MDMA as a model for determing its central actions. Using [^{3}H]MDA, Marquardt et al. [22] showed that 10% of a peripherally administered dose of the drug was present in the brain within 30 minutes of injection. For a 300 g rat given 10 mg/kg, this would amount to the 300 μg of drug we used in our studies. Since the administration of this much compound did not affect any of the neurochemical parameters measured, it is likely the drug rapidly distributed throughout the animal at a concentration too low to have any effect. Consequently, we elected to use direct i.c.v. infusions of MDMA into conscious animals to insure that behaviorally relevant brain concentrations of the drug were maintained for a period of time similar to that which might be expected following peripheral administration. Using this approach, we have been able to demonstrate significant reductions in regional tryptophan hydroxylase activity with infused doses of MDMA as low as 300 μg or a total body dose of approximately 1 mg/kg [10]; these data are shown for the cerebral cortex in Figure 8. The absence of any change in 5-HT concentrations while enzyme activity is significantly decreased is interesting in light of our observation that the loss of enzyme activity precedes the decline in transmitter concentrations. Since higher infusion doses, i.e., 600 μg, reduced both TPH activity and 5-HT concentration, the 300 μg dose may have been sufficient to elicit the loss

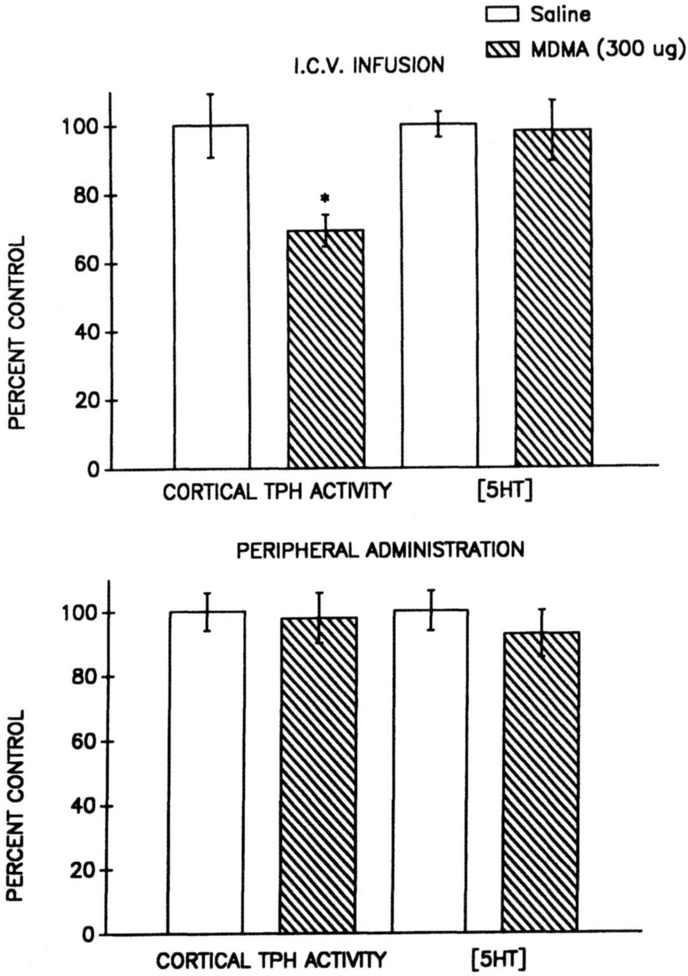

Figure 8. Comparison of the effect of a continuous i.c.v. infusion of MDMA with peripheral administration of the same dose. MDMA was infused through a stainless steel cannula for 1 hour, after which the rats were observed for an additional 2 hours prior to sacrifice. MDMA was administered peripherally by the s.c. route and the animals were sacrificed 3 hours later. $^{*}P < 0.05$ versus saline.

of enzyme activity without producing enough release to actually deplete the terminals. More important is the point that 300 μg is the same dose that had no effect when given by a bolus injection centrally, nor did it affect enzyme activity following peripheral administration, as also shown in Figure 8. These data, therefore, demonstrate that the acute effect of MDMA on serotonergic neurons is a central action of the drug, but one requiring sustained, high concentrations of the agent.

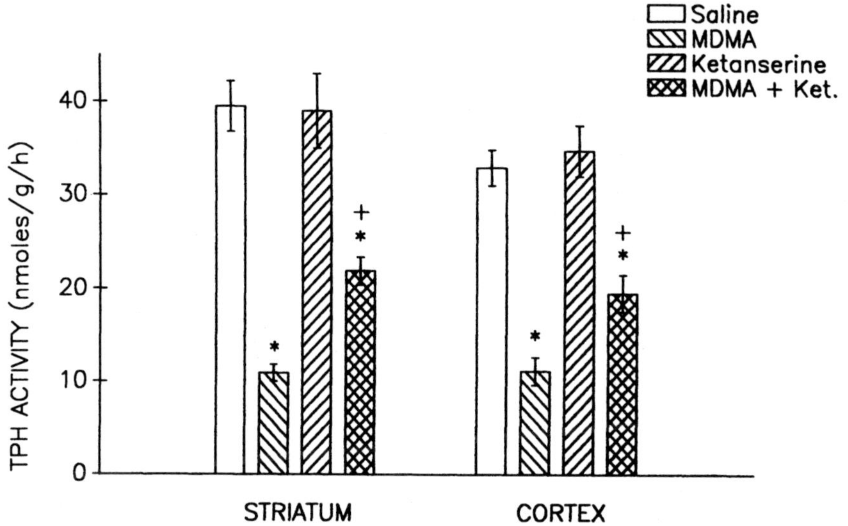

Figure 9. Partial antagonism of the acute loss of striatal and cortical TPH activity by the $5\text{-}HT_2$ receptor antagonist ketanserine. MDMA (10 mg/kg) and ketanserine (2.5 mg/kg) were administered simultaneously 3 hours prior to sacrifice. $^*P < 0.05$ versus saline; $+P < 0.05$ versus MDMA alone.

Although we believe we have excluded the possible involvement of a peripheral metabolite of MDMA in the acute effect of the drug on serotonergic neurons, the mechanism ultimately responsible for the loss of TPH activity has remained elusive. We have considered a number of possible mechanisms that would be a consequence of the release of monoaminergic transmitters produced by MDMA. Interruption of catecholamine biosynthesis by pretreatment with α-methyl-p-trysoine had no consistent effect on the loss TPH activity or the acute depletion of 5-HT produced by MDMA. Unilateral lesions of the substantia nigra with 6-hydroxydopamine were similarly without effects [Schmidt and Taylor, unpublished results]. More global depletions of monoamine stores, with reserpine [20] or the L. aromatic-amino acid decarboxylase inhibitor, α-fluoromethyldopa, also failed to block loss of TPH activity produced by MDMA [Schmidt and Taylor, unpublished, results]. With the exception of inhibitors of the 5-HT uptake carrier system, the only pharmacological manipulations that have consistently altered the acute effect of MDMA have been a modest antagonism with 5-HT receptor antagonists, such as ketanserine. The data for cortical and striatal TPH activity are shown in Figure 9. Similar results were observed with the nonselective 5-HT antagonist methiothepin [20]. Neither antagonist had any significant effect on the decrease in 5-HT concentrations after MDMA administration, however. This is not surprising in view of the very small protection provided by the drugs and the fact that receptor antagonists would not alter the carrier-mediated

Table 2. Comparison of the acute and long-term effects of MDMA and its analgoues on serotonergic neurons in the rat.

	Cortical 5-HT μ/g tissue 3 hours	7 days	[^{3}H]5-HT uptake at 7 days % control
Saline	0.323 ± 0.016	0.280 ± 0.017	100 ± 5.1
MDA (20 mg/kg)	0.056 ± 0.006*	0.093 ± 0.015*	61.2 ± 8.2*
MDMA (20 mg/kg)	0.036 ± 0.004*	0.124 ± 0.020*	64.0 ± 1.7*
MDE (20 mg/kg)	0.052 ± 0.004*	0.297 ± 0.025	99.6 ± 3.5

efflux of 5-HT from the nerve terminals. Unfortunately, the effect of the antagonists is small, and further work with other 5-HT receptor blockers is required to determine if excessive activity at 5-HT receptors is somehow involved in the acute effects of MDMA on TPH.

Although the reversibility of the acute effect of MDMA is well demonstrated by the results with the (−)-stereoisomer of MDMA (see Figure 5) and the results with MDE (see Table 2), the difference between the development of the acute and long-term neurochemical effects of MDMA is most clearly shown by the results displayed in Figure 10. In this experiment, rats were administered MDMA at time zero, with the 5-HT uptake inhibitor, fluoxetine, being administered either simultaneously or at various times after MDMA. All animals were sacrified at one week. Both cortical TPH activity and 5-HT concentrations are represented in the figure as a percent of the appropriate control group: either fluoxetine or saline at each time point. Simultaneous administration of fluoxetine with MDMA completely blocks the long-term depletion of 5-HT indicative of neurotoxicity, demonstrating that both the acute and long-term depletion of 5-HT by MDMA are sensitive to inhibitors of 5-HT uptake. Three hours after MDMA administration, 5-HT concentrations are fully depressed due to the acute effects of MDMA, yet administration of the uptake inhibitor at this time still prevents development of the neurotoxicity. Fluoxetine at six hours after MDMA provided partial protection but was without effect by 12 hours post-MDMA [12]. Similar results were observed for cortical TPH activity, although inhibition of uptake did not appear to block the loss of enzyme activity beyond three hours post-MDMA. This smaller window for protection of TPH activity may be due to the quicker response of the enzyme to MDMA, when compared to 5-HT concentrations (see Figure 3). The observation that tryptophan hydroxylase activity does recover by one week when fluoxetine is administered at three hours is further support for the hypothesis that synthesis of new enzyme is required to restore enzyme activity on the affected nerve terminals.

The ability to block the development of the neurotoxicity after the serotonergic terminal has been depleted of 5-HT suggests the massive MDMA-induced release of 5-HT is not responsible for the long-term effects of the drug. However, the results show that, in addition to the role of the 5-HT

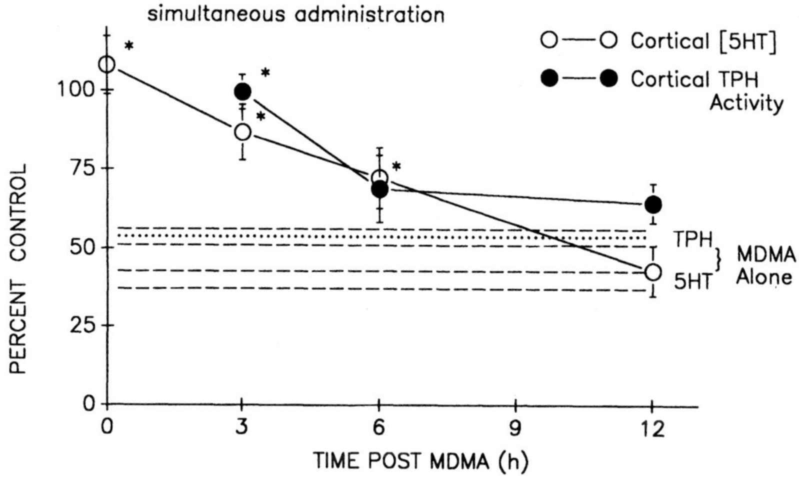

Figure 10. Time-dependency of the blockade of MDMA-induced neurotoxicity by fluoxetine. Animals were injected with MDMA (20 mg/kg) at time zero, followed by injection of saline or fluoxetin (5 mg/kg) at various times thereafter. All rats were sacrificed 1 week later. The effect of MDMA alone on TPH activity and 5-HT concentrations is represented by the horizontal bars. $^{*}P<0.05$ versus MDMA alone.

carrier in the acute depletion of 5-HT by MDMA, some late activity on the part of the carrier must be required for the development of the neurotoxic response to MDMA. By three hours the serotonergic terminal has been depleted of 5-HT; hence, fluoxetine can no longer be interfering with this activity, and by six hours, most of the behavioral effects of MDMA have abated. However, even this late interference with the activity of the uptake carrier can disrupt the development of the neurotoxicity. Because this late activity on the part of the carrier is occurring between three and 12 hours after MDMA, it is possible that a metabolite of the drug is being accumulated during this period. A similar hypothesis has been offered to explain the neurotoxicity of PCA and its sensitivity to fluoxetine for as long as 48 hours after drug administration [5]. Although MDMA is less potent than PCA as a neurotoxin, its effect apparently evolves in a shorter time period since it cannot be blocked beyond six hours after MDMA administration.

Although the generation of a hypothetical neurotoxic metabolite of MDMA is compatible with the apparent stereochemistry of the neurotoxic effect, by analogy with PCA, there are a number of problems with this hypothesis. Similar stereochemical specificity for the neurotoxic effect of PCA has reinforced the opinion that a neurotoxic metabolite of PCA may be involved in its long-term effects on serotonergic neurons. However, although there have been reports of covalent binding of a metabolite of PCA to cell macromolecules in vitro, [23, 24], this metabolite has yet to be isolated. The neurotoxi-

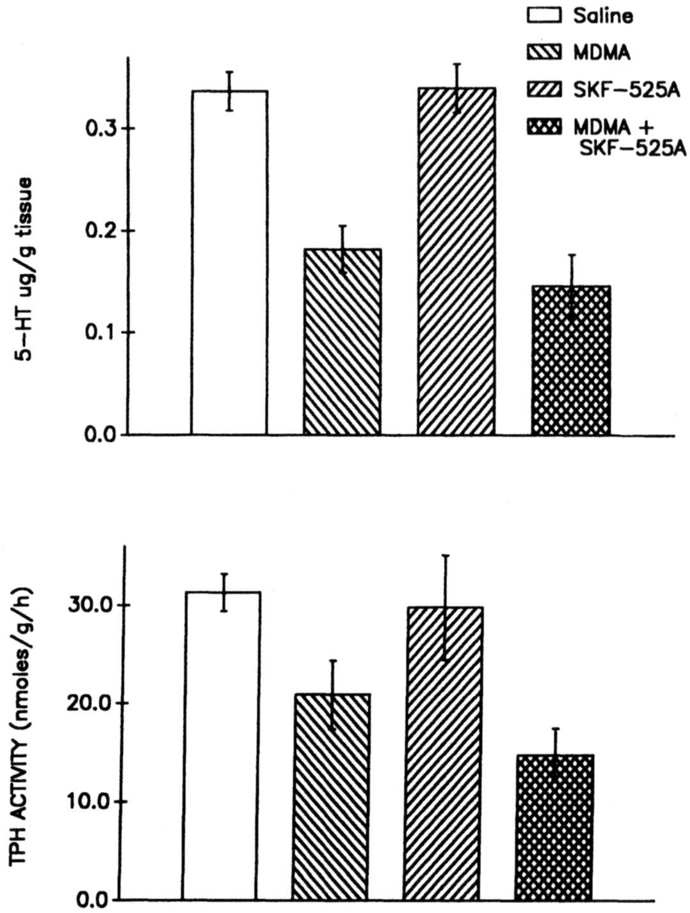

Figure 11. Lack of effect of the cytochrome P-450 inhibitor SKF-525A on the neurotoxicity of MDMA. SKF-525A (10 mg/kg, i.p.) was given 60 minutes prior to MDMA (20 mg/kg) and all rats were sacrificed 1 week later.

city of PCA is also unaffected by intracerebral glutathione [25] or pretreatment with the cytochrome P-450 inhibitor, SKF-525A [26]. Studies by Steranka and Sanders–Bush [27], using metabolic inducers such as phenobarbital and 3-methylcholanthrene, failed to find evidence of a neurotoxic hepatic metabolite of PCA. We have treated rats with SKF-525A prior to MDMA administration without any effect on either the long-term loss of TPH activity or 5-HT concentrations (see Figure 11).

In the absence of more evidence for a neurotoxic metabolite of MDMA or PCA, we are considering other possible mechanisms for the long-term effects of these drugs. These include the generation of a neurotoxic molecule from one or more of the transmitters released by these drugs, as well as direct receptor effects of the released transmitters. There is considerable evidence suggesting a role for dopamine release in the neurotoxicity of methamphetamine, including sensitivity to pretreatment with α-methly-p-tyrosine [28, 29] and 6-hydroxydopamine lesions of the substania nigra [30]. Dopamine receptor antagonists have also been found to alter methamphetamine-induced neurotoxicity [31]. Similar studies are currently under way in our laboratory, with agents selected to interfere with the function of either serotonergic or catecholaminergic systems.

In conclusion, studies in our laboratory have identified two distinct neurochemical consequences of MDMA administration to rats. The initial effects of MDMA on the serotonergic nerve terminal include an uregulated carrier-mediated release of 5-HT coupled with a rapid loss of tryptophan hydroxylase activity. Both of these events are necessary for the massive depletion of forebrain 5-HT concentrations that occur after the administration of MDMA. The loss of enzyme activity is a centrally-mediated effect of the drug, although the ultimate mechanism remains to be determined. Although tryptophan hydroxylase is apparently irreversibly inactivated, the acute effect of MDMA on the neuron is not an irrevocable event that commits the nerve terminal to degeneration, since (−)-MDMA or MDA can both produce a depletion of 5-HT quantitatively similar to (+)-MDMA or MDA, without producing evidence of neurotoxicity at later times. The ability of fluoxetine to block the process leading to the neurotoxicity after the acute depletion of 5-HT has already occurred also supports the reversibility of these early changes. In all respects, the neurochemical effects of MDMA qualitatively resemble those of PCA, suggesting the drugs may also share a common mechanism of action. This is likely to be true with regard to both their acute and neurotoxic effects. Although current opinion favors neurotoxic metabolite as the cause of the long-term effects of PCA, there is as yet no conclusive data to validate this hypothesis. MDMA has not been subject to the same degree of investigation as has PCA, however, evidence for its metabolism to a neurotoxin is also scant. In the absence of such data, we suggest that the evidence of a role for neurotransmitter release in the neurotoxicity of amphetamines is at least as strong as that for a neurotoxic metabolic. There is good evidence that dopamine release is a prerequisite for the neurotoxic effects of methamphetamine on both dopaminergic and serotonergic neurons. In the case of MDMA, the stereochemical specificity and structural requirements for the production of the neurotoxicity are also compatible with the requirements for dopamine release.

REFERENCES

1. Fuller, R.W. and Heunkick–Luecke, S.K., 1982. Further studies on the long-term depletion of striatal dopamine in iprindole-treated rats by amphetamine. Neuropharmacology 21: 433–438.
2. Hotchkiss, A.J., Morgan, M.E., and Gibb, J.W., 1979. The long-term effects of multiple

doses of methamphetamine on neostriatal tryptophan hydroxylase, tyrosine hydroxylase, choline acetyltransferase and glutamatge decarboxylase activities. Life Sci. 25:1373–1378.
3. Fuller, R.W., 1985. Persistent effects of amphetamine, p-chloroamphetamine and related compounds on central dopamine and serotonin neurons in rodents. Psychopharmacol. Bull. 21:528–532.
4. Schmidt, C.J., Wu, L., and Lovenberg, W., 1986. Methylenedioxymethamphetamine: A potentially neurotoxic amphetamine analogue. Eur. J. Pharmacol. 124:175–178.
5. Fuller, R.W., Perry, K.W., and Molloy, B.B., 1975. Reversible and irreversible phases of serotonin depletion by 4-chloroamphetamine. Eur. J. Pharmacol. 33:119–124.
6. Sanders–Bush, E. and Steranka, L., 1978. Immediate and long-term effects of p-chloroamphetamine on brain amines. In *Serotonergic Neurotoxic* (Jakoby, J.H. and Lytle, L.D. eds.). New York: Ann. NY Acad. Sci., pp. 208–221.
7. Massari, V.J., Tizabi, Y., and Sanders–Bush, E., 1978. Evaluation of the neurotoxic effects of p-chloroamphetamine: A histological and biochemical study. Neuropharmacology 17: 541–548.
8. Knapp, S., Mandell, A.J., and Geyer, M.A., 1974. Effects of amphetamines on regional tryptophan hydroxylase activity and synaptosomal conversion of tryptophan to 5-hydroxytryptamine in rat brain. J Pharmacol. Exp. Ther. 189:676–689.
9. Harvey, J.A., McMaster, S.E., and Fuller, R.W., 1977. Comparison between the neurotoxic and serotonin depleting effects of various halogenated derivatives of amphetamine in the rat. J. Pharmacol. Exp. Ther. 202:581–589.
10. Schmidt, C.J. and Taylor, V.L., 1988. Direct central effects of acute methylenedioxymethamphetamine on serotonergic neurons. Eur. J. Pharmacol., in press.
11. Schmidt, C.J., Levin, J.A., and Lovenberg, W., 1987. *In vitro* and *in vivo* neurochemical effects of methylenedioxymethamphetamine on striatal monoamingergic systems in the rat brain. Biochem. Pharmacol. 36:4095–4102.
12. Schmidt, C.J., 1986. Neurotoxicity of the psychedelic amphetamine, methylenedioxymethamphetamine. J. Pharmacol. Exp. Ther. 240:1–7.
13. Sekerke, H.J., Smith, H.E., Bushing, J.A., and Sanders–Bush, E., 1975. Correlation between brain levels and biochemical effects of the optical isomers of p-chloroamphetamine. J. Pharmacol. Exp. Ther. 193:835–844.
14. Johnson, M.P., Hoffman, A.J., and Nichols, D.E, 1986. Effects of the enantiomers of MDA, MDMA and related analogues on [^{3}H]serotonin and [^{3}H]dopamine release from superfused rat brain slices. Eur. J. Pharmacol. 132:269–276.
15. Steele, T.D., Nichols, D.E., and Yim, G.K.W., 1987. Stereochemical effects of 3,4-methylenedioxymethamphetamine (MDMA) and related amphetamine derivatives on inhibition of uptake of [^{3}H]monoamines into synaptosomes from different regions of rat brain. Biochem. Pharmacol. 36:2297–2303.
16. Schmidt, C.J., 1987. Acute administration of methylenedioxymethamphetamine: Comparison with the neurochemical effects of its N-desmethyl and N-ethyl analogs. Eur. J. Pharmacol. 136:81–88.
17. Bakhit, C. and Gibb, J.W., 1981. Methamphetamine-induced depression of tryptophan hydroxylase: Recovery following acute treatment. Eur. J. Pharmacol. 76:229–233.
18. Schmidt, C.J. and Gibb, J.W., 1985. Role of the serotonin uptake carrier in the neurochemical response to methamphetamine. Neurochem. Res. 10:637–648.
19. Hwang, E.C. and VanWoert, M.H., 1980. Comparison effects of various serotonin releasing agents in mice. Biochem. Pharmacol. 29:3163–3167.
20. Schmidt, C.J. and Taylor V.L. 1987. Depression of rat brain tryptophan hydroxylase activity following the acute administration of methylenedioxymethamphetamine. Biochem. Pharmacol. 36:4095–4102.
21. Schmidt, C.J., 1988. Acute and long-term neurochemical effects of methylenedioxymethamphetamine. In *Pharmacology and Toxicology of Amphetamine and Related Designer Drugs.* NIDA Research Monographs, in press.
22. Marquardt, G.M., DiStefano, V., and Ling, L.L., 1978. Metabolism of 3,4-methylenedioxyamphetamine in the rat. Biochem. Pharmacol. 27:1503–1505.
23. Ames, M.M., Nelson, S.D., Lovenberg, W., and Sesame, H.A., 1977. Metabolic activation of para-chloroamphetamine to a chemically reactive metabolite. Commun. Psychopharmacol. 1:455–460.

24. Miller, K.J., Anderholm, D.C., and Ames, M.M., 1986. Metabolic activation of the serotonergic neurotoxin para-chloroamphetamine to chemically reactive intermediates by hepatic and brain microsomals preparations. Biochem. Pharmacol. 35:1737–1742.
25. Sherman, A.D., Hsiao, W.C., and Gal, E.M., 1977. Cerebral metabolism of [^{3}H]-p-chloroamphetamine. Neuropharmacology 16:17–24.
26. Fuller, R.W., Snoddy, H.D., Roush, B., and Molloy, B.B., 1973. Further structure-activity studies on the lowering of brain 5-hydroxindoles by 4-chloroamphetamine. Neuropharmacology 12:33–42.
27. Steranka, L. and Sanders–Busch, E., 1978. Long-term reduction of brain serotonin by p-chloroamphetamine: Effects of inducers and inhibitors of drug metabolism. J. Pharmacol. Exp. Ther. 206:460–467.
28. Gibb, J.W. and Kogan, F.J., 1979. Influence of dopamine synthesis on methamphetamine-induced changes in striatal and adrenal tyrosine hydroxylase. Naunyn–Schmiedeberg's Arch. Pharmacol. 310:185–187.
29. Schmidt, C.J., Ritter, J.K., Sonsalla, P.K., Hanson, G.R., and Gibb, J.W., 1985. Role of dopamine in the neurotoxic effects of methamphetamine. J. Pharmacol. Exp. Ther. 233: 539–544.
30. Johnson, M., Stone, D.M., Hanson, G.R., and Gibb, J.W., 1987. Role of the dopaminergic nigrostriatal pathway in methamphetamine-induced depression of the neostriatal serotonergic system. Eur. J. Pharmacol. 135:231–234.
31. Sonsalla, P.K., Gibb, J.W., and Hanson, G.R., 1986. Roles of D_1 and D_2 dopamine receptor subtypes in mediating the methamphetamine-induced changes in monamine systems. J. Pharmacol. Exp. Ther. 238:932–937.

10. MDMA EFFECTS IN BRAIN: PHARMACOLOGIC PROFILE AND EVIDENCE OF NEUROTOXICITY FROM NEUROCHEMICAL AND AUTORADIOGRAPHIC STUDIES

GEORGE BATTAGLIA, ROBERT ZACZEK, AND ERROL B. DE SOUZA

1. INTRODUCTION

3,4-Methylenedioxymethamphetamine (MDMA), a ring-substituted derivative of methamphetamine, has been reported to exhibit both stimulant and psychotomimetic properties [1–3]. MDMA has recently attracted a great deal of attention due to its increasing abuse among certain segments of the population [4, 5] and has been the focus of a number of review articles [6, 7] and symposia [8, 9]. Recent data demonstrating that MDMA is self-administered by both rhesus monkeys [10] and baboons [11] suggest that MDMA may have high abuse potential in man. These reports are particularly disturbing, as we and others have recently demonstrated that MDMA is a potent neurotoxin that appears to cause selective degeneration of brain serotonin neurons [12–16], comparable to that reported for its structural analogue, 3,4-methylenedioxyamphetamine (MDA) [12, 17–18].

This chapter will address both the pharmacologic profile of MDMA at various brain recognition sites and the neurotoxic effects of MDMA on brain monoamine systems. We will first describe the in vitro pharmacologic profile of MDMA at a number of established brain recognition sites and receptors and the characteristics of [^{3}H]-MDA and [^{3}H]-MDMA association with brain membranes. With respect to the neurochemical consequences of in vivo administration of MDMA on brain monoamine systems, we will discuss: (1) the selective neurodegenerative effects on serotonin (5-HT) systems, (2) the effects of dose and frequency of drug administration, (3) the relative sensitivity of various animal species to MDMA, (4) the potential neuronal mechanisms involved in the neurotoxic effects of the drug, (5) the time course of recovery following neurodegenerative changes, and (6) the neuroanatomic and morphological specificity of MDMA-induced neurotoxicity. For comparative purposes, we will also describe the effects of some other designer drugs, such as MDA and N-ethyl-3,4-methylexdioxyamphetamine (MDE).

2. IN VITRO EFFECTS OF MDMA

In order to elucidate the putative sites of action of MDMA in brain, we have taken two approaches: (1) we have carried out an extensive in vitro pharmacologic profile of MDMA at known brain receptors and recognition sites to assess which neuronal systems may play a role in the central nervous system (CNS) actions of this drug, and (2) we have carried out preliminary studies investigating the characteristics and pharmacology of [$^3H^-$]MDMA binding in rat brain synaptosomes to assess whether these sites represent a novel binding site distinct from other established neurotransmitter receptors or recognition sites.

2.1. Pharmacologic profile of MDMA at various brain recognition sites

We have used radioligand binding techniques to determine the relative potencies of MDMA at various brain neurotransmitter receptors and recognition sites, in order to elucidate the sites through which MDMA might elicit its behavioral, psychotomimetic, and neurotoxic effects.

The in vitro pharmacologic profile of MDMA demonstrates a broad range of affinities of the drug for various brain recognition sites [19]. The relative potencies of MDMA at the various brain recognition sites were assessed from the inhibitory affinity constants (K_i values) derived from MDMA competition data and calculated using the non-linear curve fitting program, LIGAND [20]. These data are summarized in Table 1.

MDMA is most potent at a number of serotonin recognition sites and at α_2-adrenergic receptors with affinity constants in the high nanomolar to low micromolar range. MDMA had highest affinity for 5-HT uptake sites ($<1\mu M$), with lower but comparable affinities at 5-HT_2 serotonin, α_2-adrenergic, M-1 muscarinic cholinergic, and H-1 histamine receptors (K_i values $\leqq 5\mu M$). The rank order of affinities of MDMA at various brain receptors and uptake sites was as follows: 5-HT uptake $>$ α_2-adrenergic = 5-HT_2 serotonin = M-1 muscarinic = H-1 histamine $>$ norepinephrine (NE) uptake = M-2 muscarinic = α_1-adrenergic = β-adrenergic $\geqq$ dopamine uptake = 5-HT_1 serotonin $>>$ D-2 dopamine (DA) $>$ D-1 dopamine (Table 1). MDMA exhibited negligible affinities (>500 μM) at μ, δ, and κ opioid, central-type benzodiazepine, and corticotropin-releasing factor receptors, at choline uptake sites, and at calcium channels. MDA had a similar pharmacologic profile, with affinities comparable ($<$ twofold difference) to those of MDMA at each of the respective brain recognition sities investigated.

These data suggest that a number of the behavioral, psychotomimetic, and neurochemical effects of MDMA may be explained, in part, by interactions of this drug at multiple serotonin recognition sites in brain. MDMA may alter serotonergic transmission in brain through direct actions at post- as well as pre-synpatic 5-HT recognition sites. With respect to actions mediated at postsynaptic receptors, a number of hallucinogenic phenylisopropylamine derivatives have been shown to exhibit potent agonist-like activity at brain 5-HT_2 serotonin receptors [21, 22]. The in vitro affinities of these hallucinogens at 5-HT_2 serotonin receptors are significantly correlated with both their behavioral potencies in animals, in generalization to other hallucinogens, and with their human hallucinogenic potencies [23, 24]. As observed for other ring-substituted amphetamine derivatives, we have found that MDMA and other methylenedioxy derivatives of amphetamine (MDA and MMDA) exhibit high affinity agonist-like binding characteristics at 5-HT_2 serotonin receptors and a stereospecificity consistent with that observed for other hallucinogenic compounds at this receptor [25, 26]. Data in support of agonist-like characteristics of MDMA at 5-HT_2 serotonin receptors include interactions of MDMA with the high affinity state of 5-HT_2 serotonin receptors, which are sensitive to the effects of guanine nucleotides; similar interactions have been previously reported for 5-HT and other classical tryptamine agonists at this site [27, 28]. Furthermore, while the overall apparent affinity of MDMA for [^{3}H]-kentanserin-labeled 5-HT_2 serotonin receptors is in the low micromolar range, MDMA interactions with the high affinity state of 5-HT_2 serotonin receptors labeled directly by [3-H]-DOB [29] are markedly more potent, with an affinity in the nanomolar range ($K_i \approx 300$ nM). Since MDMA exhibits agonist-like properties at this site similar to those of other hallucinogens, it is feasible that a significant component of the "mood-altering" effects of MDMA may be mediated via direct agonist actions at 5-HT_2 serotonin receptors. A comparison of the relative affinities of MDMA and MDA at postsynaptic 5-HT_2 serotonin receptors with those of other ring-substituted amphetamine hallucinogens suggests that MDMA and MDA would be weaker hallucinogens than would be compounds such as DOM (STP) or DOI (4-iodo, 2,5 dimethoxy amphetamine). A recent study demonstrating that the 5-HT receptor antagonist methysergide can potentiate the MDMA-induced increases in locomotor activity [30] further supports the contention that some of the actions of MDMA are mediated by postsynaptic 5-HT_2 serotonin receptors. In addition to interactions with 5-HT_2 serotonin receptors, some of the effects of MDMA on serotonergic systems could be mediated via 5-HT_{1A} receptors; MDMA has a relatively high affinity for 5-HT_{1A} receptors. Direct agonist effects at this site might contribute to the "centering" and calming subjective reports, since similar effects are reported for novel, non-benzodiazepine anxiolytics, such as ipsapirone and buspirone, which interact with 5-HT_{1A} sites [31].

In addition to its relatively high affinity at postsynaptic serotonin receptors,

Table 1. Pharmacologic profile of MDMA at various brain recognition sites.

Brain recognition site	Affinity K^i (μM)	Radioligand/displacer	Brain region	Assay time, temperature	Buffer
Uptake sites					
Serotonin	0.61 ± .05	0.25nM ^{3}H-Paroxetine/1μM citalopram	1	120min, Rm T	A
Norepinephrine	15.8 ± 1.7	4.0nM ^{3}H-Mazindol/0.3μM desipramine	1	90min, 4°C	A
Dopamine	24.4 ± 1.9	1.0nM ^{3}H-GBR 12935/1μM mazindol	2	60min, Rm T	A
Choline	> 500	10nM ^{3}H-Hemicholinium-3/10μM Hemicholinium-3	2	30min, 25°C	B
Adrenoceptors					
α1	18.4 ± 1.2	0.5nM ^{3}H-Prazosin/10μM phentolamine	1	30min, 37°C	C
α2	3.6 ± 0.8	0.5nM ^{3}H-Para-aminoclonidine-10μM phentolamine	1	30min, 37°C	C
β	19.2 ± 2.1	0.5nM ^{3}H-Dihydroalprenalol/1μM propranolol	1	30min, 37°C	C
Dopamine receptors					
D-1	148 ± 14	0.2nM ^{3}H-SCH 23390/0.1μM flupenthixol	2	30min, 37°C	C
D-2	95 ± 15	0.2nM ^{3}H-Spiperone/1μM (+)butaclamol	2	30min, 37°C	C
Serotonin Receptors					
5-HT_1	23 ± 1.5	2.5nM ^{3}H-Serotonin/10μM serotonin	1	30min, 37°C	C
5-HT_2	5.1 ± 0.3	0.4nM ^{3}H-Ketanserin/0.5μM cinanserin	1	30min, 37°C	C
Cholinergic receptors					

M-1 Muscarinic	5.8 ± 0.3	0.1nM ^{3}H(−)QNB/1μM atropine	1	90min, Rm T	D
M-2 Muscarinic	15.1 ± 0.1	0.1nM ^{3}H(−)QNB/1μM atropine	3	90min, Rm T	D
Opioid receptors					
μ	> 500	2nM ^{3}H-Dihydromorphine/1μM levallorphan	4	45min, 25°C	E
δ	> 500	4nM ^{3}H-D-Ala^{2}-D=leu^{5}-enkephalin (30nM morphine)/1μM levallorphan	4	45min, 25°C	E
ϰ	> 500	1.6nM ^{3}H-Ethylketazocine (30 nM morphine + #00nM D-ala^{2}-D-leu^{5}-enkephalin)/1μM levallorphan	4	45min 25°C	E
Other sites					
H-1 Histamine receptors	5.7 ± 2.4	2nM ^{3}H-Mepyramine/1μM doxepin	1	60min, Rm T	F
Benzodiazepine receptors	> 500	0.2nM ^{3}H-Flunitrazepam/1μM clonazepam	1	60min, Rm T	G
Corticotropin-releasing factors (CRF) receptors	> 500	0.1nM ^{125}I-Tyr°-rat CRF/1μM, ovine CRF	5	120min, Rm T	H
Calcium channels	> 500	0.2nM ^{3}H-Nitredipine/0.1μM nifedipine	1	60min, Rm T	G

Affinities of MDMA at various brain recognition sites. Data represent the mean and SEM from three to five competition curves at each of the sites. K_i values were determined using the nonlinear least-squares curve fitting program LIGAND. Assay buffers were as follows: A, 50 mM TRIS-HCl, 120 mM NaCl, 5 mM KCl (pH 7.4 at Rm T); B, 50 mM glycylglycine, 200 mM NaCl (pH 7.8 at 25°C); C, 50 mM TRIS-HCl, 10 mM $MgSO_4$, 0.5 mM K_2EDTA (pH 7.4 at 37°C); D, 50 mM TRIS-HCl, 10mM $MgSO_4$ (pH 7.7 at Rm T); E, 0.17 M TRIS-HCl (pH 7.6 at 25°C); G, 50mM TRIS-HCl (pH 7.7 at Rm T); F, 50 mM $Na^{+}K^{+}$ phosphate (pH 7.4 at Rm T); H, 50 mM TRIS-HCl, 10 mM $MgCl_2$, 2 mM EGTA 0.1% Bovine Serum Albumin, 0.1 mM bacitracin, aprotinin (100 KIU/ml) (pH 7.2 at 22°C). Brain regions were as follows: 1. frontal cortex; 2. striatum; 3. brain stem; 4. whole brain; and 5. olfactory bulb.

MDMA also exhibits high affinity for 5-HT uptake sites. These data suggest that some of the actions of MDMA may be mediated at presynaptic 5-HT binding sites. MDMA has been reported to competitively inhibit [^{3}H]-5-HT uptake in vitro [3] and to increase the release of [^{3}H]-5-HT from brain synaptosomes [32] and hippocampal slices [33]. Furthermore, the neurotoxic effects of in vivo administration of MDMA on 5-HT terminals can be blocked by concomitant administration of the 5-HT uptake blockers citalopram [13, 34]. Additional evidence in support of the hypothesis that MDMA produces some of its effects through presynaptic serotonergic mechanisms is provided by data demonstrating that MDMA generalizes to a fenfluramine cue in discrimination studies [35]. As mentioned previously and as shown in Table 1, MDMA exhibits relatively high affinity for α_2-adrenergic receptors. Classic α-adrenergic receptor antagonists such as phentolamine have been reported to increase the release of ^{3}H-5-HT via effects on α_2-adrenergic receptors [36]. Thus one might speculate that the serotonin releasing effects of MDMA may be mediated, in part, by high affinity antagonist-like effects at α_2-adrenergic receptors localized to presynaptic serotonin terminals. Thus the relatively high affinity of MDMA at the 5-HT uptake site and α_2-adrenergic receptor may contribute, in part, to the neurochemical, neurotoxic, and behavioral effects mediated at presynaptic 5-HT terminals.

Interestingly, the "anxiolytic-like" effects of MDMA do not appear to be mediated through agonist actions at benzodiazepine receptors or antagonist effects at corticotropin-releasing factor receptors, as evidenced by the low affinity of MDMA (> 500 μM) at each of these receptors. In addition, neither the reinforcing, analgesic or mood-altering properties of the drug appear to be mediated through interactions with any of the opioid receptor subtypes since MDMA has relatively low affinities for these binding sites. While brain serotonin systems may play a key role in mediating some of the effects of MDMA on analgesia and body temperature, as well as in the reported anxiolytic-like, mood altering, and subjective effects of the drug, additional neurotransmitter systems may contribute to some of the unique subjective experiences reported for MDMA and other drugs in this class.

2.2. ^{3}H-MDMA binding studies

In an attempt to further elucidate the molecular interactions between MDMA and neuronal tissue that may underlie MDMA's actions in brain, we examined the incorporation of both [^{3}H]-MDMA and the related compound [^{3}H]-MDA into rat brain synaptosomes. [^{3}H]-MDMA was previously reported to bind to a "high affinity" site ($K_i = 99.2$ nM) in whole rat brain membranes [37]. A subsequent study [38] reported that the apparent "high affinity" MDMA binding was due to a "nonspecific" interaction of [^{3}H]-MDMA with glass fiber filters. In addition, the latter study did not detect any uptake of [^{3}H]-MDMA into rat brain synaptosomes.

While the previous studies of [^{3}H]-MDMA incorporation into brain membranes were performed using standard filter binding techniques, we have used centrifugation assays and have employed intact synaptosomes prepared from various regions of rat brain to investigate the incorporation of [^{3}H]-MDMA and [^{3}H]-MDA [39]. We observed that [^{3}H]-MDA was incorporated into three saturable pools. First, the radioligand was sequestered into a saturable-nonspecific site that was resistent to boiling of the membranes. Analysis of saturation data indicated that, in addition to this nonspecific site, there was also a high affinity site (K_D = 0.89 μM, B_{max} = 23 pmoles/mg synaptosomal protein) and a low affinity site (k_D = 45 μM, B_{max} = 3 nmoles/mg synaptosomal protein). The low affinity site was dependent on the presence of 0.27 M sucrose. This sucrose-dependence was not due to the maintenance of iso-osmotic conditions, since [^{3}H]-MDA incorporation was reduced by 74% when synaptosomes were incubated in iso-osmotic saline. [^{3}H]-MDMA interacts with sites similar to those characterized for [^{3}H]-MDA. In addition to saturable-nonspecific sites (i.e., resistant to boiling), two specific [^{3}H]-MDMA sites were observed on analysis of saturation data (K_D high: 2.9 μM, B_{max}: 79 pmole/mg protein; K_D low: 128 μM, B_{max}: 7.4 nmole/mg protein). The pharmacological profiles of [^{3}H]-MDA and [^{3}H]-MDMA binding were also similar. The order of potency of inhibition of [^{3}H]-MDA or [^{3}H]-MDMA incorporation was paroxetine = desipramine > mazindol > serotonin.

The high binding capacity of the MDA/MDMA site suggests that the binding does not represent a bimolecular ligand-protein interaction. In addition, [^{3}H]-MDA and [^{3}H]-MDMA do not appear to be internalized or sequestered in a intrasynaptic pool, since the binding is not dependent on temperature and is relatively insensitive to the effects of detergents. However, the heterogenous distribution of [^{3}H]-MDA binding in different brain regions indicates that this site is not a nonspecific interaction of MDA or MDMA with brain lipid, as has been suggested for an apparent low affinity component of [^{3}H]-imipramine binding [40]. Further studies are required to determine the exact nature of the interactions of [^{3}H]-MDA and [^{3}H]-MDMA with novel brain recognition sites and the possible relevance of these interactions to the clinical, biochemical, and toxic actions of these compounds.

In order to address the question of pharmacologic relevance of micromolar affinities of MDMA and MDA at various brain recognition sites and the micromolar affinites of [^{3}H]-MDMA and [^{3}H]-MDA binding sites, we have carried out preliminary studies to assess brain concentrations of drug following systemic administration of MDA and MDMA. Concentrations of drug were measured at 45 minutes after systemic administration of the compounds, as peak locomotor activity as well as peak levels of [^{3}H]-MDA and [^{3}H]-MDMA in brain were present during this period. As shown in Table 2, following a single subcutaneous injection of 20mg/kg [^{3}H]-MDMA or [^{3}H]-MDA, fairly comparable concentrations of each drug were observed in all brain regions, with slightly higher levels of drug measured in liver. Assuming a conversion factor of 1 gm of tissue being equivalent of 1 ml, then the values

Table 2. Regional distribution of [^{3}H]-MDA and [^{3}H]-MDMA in rat brain and peripheral tissue.

Region	*μmol/g tissue* [^{3}H]-MDMA	[^{3}H]-MDA
Frontal cortex	0.22	0.42
Rest of cortex	0.19	0.32
Striatum	0.22	0.42
Hippocampus	0.22	0.44
Thalamus	0.21	0.44
Hypothalamus	0.18	0.37
Midbrain	0.17	0.36
Cerebellum	0.15	0.39
Brainstem	0.15	0.23
Pituitary	0.31	—
Liver	0.48	1.26
Spleen	0.25	0.56

Rats were injected with 20 mg/kg of [^{3}H]-MDMA or [^{3}H]-MDA, sacrificed at 45 minutes, and brain regions and peripheral tissues were dissected. A portion of the respective tissues was weighed, solubilized overnight in protosol, and counted by liquid scintillation spectrometry.

of [^{3}H]-MDMA and [^{3}H]-MDA found in all brain regions would convert to concentrations of MDMA and MDA in the high micromolar range for a 20 mg/kg dose of drug. This dose of MDA and MDMA has previously been shown to produce a variety of behavioral and neurochemical effects [12, 14, 17, 31]. These data suggest that the micromolar affinity interactions of MDMA and MDA with binding sites and receptors in brain appear to be pharmacologically relevant with respect to the behavioral and/or neurotoxic effects of the drugs.

3. IN VIVO EFFECTS OF MDMA: NEUROCHEMICAL STUDIES

Typically, neurotoxic effects of drugs on 5-HT neurons have been assessed from changes in a number of serotonergic parameters including (1) reductions in brain levels of 5-HT and its metabolite 5-hydroxyindoleacetic acid (5-HIAA), (2) decreases in the maximal activity of tryptophan hydroxylase (TPH), and (3) decreases in the activity of the serotonin active uptake carrier. Since MDMA can inhibit the activity tryptophan hydroxylase, the rate-limiting enzyme in 5-HT synthesis [18, 41], it is unclear whether MDMA-induced reductions in 5-HT and 5-HIAA may be due to suppressed neurotransmission in otherwise structurally intact 5-HT neurons or whether these changes may represent the consequence of the destruction of 5-HT axons and terminals.

The following studies were designed to assess and quantify both the neurochemical and neurodegenerative effects of short-term administration of MDMA on monoamine neurons in rat brain. In some studies we have also investigated the changes induced by MDA and MDE. Since monoamine uptake sites are highly concentrated on their respective nerve terminals in brain [42], we can directly quantify the degree of neurodegeneration of particular monoamine terminals by measuring reductions in the density of their respective uptake

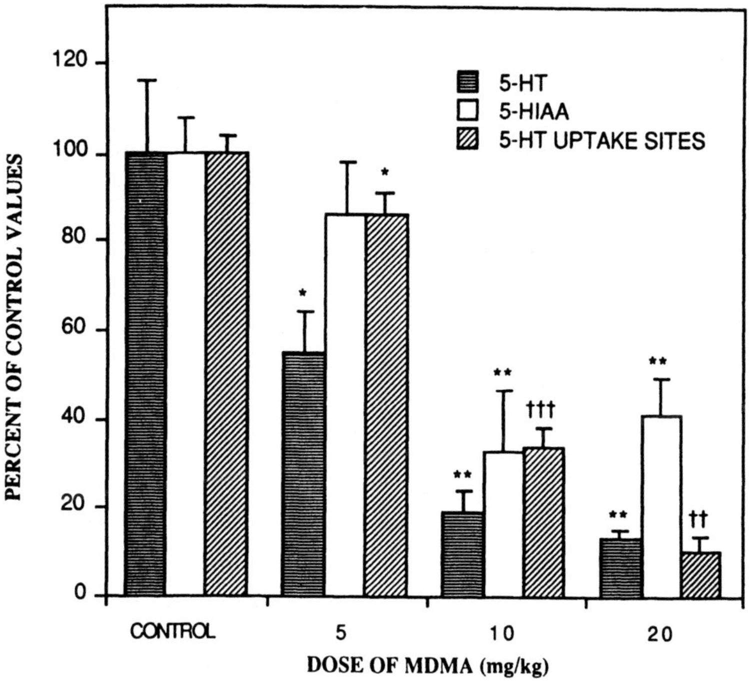

Figure 1. The effect of repeated systemic administration of various doses of MDMA on the content of serotonin (5-HT) and 5-hydroxyindoleacetic acid (5-HIAA) and on the density of 5-HT uptake sites in rat frontal cerebral cortex. Rats were administered either saline or MDMA twice a day for 4 consecutive days and sacrificed 18 hours after the last injection. Data represent the mean ± SEM from 3–5 rats and are expressed as a percent of values in control, saline-injected rats. Control values for 5-HT and 5-HIAA levels were 387 ± 61 and 251 ± 20 pg/mg tissue, respectively. The density of 5-HT uptake sites in the frontal cerebral cortex in controls was 396 ± 15 fmol/mg protein. Data were analyzed by one-way ANOVA and Duncan's multiple range test. * and ** indicate significant differences at $p < 0.05$ and $p < 0.01$, respectively, from control saline-treated rats. †† and ††† indicate significant differences at $p < 0.01$ and $p < 0.001$, respectively, from all other MDMA-treated groups. (From Battaglia et al., 1988.)

sites. We have recently reported the feasibility of using radioligand binding techniques and high affinity ligands for 5-HT, dopamine, and norepinephrine uptake sites to monitor the loss of monoamine axons and terminals [12, 13]. We report here that MDMA, MDA, and MDE cause marked reductions in various serotonergic markers (i.e., 5-HT, 5-HIAA, and 5-HT uptake sites). Marked long-term reductions in the density of 5-HT uptake sites in brain, which reflect the loss 5-HT axons and/or terminals, indicate that both MDMA and MDA are potent, long-lasting 5-HT neurotoxins.

3.1. Dose dependence

As shown in Figure 1, a repetitive dosing regimen (i.e., subcutaneous administration twice a day for four consecutive days) of MDMA at various doses up

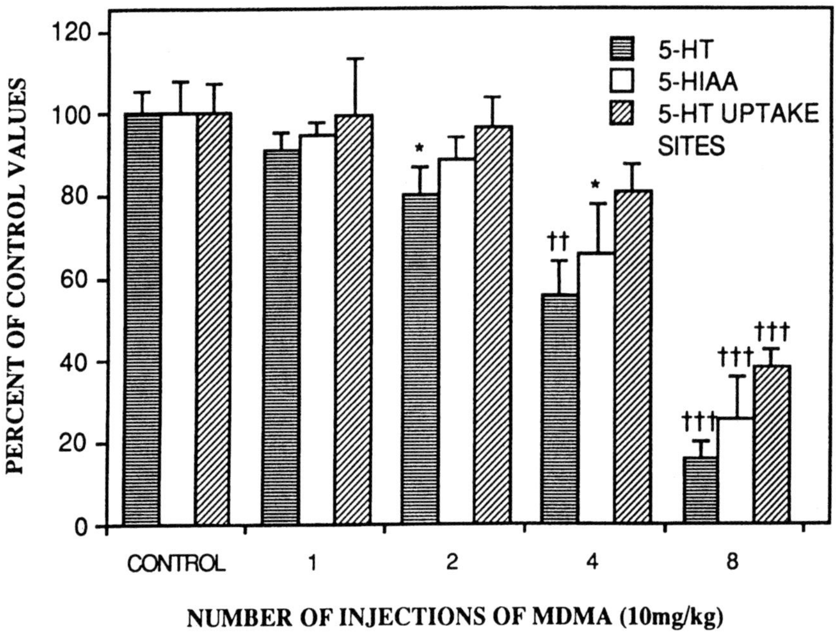

Figure 2. The effects of single and multiple injections of MDMA on the content of serotonin (5-HT) and 5-hydroxyindoleacetic acid (5-HIAA) and on the density of 5-HT uptake sites in rat frontal cerebral cortex. Rats were injected subcutaneously the specificed number of times with either saline or 10 mg/kg MDMA and sacrified 18 hours after the last injection. Data that represent the mean and SEM from three to five animals are plotted as a percent of respective values for each of the markers in control, saline-injected rats. Control levels of 5-HT and 5-HIAA were 475 ± 24 and 332 ± 24 pmol/mg tissue, respectively. The density of 5-HT uptake sites was 349 ± 24 fmol/mg protein in controls. Data were analyzed by one-way ANOVA and Duncan's multiple range test. * indicates a significant difference at $p < 0.05$ from corresponding control saline-injected rats; †† and ††† indicate significant differences at $p < 0.01$ and $P < 0.001$, respectively, from all other groups. (From Battaglia et al., 1988.)

to 20 mg/kg resulted in dose-dependent decreases in the content of 5-HT and 5-HIAA and in the density of 5-HT uptake sites in rat frontal cerebral cortex at 18 hours following the last injection. At the lowest dose of MDMA tested (5 mg/kg), 5-HT content was markedly reduced (45%), while only a small (14%) but statistically significant decrease in the density of 5-HT uptake sites was observed; a small decrease in 5-HIAA content was also observed at this dose, although this change was not statistically significant. Higher doses of MDMA (10 and 20 mg/kg) resulted in comparable reductions in 5-HIAA levels (60–70%), while the decrease in 5-HT content was significantly greater at 20 mg/kg (90%) than at 10 mg/kg (80%). The density of 5-HT uptake sites decreased progressively as the dose of MDMA was increased, with a maximal reduction of 90% observed at 18 hours following repetitive administration of 20 mg/kg MDMA. In contrast, following the identical treatment regimen of 20 mg/kg MDMA, there were no significant differences in the density of [^{3}H]-mazindol-labeled norepinephrine uptake sites (fmol/mg protein) in the frontal cerebral cortex between saline-treated (159 ± 17) and drug-treated (152 ± 5) animals.

3.2. Effect of single versus multiple injections

Since repeated systemic administration of 10 mg/kg MDMA caused marked caused marked neurodegeneration of cerebral cortical 5-HT neurons, we chose to investigate the neurodegenerative effects of single versus multiple subcutaneous injections of MDMA at this dose. Although neurotoxicity studies for MDMA typically employ a subcutaneous route of drug administration, whereas humans self-administer this drug orally, recent reports indicate comparable neurotoxic effects with either type of drug administration [43]. As shown in Figure 2, increasing the number of injections of MDMA (10 mg/kg s.c.) resulted in significant and progressively greater reductions in 5-HT and 5-HIAA content. While one injection of MDMA was without effect on any of the serotonergic markers examined, two doses were sufficient to elicit a significant reduction (~20%) in 5-HT content. A significant reduction (~34%) in 5-HIAA content was observed only after four injections of MDMA. Marked reductions of 84% and 75% in 5-HT and 5-HIAA, respectively, were observed following an eight injection treatment regimen of 10 mg/kg MDMA. The density of [^{3}H]-paroxetine-labeled 5-HT uptake sites was only significantly decreased (~64%) following eight injections of MDMA at this dose. While there was no change in dopamine content in any of the groups examined, a small and consistent decrease in norepinephrine content (~20%) was observed in all MDMA-treated rats. This small change in norepinephrine following MDMA treatment was not accompanied by a reduction in the density of [^{3}H]-mazindol-labeled norepinephrine uptake sites (data not shown).

3.3. Potential mechanisms

As the neurotoxic effects of drugs, such as para-chloroamphetamine, on 5-HT neurons can be prevented by 5-HT uptake blockers [44, 45], we investigated if the 5-HT uptake carrier protein was likewise involved in the neurotoxic effects of MDMA. As shown in Figure 3, pretreatment of rats with the selective 5-HT uptake blocker citalopram (10 mg/kg) prior to each injection of 10 mg/kg MDMA resulted in nearly complete protection against the neurotoxic and neurodegenerative effects of MDMA. Citalopram-pretreated rats showed only a small (15%) decrease in 5-HT uptake sites following MDMA treatment, in comparison with a 60% reduction in 5-HT uptake sites observed in rats treated with an identical dose of MDMA alone. In addition, citalopram pretreatment completely prevented the MDMA-induced decreases in the concentration of both 5-HT and 5-HIAA. Similar protective effects of serotonin uptake blockers, such as citalopram and fluoxetine, on MDMA-induced neurotoxicity have previously been demonstrated [15, 34]. These data would suggest that the

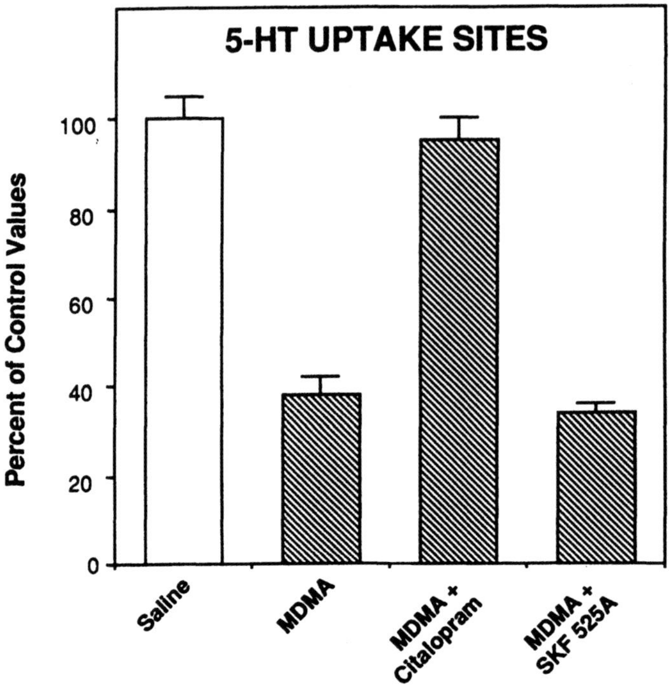

Figure 3. The effect of repeated systemic administration of 10 mg/kg MDMA, MDMA plus 10 mg/kg citalopram, and MDMA plus 25 mg/kg SKF 525A on the density of serotonin (5-HT) uptake sites in homogenates of rat frontal cerebral cortex. Data are expressed as a percent of values in control saline-treated rats and represent the mean and SEM from 4–6 animals. Control levels of 5-HT uptake sites were 356 ± 15 fmol/mg protein.

active uptake of MDMA, some related metabolite, or endogenous neurotoxin is required for MDMA-induced degeneration of serotonin neurons.

We have previously reported that in contrast to the marked neurodegenerative effects following systemic administration of MDMA or MDA on brain 5-HT neurons, single, direct intracerebral injections of MDMA or MDA had no effect on cerebral cortical 5-HT neurons, as visualized directly using serotonin immunocytochemistry [46]. Our observation of marked differences in 5-HT neurotoxicity to central versus systemically-administered MDMA would be consistent with the hypothesis of a peripherally produced neurotoxic metabolite of MDMA. Since MDMA has been reported to interact with the hepatic microsomal enzyme cytochrome P-450 [47], we investigated whether inhibition of this enzyme could alter the neurotoxic effects of MDMA. In preliminary studies, we found that in rats pretreated with the cytochrome P-450 enzyme inhibitor, SKF 525A, 45 minutes prior to each administration of MDMA, there was neither potentiation nor attenuation of the neurodegeneration found following repeated administration of 10 mg/kg of MDMA. As shown in Figure 3, no changes in the density of 5-HT uptake sites were observed between MDMA and MDMA plus SKF 525A-treated rats. While these data suggest that it is unlikely that the putative neurotoxic species is a cytochrome P–450-dependent metabolite of MDMA, the involvement of some other peripheral and/or central metabolite of MDMA cannot be ruled out.

3.4. Regeneration of serotonin neurons

In order to assess whether 5-HT neurons regenerate in MDMA-treated rats, we recently investigated the time course of changes in 5-HT content and 5-HT uptake sites in frontal cerebral cortex for up to one year following a repeated systemic administration (i.e., twice daily s.c. injections for four days) of 20 mg/kg MDMA [13]. As shown in Figure 4, at all time points up to six months during the recovery time course, the density of 5-HT uptake sites was significantly below the corresponding values in age-matched, saline-treated controls. At the six-month time point, the density of 5-HT uptake sites was only 75% of the values of saline-treated controls, whereas by 12 months after MDMA treatment, the density of 5-HT uptake sites returned to control levels. The shape of the recovery curve suggests that there may be a faster initial rate of recovery of 5-HT uptake sites that occurs between 18 hours and four weeks, which is followed by a slower rate of recovery that occurs between four weeks and 12 months. These data indicate that more than six months are required for a complete recovery of 5-HT uptake sites to control levels.

It was of interest that despite the recovery of 5-HT uptakes sites to control levels, the content of 5-HT in the same brain region remained markedly (40–50%) below age-matched controls for as long as one year after MDMA administration. It is unclear from these data whether there is a regeneration of axons that have previously degenerated or whether the increased density of 5-HT uptake sites is a consequence of increased collateral sprouting of neurons unaffected by the drug treatment. It is also possible that axonal regeneration or collateral sprouting is associated with considerably greater densities of uptake sites per neuron, thereby making it more difficult to assess neuronal recovery from this index. The observation that 5-HT levels remain 40–50% below age-matched controls for up to one year despite control-level densities of 5-HT uptake sites indicates that following lesion by MDMA, the 5-HT neurons that "recover" may not be functionally identical to those present in age-matched control brains.

3.5. Species differences

Since amphetamines have been shown to be metabolized by different pathways in rat, mouse, and guinea pig [48], studies were carried out to investigate whether MDMA neurotoxicity can be demonstrated in other species, such as mouse and guinea pig. Animals in these studies were treated repeatedly (i.e.,

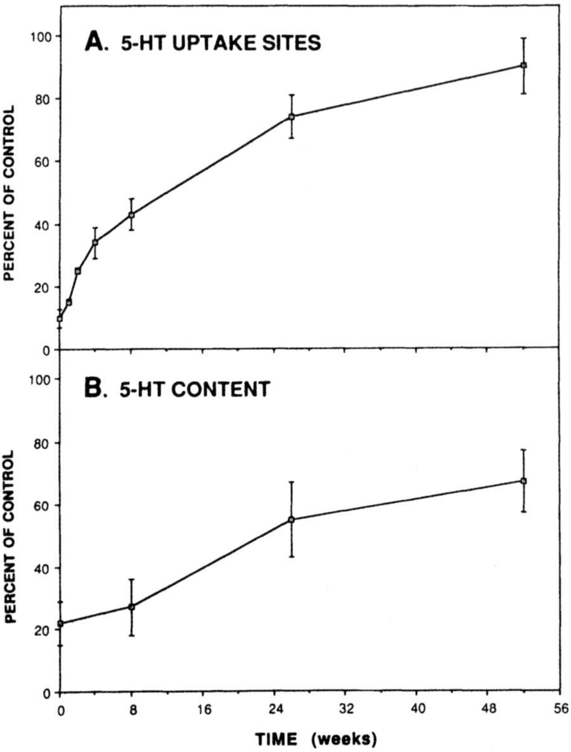

Figure 4. Time course of recovery of (A) serotonin (5-HT) uptake sites and (B) 5-HT content in rat cerebral cortex following repeated systemic administration of MDMA. Rats were injected subcutaneously with either saline or 20 mg/kg MDMA twice a day for 4 consecutive days and then sacrificed at various times up to 12 months later following the last injection of the drug. Saline-injected control rats were killed at each of the time points, and the data that represent the mean ± SEM of five rats per group are plotted as a percent of the value of age-matched saline-injected control rats. (Adapted from Battaglia et al., 1988.)

twice a day for four consecutive days) with 20 mg/kg MDMA, and levels of 5-HT, 5-HIAA, and 5-HT uptake sites were measured seven days later to assess the long-term effects of the treatment [13]. As shown in Figure 5, MDMA caused comparable and marked decreases in 5-HT and 5-HIAA content and in the density of 5-HT uptake sites in rat and guinea pig cerebral cortex but appeared to be without effect on any of these serotonergic markers in the mouse. Other studies [49] have also suggested that mice are less susceptible to the neurotoxic effects of MDMA. In more recent studies, we have also demonstrated that administration of 2.5 or 10 mg/kg MDMA for four consecutive days results in neurotoxic effects in primates, with decreases in the density of 5-HT uptake sites observed following the higher dose [50]. The

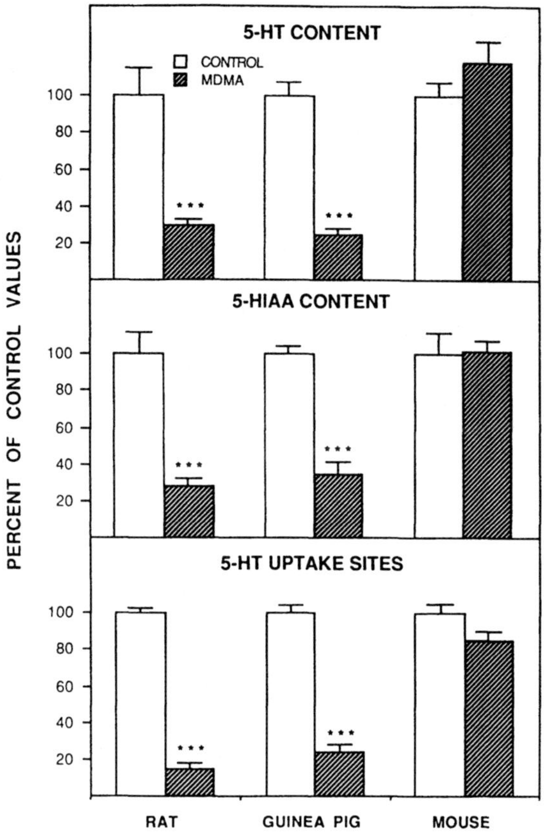

Figure 5. The effects of repeated systemic administration of MDMA on (A) the content of serotonin (5-HT), (B) the content of 5-hydroxyindoleacetic acid (5-HIAA), and (C) the density of 5-HT uptake sites in rat, guinea pig, and mouse frontal cerebral cortex. Animals were treated with saline or 20 mg/kg MDMA twice a day for 4 consecutive days and sacrificed 7 days after the last injection. Data represent mean ± SEM of five animals per group and are expressed as a percent of saline-injected control values in the respective species. In rat, guinea pig, and mouse, control values of 5-HT were 275 ± 41, 296 ± 14, and 449 ± 36 pg/mg tissue, respectively; control values of 5-HIAA were 345 ± 40, 92 ± 4, and 319 ± 34 pg/mg tissue, respectively; control values of 5-HT uptake sites were 397 ± 10, 210 ± 6, and 233 ± 12 fmol/mg protein, respectively. Data were analyzed by Students t-test. *** indicates a significant difference at $p < 0.001$ from respective control values. (From Battaglia et al., 1988.)

neurotoxic effects of MDMA observed in all species included reductions in the content of 5-HT and 5-HIAA and decreases in the density of 5-HT uptake sites. In primates, marked reductions in CSF 5-HIAA levels were also observed after drug administration [50]. These findings and other reports of neurotoxic effects of MDMA in primates [51, 52] raise serious concern for its hazards in humans.

4. NEUROTOXIC EFFECTS OF MDMA IN VARIOUS BRAIN REGIONS

The results of the studies described previously demonstrate the specific and marked neurodegenerative effects of MDMA on 5-HT axons and terminals in cerebral cortex. The following studies will address the neuroanatomical and morphological specificity of the neurotoxic actions of MDMA. Specifically we will describe (1) the neurotoxic and neurodegenerative effects of MDMA and MDA in homogenates of gross brain regions and (2) the neuroanatomic and morphological specificity of MDMA-induced degeneration of serotonergic pathways in rat brain. We have used radioligand binding studies and in vitro autoradiography of [^{3}H]-paroxetine-labeled 5-HT uptake sites to address the latter point.

4.1. Effects of MDMA on monoamine content

The effects of repeated systemic administration of MDMA and MDA on brain monoamine and monoamine metabolite levels were investigated at two weeks following the last injection. As shown in Figure 6, MDMA and MDA produced marked decreases in the content of 5-HT and 5-HIAA in various brain regions. Both MDMA and MDA caused dramatic decreases in 5-HIAA levels in cerebral cortex, hippocampus, striatum, and hypothalamus (figure 6B). In hypothalamus, the reduction in 5-HIAA levels elicited by MDA was significantly greater ($p < 0.05$) than that observed with MDMA (Figure 6B). When plotted as a percent of control values in the respective brain regions, 5-HIAA content was observed to decrease in all the brain regions examined, but the reductions in cerebral cortex and hippocampus (40–60%) were greater than those observed in striatum and hypothalamus (30–40%). With respect to 5-HT levels, marked decreases were observed in cerebral cortex and hypothalamus in both MDMA-treated and MDA-treated rats (Figure 6A). While small decreases were observed in hippocampal and striatal 5-HT content following either MDA or MDMA treatment, these reductions were found to be statistically significant only in striatum ($p < 0.01$) of MDMA-treated rats. When calculated as a percent of control 5-HT levels in the respective brain regions (Figure 6A), the data indicate a more marked reduction in 5-HT in cerebral cortex (40–60%) than in hypothalamus (18–33%).

In contrast with the marked and consistent effects of these compounds on serotonergic systems, neither MDMA nor MDA administration produced any widespread or consistent changes in levels of norepinephrine, dopamine, 3,4-dihydroxyphenylacetic acid (DOPAC), or homovanillic acid (HVA) in the various brain regions examined; however, small changes were observed in some brain regions. Both MDMA and MDA produced statistically significant increases in striatal DOPAC and cerebral cortical HVA content, whereas only MDMA treatment resulted in an increase in hippocampal DOPAC levels (Table 3). Since the changes in catecholamine levels were not accompanied by corresponding changes in the density of dopamine or norepinephrine uptake

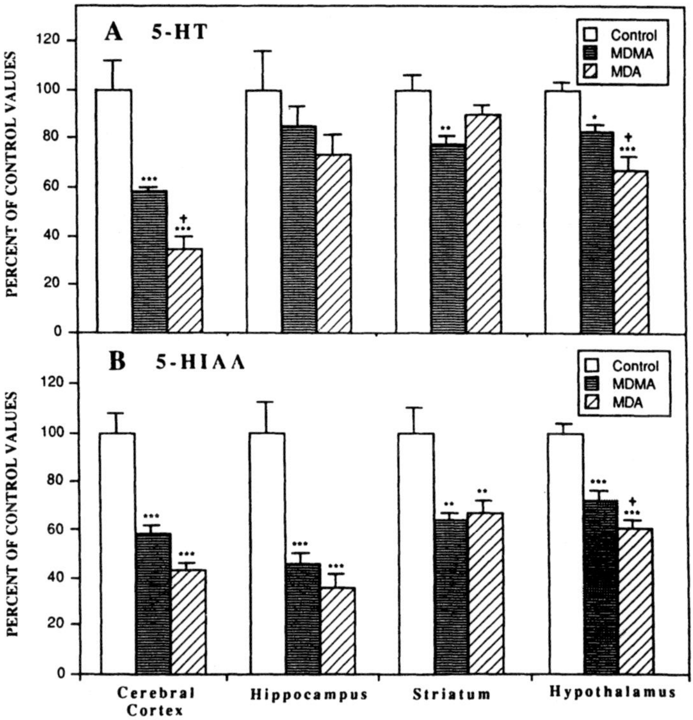

Figure 6. Effect of repeated systemic admininstration of MDMA and MDA on the concentration of (A) serotonin (5-HT) and (B) its metabolite 5-hydroxyindoleacetic acid (5-HIAA) in various brain regions. Rats were injected subcutaneously twice daily for 4 days with drug (20 mg/kg) or saline vehicle (1 ml/kg) and sacrificed at 2 weeks after the last injection. 5-HT and 5-HIAA levels were measured using reversed phase HPLC. Data are plotted as a percent of control values in each brain region and represent the mean and SEM from four to six control and drug-treated rats. Control values for 5-HT and 5-HIAA in each of the regions were as follows: cerebral cortex, 504 ± 58 and 422 ± 32; hippocampus, 410 ± 67 and 684 ± 89; striatum, 363 ± 22 and 492 ± 50; hypothalamus, 1605 ± 55 and 997 ± 42 pg/mg tissue, respectively. Data were analyzed by one-way ANOVA and Duncan's multiple range test. *, **, and *** indicate significant differences at $p < 0.05$, $p < 0.01$, and $p < 0.001$, respectively, from control saline-treated rats. † indicates a significant difference at $p < 0.05$ from MDMA-treated rats. (From Battaglia et al., 1987.)

sites, these effects may be mediated indirectly via MDMA- or MDA-induced alterations in serotonergic pathways.

4.2. Effects of MDMA on monoamine uptake sites

To determine whether changes in 5-HT and/or 5-HIAA were the consequence of long-term suppression of serotonergic function in structurally intact neurons or whether MDMA or MDA may be affecting a neurodegenerative process in each of the brain regions, we measured the density of monoamine uptake sites in these brain regions. Densities of uptake sites for 5-HT [12],

Table 3. Effect of repeated systemic administration of MDMA and MDA on norepinephrine, dopamine, and dopamine metabolite level in various regions of rat brain.

			Concentration (pg/mg tissue)	
Brain Region	NE	DA	DOPAC	HVA
Cerebral cortex				
Control	447 ± 53	59 ± 12	96 ± 14	19 ± 4
MDMA	424 ± 13	72 ± 3	73 ± 5	32 ± 4*
MDA	404 ± 26	63 ± 4	94 ± 19	36 ± 5*
Hippocampus				
Control	528 ± 62	31 ± 9	39 ± 6	6 ± 2
MDMA	572 ± 34	13 ± 5	65 ± 12*	15 ± 5
MDA	608 ± 46	16 ± 4	32 ± 13	8 ± 5
Striatum				
Control	N.D.	6091 ± 596	3212 ± 159	788 ± 58
MDMA	N.D.	6974 ± 228	3954 ± 320*	767 ± 46
MDA	N.D.	6168 ± 569	3669 ± 189*	890 ± 48
Hypothalamus	3320 ± 209	569 ± 40	228 ± 43	54 ± 4
Control	3052 ± 159	475 ± 38	200 ± 30	54 ± 5
MDMA	3577 ± 148	585 ± 67	229 ± 26	58 ± 3
MDA				

Regional brain levels of norepinephrine (NE), dopamine (DA), 3,4-dihydroxy phenylacetic acid (DOPAC), and homovanillic acid (HVA) in rats 2 weeks after administration of 20 mg/kg 3,4-methylenedioxymethamphetamine (MDMA) or 3,4-methylenedioxyamphetamine (MDA). Drugs were administered subcutaneously every 12 hours for 4 consecutive days. Values (in picograms per milligram of tissue) represent the mean and standard error of the mean (SEM.) of determinations in four to six individual rats. N.D. indicates that levels were below the sensitivity of the assay. Data were analyzed by one-way ANOVA and Duncan's multiple range test. * indicates a significant difference from saline-treated control rats at $p < 0.05$.

dopamine and norepinephrine [53] were carried out as previously described. Both MDMA and MDA caused substantial reductions in the densities of [^{3}H]-paroxetine-labeled 5-HT uptake sites in all brain regions examined. As shown in Figure 7, the densities of 5-HT uptake sites are represented as a percent of the respective control values in cerebral cortex, hippocampus, striatum, hypothalamus, and midbrain. Significant reductions (all $p < 0.001$) were observed in cerebral cortex (60–70%). hippocampus (70–75%), striatum (50%), hypothalamus (40–50%), and midbrain (50–60%). Interestingly, MDA produced a significantly greater reduction in the density of 5-HT uptake sites in cerebral cortex than that observed with MDMA. We have also observed that MDE, the N-ethyl derivative of MDA, with a comparable treatment regimen, causes a 40% reduction in 5-HT uptake sites in cerebral cortex, suggesting that this compound may be less toxic than MDA or MDMA. Scatchard analyses of [^{3}H]-paroxetine saturation data in control and drug-treated rats indicated that, in cerebral cortex, [^{3}H]-paroxetine binding was to a single population of sites (Hill coefficient values = 1.02, 1.01, and 1.03 in control, MDMA-treated and MDA-treated animals, respectively) and that there were no significant differences in the K_D values between control and drug-treated rats (18.8, 20.8, and 17.9 pM in control, MDMA-treated, and MDA-treated rats, respectively).

In contrast with the effects of these drugs on the density of 5-HT uptake

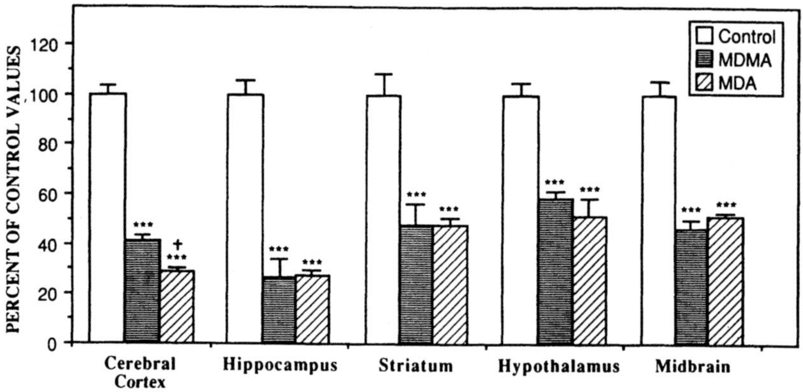

Figure 7. Effect of repeated systemic administration of MDMA and MDA on the density of serotonin (5-HT) uptake sites in various brain regions. Rats were injected subcutaneously twice daily for 4 days with MDMA and MDA (20 mg/kg) or saline vehicle (1 ml/kg) and sacrificed at 2 weeks after the last injection. Values were determined from saturation studies in each of the regions except striatum and hypothalamus, where the density of 5-HT uptake sites was assessed using a saturating concentration (0.25 nM) of [^{3}H]-paroxetine. No significant differences from control K_D values (10–20 pM) were observed in either MDMA- or MDA-treated rats. Data are plotted as a percent of the 5-HT uptake site density observed in controls in each brain region and represent the mean and SEM from 3–6 rats per group. Control values were as follows: cerebral cortex, 338 ± 10; hippocampus, 360 ± 17; striatum, 344 ± 30; hypothalamus, 775 ± 36; and midbrain, 570 ± 16 fmol/mg protein. Data were analyzed by one-way ANOVA and Duncan's multiple range test. Significant differences at $p < 0.001$ from control values are denoted by ***, while differences at $p < 0.001$ between MDA and MDMA treatments are denoted by †. (From Battaglia et al., 1987.)

sites, neither MDMA nor MDA treatment caused any significant reduction in the levels of [^{3}H]-mazindol-labeled norepinephrine uptake sites in cerebral cortex, hippocampus, or midbrain, when compared with the respective saline-treated controls (Figure 8). Although a small reduction was noted in norepinephrine uptake sites in hippocampus, this change was not statistically significant. Similarly, no significant decreases were observed in the density of [^{3}H]-mazindol-labeled dopamine uptake sites in cerebral cortex, hippocampus, striatum, and midbrain, following treatment with MDA. MDMA caused a statistically significant reduction (37%) in the density of dopamine uptake sites only in midbrain. These findings are consistent with the results from our measurements of the content of catecholamines and catecholamine metabolites in various brain regions and support the contention that neither MDMA nor MDA cause any marked widespread alterations in the integrity of catecholaminergic neurons.

4.3. Autoradiographic studies of 5-HT uptake sites

In vitro autoradiographic studies of [^{3}H]-paroxetine-labeled 5-HT uptake sites in brains of saline-treated (control) and MDMA-treated rats were carried out as previously described [54] in order to assess the neuroanatomic localization

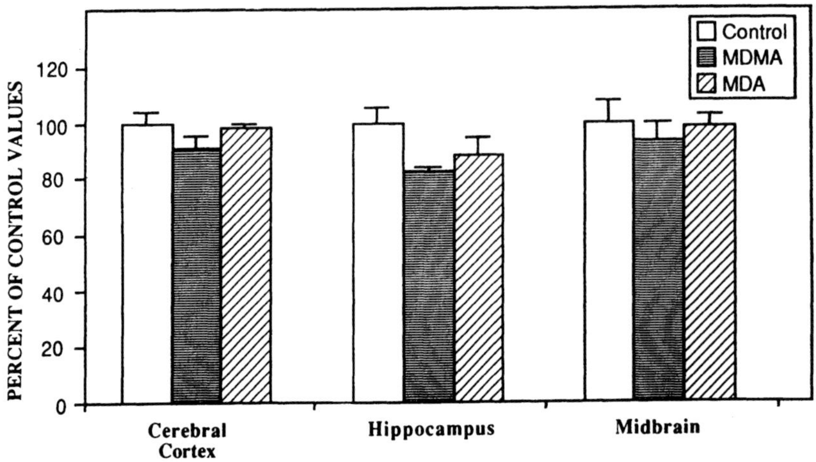

Figure 8. Effect of repeated systemic administration of MDMA and MDA on the density of norepinephrine (NE) uptake sites in various brain regions. Rats were injected subcutaneously twice daily for 4 days with MDMA or MDA (20 mg/kg) or saline vehicle (1 ml/kg) and sacrificed at 2 weeks after the last injection. NE uptake sites were measured using 6 nM [^{3}H]-mazindol in the presence of selective blockers as previously described [53]. Data are plotted as a percent of control values in each brain region and represent the mean and SEM from six control, MDMA-treated, and MDA-treated animals. Control values of NE uptake sites were as follows: cerebral cortex, 164 ± 6; hippocampus, 176 ± 9; midbrain, 157 ± 13 fmol/mg protein. (From Battaglia et al., 1987.)

of lesions induced by MDMA. In a detailed quantitative study, substantial reductions in 5-HT uptake sites (50–100% decreases) were observed in all areas of cerebral cortex as early as 18 hours after a four-day treatment regimen (20 mg/kg MDMA twice daily) and lasted for at least two weeks. As shown in Table 4, cerebral cortical regions that showed the most extensive destruction of 5-HT uptake sites (i.e., >90%) were the prefrontal (area 32), anterior cingulate (area 24), entorhinal, and parietal cortex. Comparable decreases in 5-HT uptake sites were observed between Day 0 (18 hours after last injection) and Day 14 (14 days after last injection) in several regions of cerebral cortex, such as prefrontal, pyriform, frontal area 8 and 10, entorhinal, and primary auditory regions. In other areas of cerebral cortex, such as the sensory motor regions, significant reductions in 5-HT uptake sites were observed only two weeks after the treatment.

As shown in Table 4 and Figure 9, marked decreases in 5-HT uptake sites were observed in all regions of caudate-putamen, olfactory tubercle, endopiriform nucleus, islands of calleja, and nucleus accumbens, following MDMA administration. Within the caudate-putamen, some time-dependent reductions were observed. For example, equivalent decreases in 5-HT uptake sites were observed between 18 hours and two weeks in the ventrolateral region, while in dorsolateral and dorsomedial areas, a significantly greater reduction in 5-HT

Table 4. Effects of repeated systemic administration of MDMA on the regional decreases in [^{3}H]-paroxetine-labeled serotonin uptake sites.

Brain region	Control	MDMA
Cerebral Cortex		
Prefrontal area 32	2−	1
Cingulate area 24	2	1
Indusium griseum	1	2
Pyriform	2	1
Frontal area 8	2	1+
Frontal area 10	1+	1−
Sensorymotor	2	1−
Parietal	1+	0
Entorhinal	4	1+
Primary auditory	1	1−
Primary visual	2+	2
Olfactory tubercle	4	2
Endopiriform nucleus	3	2
Island of calleja	3+	1+
Basal ganglia		
Caudate putamen:		
Dorsolateral	2	1−
Dorsomedial	1+	1−
Ventrolateral	3	2
Ventromedial	2+	1+
Nucleus accumbens	2	1−
Septal area		
Medial septal nucleus	4−	3+
Lateral septal nucleus	3	2
Amygdala basolateral nucleus	4−	3+
Thalamus and epithalamus		
Anteroventral nucleus	3−	0
Anteromedial nucleus	3	0
Anteroventral dorsomedial nucleus	3−	0
Reuniens	4+	1
Lateroposterior nucleus	3	1
Posterior nucleus	1	1
Posterioventromedial nucleus	1	1
Parafascicular nucleus	2	1+
Lateral geniculate body	5−	1+
Medial geniculate body	2	1
Lateral habenula	3	1−
Hypothalamus		
Lateral nucleus	4	4−
Hippocampus		
CA3 region	2+	1
Dentate gyrus	2	1
Molecular layer	2+	1
Parasubiculum	3+	2
Presubiculum	3	2
Midbrain		
Inferior colliculus	3	1−
Interpeduncular nucleus	3+	3
Central gray	5+	5+

Table 4. Cont.

Brain region	Control	MDMA
Superior colliculus:		
Superficial layers	3	1
Profundum	2+	1
Substantia nigra:		
Pars compacta	3+	2
Pars reticulata	3	2+
Paranigral nucleus	2	3
Ventral tegmental area	2+	2
Dorsal raphe nuclei	5+	5+
Median raphe nuclei	5+	5+
Pons-medulla		
Locus coeruleus	5+	5
Pontine reticular formation	2	2
Cerebellum (all lobules)	2	2

The data are based on observations from three animals per group. Rats were injected twice daily subcutaneously for four days with MDMA (20 mg/kg) or saline (1 ml/kg) (control) and sacrificed 14 days after the last injection. The anatomical terminology is derived from Paxinos and Watson [56]. [^{3}H]-Paroxetine binding sites were visualized by using a saturating concentration (0.25 nM) of [^{3}H]-paroxetine. Autoradiograms of rat brain were generated using [^{3}H]-Ultrofilm. Analysis of [^{3}H]-paroxetine-labeled serotonin uptake site densities in the various brain regions was performed by computerized image analysis densitometry. The relative density of [^{3}H]-paroxetine binding sites corresponds to the following range: 1=0–50 fmol/mg tissue; 2=50–150 fmol/mg tissue; 3=150–250 fmol/mg tissue; 4=250–400 fmol/mg tissue; and 5= >400 fmol/mg tissue. + amd − values indicate the upper and lower limits, respectively, of each range. No correction for "grey-white" quenching of tritium was used.

uptake sites was observed only at the later time point. Other brain regions that were sensitive to the neurodegenerative effects of MDMA included various thalamic nuclei and regions of hippocampus. In contrast, the dorsal and medial septal nuclei appeared to be less sensitive to the neurotoxic effects of MDMA, as the reductions (~25%) in 5-HT uptake sites in these regions were not statistically significant. Likewise, no significant reductions were observed in the indusium griseum, which contains primarily 5-HT axons of passage.

Within midbrain structures, regions containing 5-HT projections appeared to be more dramatically affected by MDMA than those containing 5-HT cell bodies (see Figure 10). For example, in both the superficial layers of superior colliculus and profundum, 5-HT uptake sites were reduced 85–90%, while in dorsal and median raphe, central grey, and the ventral tegmental region, there was little or no change after MDMA. Likewise, 5-HT projections to substantia nigra pars compacta and reticulata were markedly affected, whereas no changes in 5-HT uptake sites were observed in the interpreduncular nucleus and pontine reticular formation up to 14 days following MDMA administration.

In order to assess the serotonergic selectivity of the neurodegenerative effects of MDMA in brain, we have carried out additional autoradiographic studies of norepinephrine and dopamine uptake sites in brain regions containing catecholamine terminals and cell bodies. Norepinephrine and dopamine uptake sites were labeled using [^{3}H]-mazindol in the presence of specific

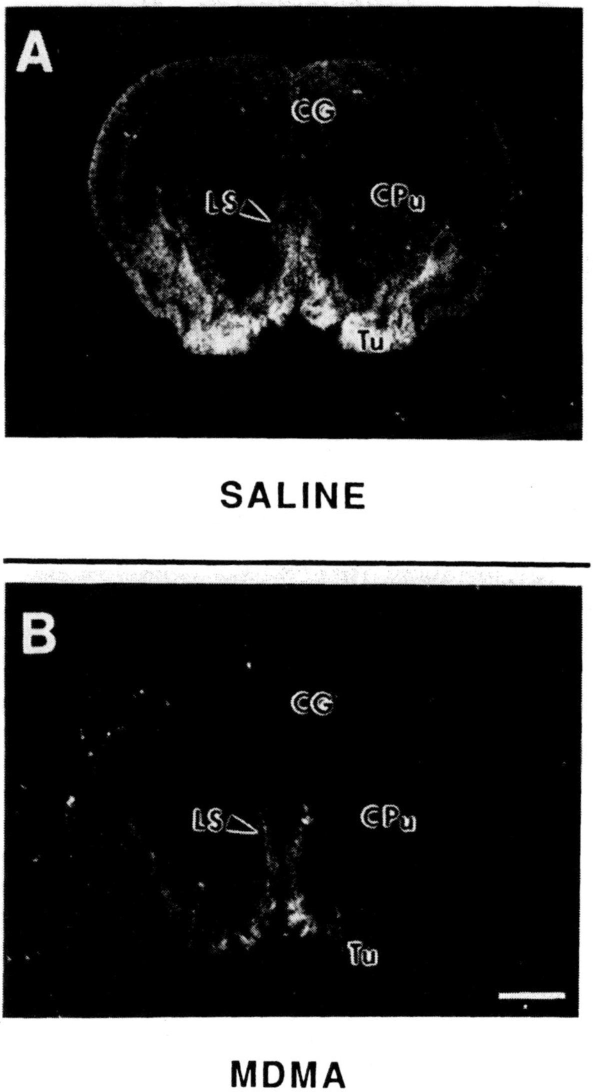

Figure 9. Autoradiographic distribution of [^{3}H]-paroxetine-labeled serotonin uptake sites in coronal sections at the level of the caudateputamen from (A) saline-treated and (B) MDMA-treated rats. These are darkfield photomicrographs (Tritium-sensitive Ultrofilm) in which the autoradiographic grains (i.e., binding sites) appear as white spots and the tissue is not visible. The degree of nonspecific binding defined in the presence of 2 μM citalopram was comparable for both treatments. In A, note the high density of serotonin uptake sites in cingulate cortex (CG), caudate putamen (CPu), olfactory tubercle (Tu), islands of calleja, and lateral septal nuclei (LS) in control brains. In MDMA-treated animals (B), marked reductions were observed in most regions except for the septal nuclei, which were relatively unaffected.

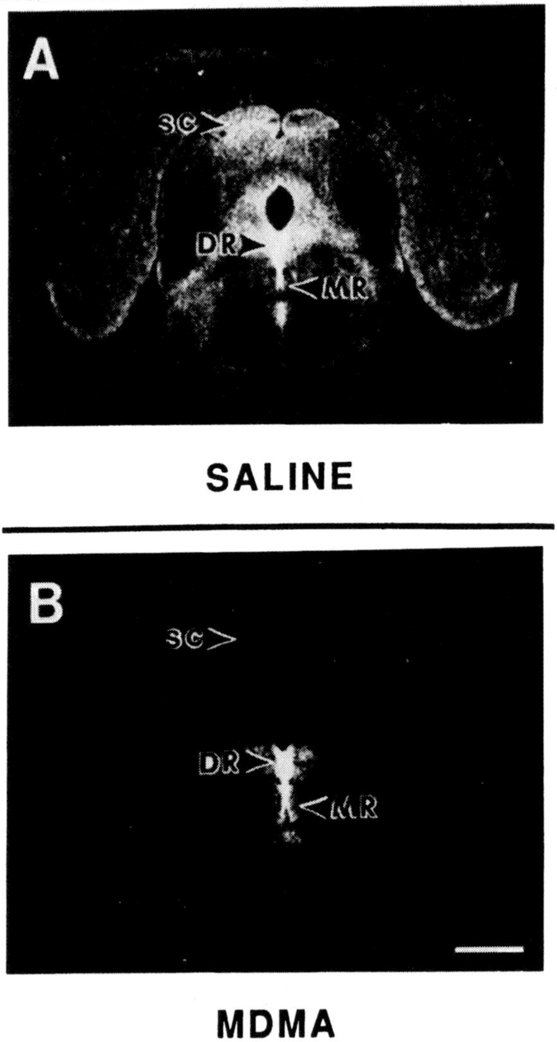

Figure 10. Autoradiographic distribution of [^{3}H]-paroxetine-labeled serotonin uptake sites in coronal sections at the level of midbrain in (A) saline-treated and (B) MDMA-treated rats. In A, note the high density of serotonin uptake sites in control brain in regions containing serotonin projections, such as entorhinal cortex, superior colliculus (SC), presubiculum, and parasubiculum, as well as cell body regions, such as dorsal (DR) and median (MR) raphe and central grey. MDMA-treated animals exhibited marked reductions in [^{3}H]-paroxetine binding sites in pre- and para-subiculum, entorhinal cortex, and superior colliculus (SC), while no change in densities were observed in areas containing primarily serotonin perikarya, such as the raphe nuclei (DR and MR).

blockers, as previously published [55]. With respect to norepinephrine uptake sites, we observed no change from control levels of binding sites in midbrain regions, such as locus coeruleus, interpeduncular nucleus, or substantia nigra pars compacta or reticulata, up to 14 days following MDMA treatement. In a number of cerebral cortical regions that receive norepinephrine projections, norepinephrine uptake sites were not decreased at 18 hours after treatment but appeared to be slightly increased at Day 14. Consistent with the minimal effect of MDMA on parameters associated with catecholamine neurons, there were no changes in the density of dopamine uptake sites when compared to levels in saline-treated rats in either cell body regions, such as substantia nigra pars compacta and reticulata, or terminal regions, such as caudate-putamen, nucleus accumbens, and olfactory tubercle. These results are, therefore, consistent with what has been observed in homogenate studies and indicate that the neurodegenerative effects of MDMA appear to be confined primarily to serotonergic neurons since this treatment regimen did not reduce the density of uptake sites associated with catecholamine-containing neurons. Additional studies are necessary to further access the neuroanatomic localization of any long-term compensatory changes in norepinephine projections or other neurotransmitter recognition sites that may occur as a consequence of MDMA lesion of 5-HT pathways.

5. SUMMARY AND CONCLUSIONS

The present chapter attempts to elucidate some of the in vitro and in vivo properties of the drug 3,4-methylenedioxymethamphetamine (MDMA). We have utilized in vitro radioligand binding studies to obtain a detailed pharmacologic profile of MDMA at various established brain receptors and recognition sites, as well as to assess the putative role of novel [^{3}H]-MDMA binding sites in brain. With respect to established neurotransmitter receptors and recognition sites, the relatively high affinities of MDMA at pre-synaptic and post-synaptic serotonin recognition sites (i.e., 5-HT uptake sites, 5-HT$_2$ and 5-HT$_{1A}$ serotonin receptors) and α_2-adrenergic receptors suggest that these sites may play a significant role in both the psychotomimetic and neurodegenerative properties of this drug. In addition, some preliminary data from [^{3}H]-MDMA binding studies suggest that MDMA may interact with a novel recognition site in brain, which may play some role in the actions of this compound. It is unlikely that the anxiolytic-like or reinforcing properties of MDMA are mediated via benzodiazepine or opioid receptors since the drug exhibits much lower affinities for these binding sites.

We have also addressed the neurochemical effects of MDMA following in vivo administration of the drug and have provided substantial evidence that this drug causes widespread and long-lasting degeneration of serotonin neurons in brain. With respect to the parameters involved in the neurotoxic and neurodegenerative actions of MDMA, the data indicate that: (1) the severity of lesion by MDMA is dependent on both the dose and the frequency of drug administration; (2) MDMA-induced neurodegenerative effects are preferentially localized to serotonergic neurons while sparing catecholamine neurons; (3) the neurodegenerative effects of MDMA can be elicited in a number of animal species, including primates; (4) the neurodegenerative effects of MDMA on 5-HT neurons can be prevented by 5-HT uptake blockers, suggesting a role for the active uptake of the drug or a neurotoxic metabolite of MDMA; and (5) the neurodegenerative effects are long-lasting (up to one year) with respect to structural recovery, while functional recovery may be permanently impaired.

In addition, we have reported that the neurotoxic and neurodegenerative effects of MDMA are widespread, with marked reductions in all 5-HT parameters observed in a number of brain regions. In vitro autoradiographic studies have revealed that there is neuroanatomic and morphologic specificity to the neurodegenerative effects of MDMA, as evidenced from the predominant reductions in 5-HT uptake sites in brain regions containing primarily 5-HT terminals whereas regions containing 5-HT axons of passage and perikarya are relatively unaffected. These findings have been confirmed by additional data from 5-HT immunocytochemical studies [16] that demonstrate the specificity of MDMA for neurodegeneration of 5-HT axons and terminals, while 5-HT perikaya are spared.

ACKNOWLEDGEMENTS

We would like to thank Drs. S.Y. Yeh, Thomas R. Insel, and Michael J. Kuhar for their contributions to various aspects of the work described. We also thank Theresa Kopajtic, Chai Kulsakdinun, and Brian Brooks for technical assistance and Mary Flutka and Sharon Amos for manuscript preparation. Certain aspects of this work were supported in part from funds provided by the U.S. Food and Drug Administration.

REFERENCES

1. Anderson, G.M., Braun, G., Braun, U., Nichols, D.E., and Shulgin, A.T., 1978. Absolute configuration and psychotomimetic activity. NIDA Research Monograph 22:8–15.
2. Braun, U., Shulgin, A.T., and Braun, G., 1980. Study of the central nervous activity and analgesia of the N-substituted analogs of the amphetamine derivative 3,4-methylenedioxyphenylisopropylamine. Drug Res. 30:825–830.
3. Shulgin, A.T., 1986. The background and chemistry of MDMA. J. Psychoactive Drugs 18:291–304.
4. Adler, J., Abramson, B., Katz, S., and Hager, H., 1985. Getting high on "Ecstasy." Newsweek, April 15, p. 96.
5. Gertz, K.R., 1986. The agony of ecstasy. Science Digest, February, pp. 27.
6. Battaglia, G. and De Souza, E.B., 1987. New perspectives on MDMA (3,4-methylenedioxymethamphetamine). Substance Abuse 8:31–42.
7. Barnes, D.M., 1988. New data intensify the agony over ecstasy. Science 239:864–866.
8. MDMA Conference Proceedings, 1986. J. Psychoactive Drugs. Vol. 18.
9. Pharmacology and toxicology of amphetamines and related designer drugs, 1989. NIDA Research Monograph, in press.
10. Beardsley, P.M., Balster, R.L., and Harris, L.S., 1986. Self-administration of methylenedioxymethamphetamine (MDMA) by rhesus monkeys. Drug Alc. Depend. 18:149–157.
11. Lamb, R.J. and Griffiths, R.R., 1987. Self-administration of d,1-methylenedioxymetham-

phetamine (MDMA) in the baboon. Psychopharmacology 91: 268–272.
12. Battaglia, G., Yeh, S.Y., O'Hearn, E., Kuhar, M.J., Molliver, M.E., and De Souza, E.B., 1987. 3,4-methylenedioxymethamphetamine and 3,4-methylenedioxyamphetamine destroy serotonin terminals in rat brain: Quantification of neurodegeneration by measurement of ^{3}H-paroxetine-labeled serotonin uptake sites. J. Pharmacol. Exp. Ther. 242:911–916.
13. Battaglia, G., Yeh, S.Y., and De Souza, E.B., 1988. MDMA-induced neurotoxicity: Degeneration and recovery of brain serotonin neurons. Pharmacol. Biochem. Behav. 29:269–274.
14. Commins, D.L., Mosmer, G., Virus, R.M., Woolverton, W.L., Schuster, C.R., and Seiden, L.S., 1987. Biochemical and histological evidence that methylenedioxymethamphetamine (MDMA) is toxic to neurons in the rat brain. J. Pharmacol. Exp. Ther. 241:338–345.
15. Schmidt, C.J., 1986. Neurotoxicity of the psychedelic amphetamine, methylenedioxymethamphetamine. J. Pharmacol. Exp. Ther. 240:1–7.
16. O'Hearn, E., Battaglia, G., De Souza, E.B., Kuhar, M.J., and Molliver, M.E., 1988. Methylenedioxyamphetamine (MDA) and methylenedioxymethamphetamine (MDMA) cause ablation of serotonin axon terminals in forebrain: Immunocytochemical evidence. J. Neurosci., 8:2788–2803.
17. Ricaurte, G., Bryan, G., Strauss, L., Seiden, L., and Schuster, C., 1985. Hallucinogenic amphetamine selectively destroys brain serotonin nerve terminals. Science 229:986–988.
18. Stone, D.M., Stahl, D.C., Hansen, G.R., and Gibb, J.W., 1986. The effects of 3,4 methylenedioxymethamphetamine (MDMA) and 3,4 methylenedioxyamphetamine (MDA) on monoaminergic systems in the rat brain. Eur. J. Pharmacol. 128:41–48.
19. Battaglia, G., Brooks, B., Kulsakdinun, C., and De Souza, E.B., 1988. Pharmacologic profile of MDMA (3,4-methylenedioxymethamphtamine) at various brain recognition sites. Eur. J. Pharmacol. 149:159–163.
20. Munson, P. and Rodbard, 1980. LIGAND: a versatile approach for characterization of ligand-binding systems. Anal. Biochem. 107:220–239.
21. Lyon, R.A., Davis, K.A., and Titeler. M., 1987. ^{3}H-DOB (4-bromo-2, 5-methoxyphenylisopropylamine) labels a guanine nucleotide-sensitive state of corticol 5-HT_2 receptors. J. Pharmacol. Exp. Ther. 31:194–199.
22. Shannon, M., Battaglia, G., Glennon, R.A., and Titeler, M., 1984. 5-HT_2 and 5-HT_1 serotonin receptor binding properties of derivatives of the hallucinogen 1-(2,5-dimethoxyphenyl)-2-aminopropane (2,5 DMA). Eur. J. Pharmacol. 102:23–29.
23. Glennon, R.A., Titeler, M., and McKenney, J.D., 1984. Evidence for 5-HT_2 involvement in the mechanism of action of hallucinogenic agents. Life Sci. 35:2505–2511.
24. Titeler, M., Lyon, R.A., and Glennon, R.A., 1988. Radioligand binding evidence implicates the brain 5-HT_2 receptor as a site of action for LSD and phenylisopropylamine hallucinogens. Psychopharmacol. 94:213–216.
25. Battaglia, G., Kuhar, M.J., and De Souza, E.B., 1986. MDA and MDMA (ecstasy) interactions with brain serotonin receptors and uptake sites: *In vitro* studies. Soc. Neurosci. Abstr. 12:1234.
26. Lyon, R.A., Glennon, R.A., and Titeler, M., 1986. 3,4 methylenedioxymethamphetamine (MDMA): Stereoselective interactions at 5-HT_1 and 5-HT_2 receptors. Psychopharmacol. 88:525–526.
27. Battaglia, G., Shannon, M., and Titeler, M., 1984. Guanyl nucleotide and divalent cation regulation of cortical S_2 serotonin receptors. J. Neurochem. 43:1213–1219.
28. Titeler, M., Battaglia, G., and Shannon, M., 1984. Guanine nucleotides modulate cortical S_2 serotonin receptors. J. Rec. Res. 4: 705–712.
29. Titeler, M., Herrick, K., Lyon, R.A., and Glennon, R.A., 1985. ^{3}H-DOB: A specific radioligand for 5-HT_2 sites. Eur. J. Pharmacol. 117:145–147.
30. Gold, L.H. and Kobb, G.F., 1988. Methysergide potentiates the hyperactivity produced by MDMA in rats. Pharmacol. Biochem. Behav. 29: 645–648.
31. Taylor, D.P., 1988. Buspirone, A new approach to the treatment of anxiety, FASEB. 2:2445–2452.
32. Nichols, D.E., Lloyd, D.H., Hoffman, A.J., Nichols, M.B., and Yim, G.K., 1982. Effect of certain hallucinogenic amphetamine analogues on the release of ^{3}H-serotonin from rat brain synaptosomes. J. Med. Chem. 25:530–535.
33. Johnson, M.P., Hoffman, A.J., and Nichols, D.E., 1986. Effects of the enantiomers of MDA, MDMA and related analogues on [^{3}H]serotonin and [^{3}H]dopamine release from superfused

rat brain slices. Eur. J. Pharmacol. 132:269–276.
34. Schmidt, C.J. and Taylor, V.L., 1987. Depression of rat brain tryptophan hydroxylase activity following the acute administration of methylenedioxymethamphetamine. Biochem. Pharmacol. 36:4095–4102.
35. Schechter, M.D., 1986. Discriminative profile of MDMA. Pharmacol. Biochem. Behav. 24:1533–1536.
36. Timmermans, P.B.M.W.M. and van Zwieten, P.A., 1982. Alpha-2 adrenoreceptors: Classification, localization mechanisms and targets for drugs. J. Med. Chem. 25:1389–1401.
37. Gehlert, D.R., Schmidt, C.J., Wu, L., and Lovenberg, W., 1985. Evidence for specific methylenedioxymethamphetamine (ecstasy) binding sites in the rat brain. Eur. J. Pharmacol. 119:135–136.
38. Wang, S.S., Ricaurte, G.A., and Peroutka, S.J., 1987. [^{3}]3,4-Methylenedioxymethamphetamine (MDMA) interactions with brain membranes and glass fiber filter paper. Eur. J. Pharmacol. 138:439–443.
39. Culp, S., Zaczek, R., and De Souza, E.B., 1988. Incorporation of (MDA) into rat brain synaptosomes. Soc. Neurosci. Abstr., Vol. 14.
40. Marcusson, J., Fowler, C.J., Nall, H., Ross, S.B., and Winblad, B., 1985. "Specific" binding of [^{3}H] imipramine to protease-sensitive and protease-resistant sites. J. Neurochem. 44: 705–711.
41. Stone, D.M., Johnson, M., Hanson, G.R., and Gibb, J.W., 1987. A comparison of the neurotoxic potential of methylenedioxyamphetamine (MDA) and its N-methylated and N-ethylated derivatives. Eur. J. Pharmacol. 134:245–248.
42. Kuhar, M.J. and Aghajanian, G.K., 1973. Selective accumulation of ^{3}H-serotonin by nerve terminals of raphe neurons: An autoradiographic study. Nature 241:187–189.
43. Finnegan, K.T., Ricaurte, G.A., Ritchie, L.D., Irwin, I., Peroutka, S.J., and Langston, J.W., 1988. Orally administered MDMA causes a long-term depletion of serotonin in rat brain. Brain Res. 447:141–144.
44. Ross, S.B., Ogren, S.O., and Renyi, L., 1976. Antagonism of the acute and long-term biochemical effects of 4-chloramphetamine on the 5-HT neurons in rat brain by inhibitors of the 5-hydroxytryptamine uptake. Acta Pharmacol. Toxicol. 39:456–476.
45. Sanders-Bush, E. and Steranka, L.A., 1978. Immediate and long-term effects of p-chloroamphetamine on brain amines. Ann. N.Y. Acad. Sci. 305:208–221.
46. Moliver, M.E., O'Hearn, E., Battaglia, G., and De Souza, E.B., 1986. Direct intracerebral administration of MDA and MDMA does not produce serotonin neurotoxicity. Soc. Neurosci. Abstr. 12:1234.
47. Brady, J.F., DiStefano, E.W., and Cho, A.K., 1986. Spectral and inhibitory interactions of (±)3,4-methylenedioxymethamphetamine (MDA) and (±)3,4-methylenedioxymethamphetamine (MDMA) with rat hepatic microsomes. Life Sci. 39:1457–1464.
48. Caldwell, J., 1980. The metabolism of amphetamines and related stimulants in animals and man. In *Amphetamines and Related Stimulants: Chemical Biological Clinical and Sociological Aspects* (Caldwell, J., ed.) Boca Raton, FL: CRC Press, pp. 29–46.
49. Stone, D.M., Hanson, G.R., and Gibb, J.W., 1987. Differences in the central serotonergic effects of methylenedioxymethamphetamine (MDMA) in mice and rats. Neuropharmacology 26: 1657–1661.
50. Johannessen, J.N., Insel, T.R., Battaglia, G., Kuhar, M.J., and De Souza, E.B., 1988. MDMA selectively destroys brain serotonin terminals in rhesus monkeys. Soc. Neurosci. Abstr., Vol. 14.
51. Ricaurte, G.A., DeLanny, L.E., Irwin, I., and Langstrom, J.W., 1988. Toxic effects of MDMA on central serotonergic neurons in the primate: Importance of route and frequency of administration. Brain Res. 446:165–168.
52. Ricaurte, G.A., Forno, L.S., Wilson, M.A., DeLanny, L.E., Irwin, I., Molliver, M., and Langston, J.W., 1988. (±)3,4 Methylenedioxymethamphetamine selectively damages central serotonergic neurons in non-human primates. JAMA 260:51–55.
53. Javitch, J.A., Blaustein, R.O., and Snyder, S.H., 1984. [^{3}H]Mazindol binding associated with neuronal dopamine and norepinephrine uptake sites. Mol. Pharmacol. 26: 35–44.
54. De Souza, E.B. and Kuyatt, B.L., 1987. Autoradiographic localization of ^{3}H-paroxetine-labeled serotonin uptake sites in rat. Synapse 1:488–496.

55. Javitch, J.A., Strittmatter, S.M., and Snyder, S.H., 1985. Differential visualization of dopamine and norepinephrine uptake sites in rat brain using ^{3}H-mazindol autoradiography. J. Neurosci. 5:1513–1521.
56. Paxinos, G. and Watson, C., 1982. The rat brain and sterotaxic coordinates. Sydney: Academic Press.

11. A TISSUE CULTURE MODEL OF MDMA TOXICITY

PATRICA M. WHITAKER-AZMITIA AND EFRAIN C. AZMITIA

1. INTRODUCTION

Tissue culture of central nervous system neurons gives a unique opportunity to study both toxic and trophic effects of drugs and other substances on the growth and development of these cells. The substances under examination can simply be added to the culture media, and detailed dose-response and time-course relationships can be studied in a way that no other system affords. Moreover, by altering the cellular composition of the cultures, it can be determined if the drug is acting on a presynaptic or postsynaptic site or even if the effect is mediated through a non-neuronal cell, such as astrocytes [1].

Admittedly, there are shortcomings with the method. For example, direct application of the drugs under investigation to the cultures might result in concentrations that would never be attained in vivo. However, for studying the toxic effects of the isomers of MDMA, tissue culture is a useful system because it can be used to explore the mechanism, on a cellular level, of the toxicity that has already been observed in vivo [2].

We have approached the problem of MDMA toxicity to serotonergic neurons by first observing the effects of various concentrations of each isomer on the growth of these neurons in primary cultures derived from fetal rat brains. From there we have determined the cellular mediators of the toxicity by removing astrocytes (using an anti-mitotic agent, floxuridine, FUDR) or by adding a target tissue to the cultures. In this case, we have chosen to use hippocampus.

After determining the concentration of MDMA that is toxic to the cultures, we have used information from animal studies to predict which subcellular constituents of the cultures would be affected at that concentration and would thus produce the toxicity. For example, MDMA can affect the serotonergic system through actions on the release [3] or re-uptake [4] of the neurotransmitter. There have also been several monoamine receptor interactions reported [5]. To test the individual components of MDMA action, therefore, we devised experiments that were either pharmacologically mimicking or pharmacologically antagonistic. To test the effects of releasing agents, we used p-chloroamphetamine (PCA), and for uptake inhibition, we used fluoxetine. Finally, to test for specific receptor interactions, we observed the effects of MDMA in the presence of mianserin (a 5-HT_2 antagonist), chlorpheniramine or terfenadine (histamine antagonists), or BHT 920 (an alpha2 agonist).

2. TISSUE CULTURE SYSTEM

2.1. Culture of neurons

Embryos at 15 days of gestation were removed by Caesarean section from time-mated pregnant rats (Hilltop Breeding Laboratories). The brains were removed and placed in Eagle's minimum essential media with 1% glucose. A raphe slice that contains the majority of the midbrain serotonergic cell bodies was dissected [6].

Briefly, two scalpels were used to transect the fetal brain at the pontine and mesencephalic flexures. A cut was made through the cerebral aqueduct to expose the tegmentum. Two longitudinal cuts were made approximately 0.5 mm off the midline. The underlying meninges were removed and the strip transferred to Ca^{++}/Mg^{++} free MEM containing 1% glucose.

The strips were transferred to 5–10 ml of Versene (Gibco brand EDTA) solution and gently agitated by repeated trituration using a fire-polished Pasteur pipette. The cell suspension was spun at 500 g for five minutes and the pellet resuspended in complete media (MEM with non-essential amino acids, 1% glucose, and 5% fetal calf serum) before being spun again. The final pellet was resuspended in complete media (approximately 0.5 cc/raphe strip). The complete media consisted of MEM with non-essential amino acids, 1% glucose, and 10% fetal calf serum.

The cells were plated onto 96-well Linbro plates (Falcon Labware) previously coated with poly-l-lysine (25 mcg/ml) for quantitation by specific uptake of serotonin, or onto 8-well slides (Titer Tek) previously coated with poly-l-lysine for morphological assessment using immunocytochemistry (see below).

The cells were plated at initial plating densities averaging 1.2×10^6 cells/cm^2. In most experiments, the cells were maintained in an incubator at 37° in an atmosphere of 95%/5% CO_2/O_2 for five days. Drugs to be tested were added aseptically either at the time of plating or at various timepoints throughout the five days. Three to four wells were used for each drug condition in at least two separate experiments. In any one experiment, up to twenty-three concentrations of a drug could be tested, generally in a range from 10^{-15} to 10^{-3} M.

2.2. Cellular composition

As a model of a complete serotonergic synapse, cultures of raphe neurons were co-cultured with cells from the hippocampus. Briefly, the hippocampal regions from 18 day rat fetuses were dissected out and freed of meninges and choroid plexus. The tissue was minced and dissociated as described for raphe cells. Using this method, we have shown that development of the serotonergic neurons is enhanced by presence of target regions [1].

Since the toxicity of the dopaminergic toxin MPTP is mediated via astroglial cells, we tested the role of astroglial cells in MDMA toxicity by incubating cultures with an anti-mitotic, floxuridine (FUDR). Astroglial cells are actively dividing in our cultures and are therefore inhibited or destroyed by the addition of 20 microgm/ml FUDR to the culture media. We have found that this addition has little if any effect on the serotonergic neurons during the short period of time we have been growing the cultures.

2.3. Quantitation of growth/toxicity

Immunocytochemistry was performed using a specific antibody raised against serotonin conjugated to hemocyanin. The cultures were pretreated with 10^{-4} pargyline for 30 minutes, followed by 10^{-5} L-tryptophan for one hour. Cultures were washed with tris-buffered saline (TBS) and fixed with 4% paraformaldehyde containing 0.05% $MgCl_2$ and 5% glucose. After two hours fixation, the cells were rinsed with TBS and processed using the ABC kit (Vectastain) with DAB as substrate. Morphological analysis was performed on ten randomly selected cells from coded cultures for each culture condition. The somal area and cumulative process length were calculated using a Bioquant Computer System IV (R&M Biometrics, Inc.) with a Leitz Diaplan microscope with a 25X Pl-apo objective.

Cultures (in 96-well plates) were washed with MEM, and fresh MEM with 0.5% glucose and 50 nM ^{3}H-5-HT (NEN, 26 Ci/mmole) was added for twenty minutes at 37%. Non-specific uptake was determined in the presence of 10^{-5} M fluoxetine. After incubation, cultures were washed twice with MEM and retained radiolabel extracted with absolute ethanol and counted in a Beckman liquid scintillation counter. Results (expressed as cpm/well) were statistically analyzed and plotted using Sigma Plot (Version 3; Jandel Scientific).

3. MDMA EFFECTS ON CULTURED NEURONS

Figure 1 shows the specific uptake of serotonin in cultures of neurons exposed to the isomers of MDMA. The approximate K_i for toxicity is 8 micromolar for *S*-(+)-MDMA and 90 micromolar for *R*-(−)-MDMA. Morphometric

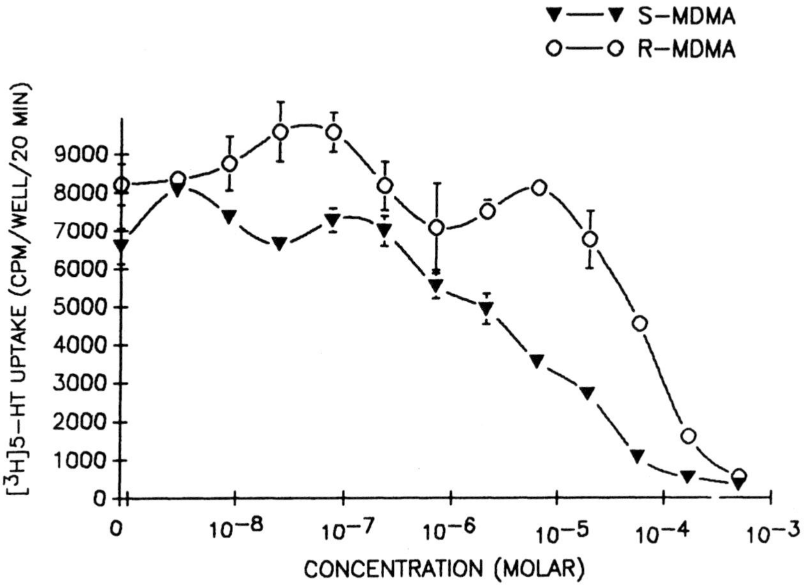

Figure 1. Effects of various concentrations of the isomers of MDMA on specific uptake of [^{3}H]-serotonin into primary cultures of serotonergic neurons maintained in culture for four days. Each point represents the mean and SEM from four cultures.

analysis (Table 1) showed significant degeneration of serotonergic cell bodies at an MDMA concentration of 5×10^{-4} but not 5×10^{-7} M. Moreover, exposure of cultures to MDMA not only reduced cell number but also decreased the process outgrowth from those neurons which remained (Figure 2).

In studies pharmacologically mimicking the known actions of MDMA, the releasing agent p-chloroamphetamine (PCA) was examined (Figure 3) and found to have an approximate K_i of 1×10^{-5} M. The specific serotonin uptake inhibitor fluoxetine, was also found to be toxic (Figure 3), with an approximate K_i of 100 nM. However, at lower (non-toxic) doses (10^{-7} M), fluoxetine attenuated the toxicity of 10^{-4} MDMA (Table 2).

Table 1. Effects of the stereoisomers of MDMA on survival of serotonergic neurons in tissue culture.

Number of cells/unit area		
Control		8.7 ± 1.5
S(+)-MDMA:	10^{-10} M	8.7 ± 1.5
	5×10^{-7} M	9.8 ± .15
	5×10^{-4} M	2.3 ± .8
R(−)-MDMA:	10^{-10} M	8.5 ± .2
	5×10^{-7} M	9.5 ± 2
	5×10^{-4} M	3.5 ± .5

In studies pharmacologically blocking the known effects of MDMA, we examined antagonists for the following receptors: $serotonin_2$ (mianserin), histamine (chlorpheniramine and terfenadine), and the $alpha_2$ agonist BHT 920. Since there has also been a report on MDMA causing release of dopamine, we studied the effects of the dopamine uptake inhibitor nomifensine. Toxicity was not attenuated by any of these agents, with the exception of BHT 920. At a concentration of 10^{-8} M, BHT 929 inhibited the toxicity of *S*-(+)-MDMA by 49% and *R*-(−)-MDMA by 34%. All of these results are given in Table 3.

The final series of experiments aimed at localizing the cellular site of toxicity showed that toxicity is attenuated by either postsynaptic tissues or astroglial cells. In co-cultures of raphe cells with hippocampus, the toxicity curves are shifted such that the two isomers have identical profiles; that is, the *S*(+) isomer becomes 10X less toxic (Figure 4). In cultures that have been treated with FUDR, MDMA toxicity is greatly increased (Figure 5).

4. CONCLUSIONS

Using a tissue culture model of serotonergic neurons, we have studied the mechanism of MDMA toxicity. Although our work is still in progress, we can draw some preliminary conclusions. First, the *S*(+) isomer of MDMA is approximately ten times more potent in producing toxicity than is the *R*(−) isomer. This is comparable to results in vivo, which have also indicated that this is the more toxic isomer [3]. However, the in vivo studies indicate that the *R*(−) isomer is only toxic (ie., causes reduction of serotonin) acutely and that one week after treatment, control levels of serotonin are found. Conversely, the *S*(+) isomer produces a long-lasting depletion. This discrepancy could be related to the fact that our model allows the neurons to be left in contact with the MDMA for four days without a washout period, before measuring serotonin uptake. Also, it is possible that the recovery of serotonin levels would occur in our cultures, as well, if they were maintained for one week. Both of these issues will be addressed in future experiments, as they will give important information on the two forms of toxicity that appear to occur.

One notable difference between the isomers of MDMA, is that *S*(+)-MDMA is approximately tenfold more potent on the release of serotonin. This then seems a likely candidate to at least partially explain the toxicity of MDMA. There is further evidence from other drugs we have tested.

The only receptor-active drug that we have tried and that did have an effect is the $alpha_2$ agonist BHT 920. This drug inhibited the toxicity produced by either isomer with an approximate K_i value of 10 nM. $Alpha_2$ adrenergic receptors are known to occur on serotonergic neurons and to be inhibitory to release to serotonin [7]. Thus inhibition of release of serotonin appears to be one mechanism by which toxicity is attenuated. This would be consistent with the observation that the serotonin releaser PCA is also toxic. Whether or not release is mediated by MDMA through an action on the same receptor (for

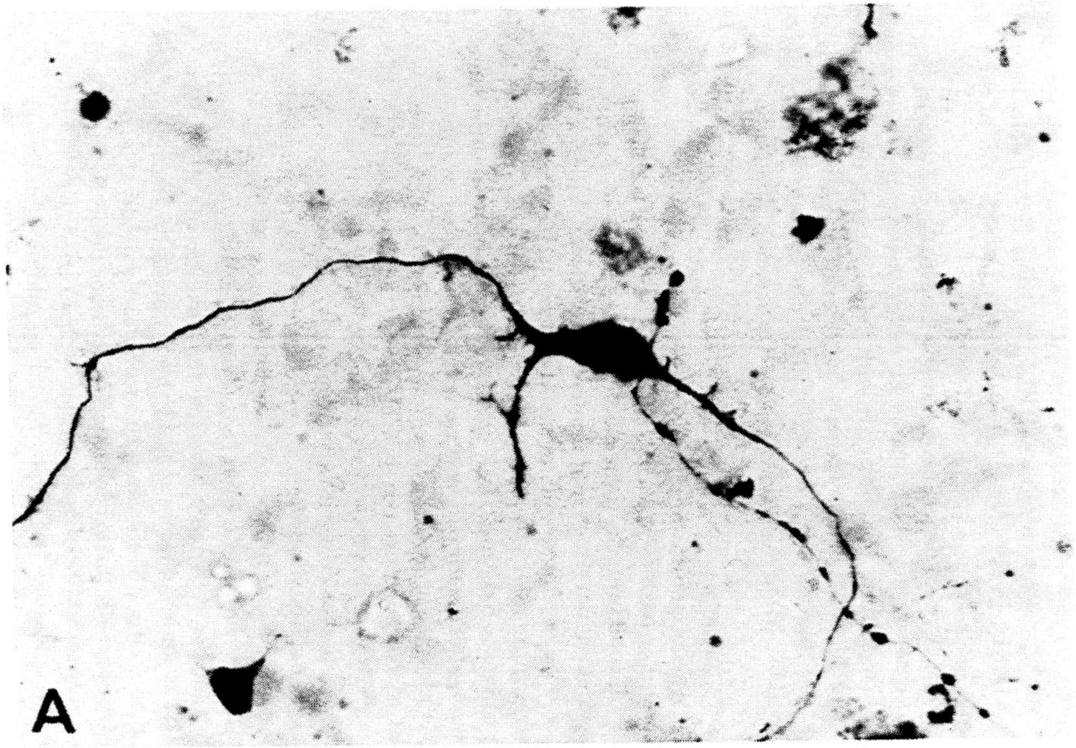
A

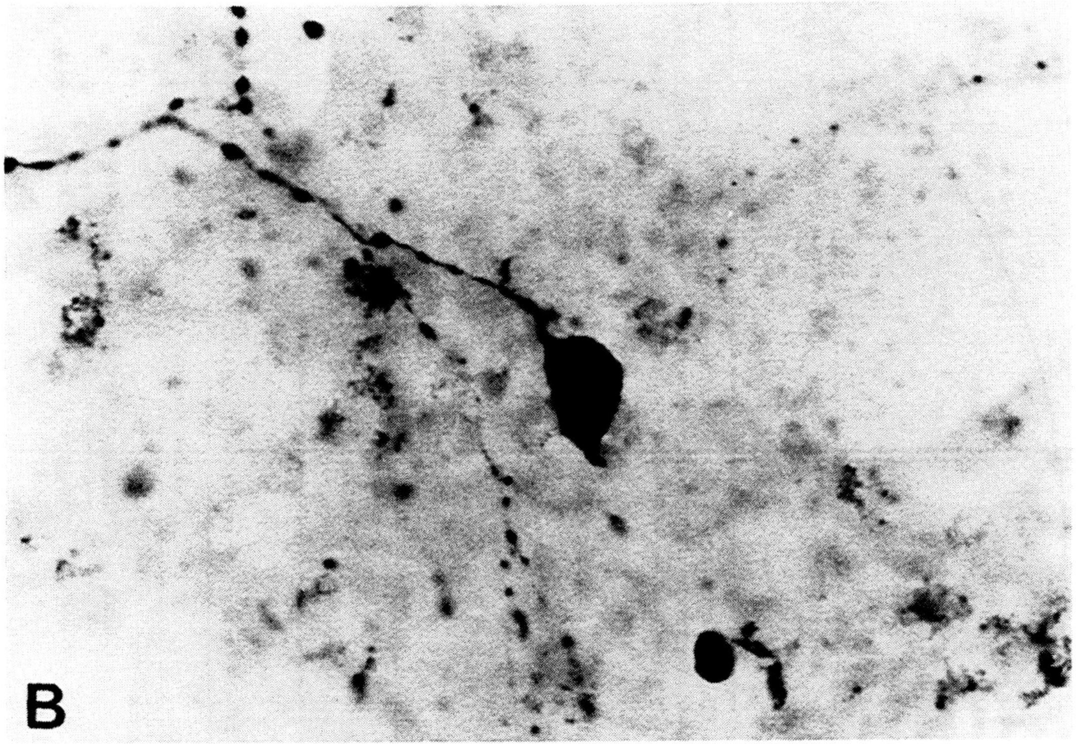

Figure 2. Immunocytochemically stained serotonergic neurons grown in culture for four days without (A) or with (B) 10^{-4} M MDMA.

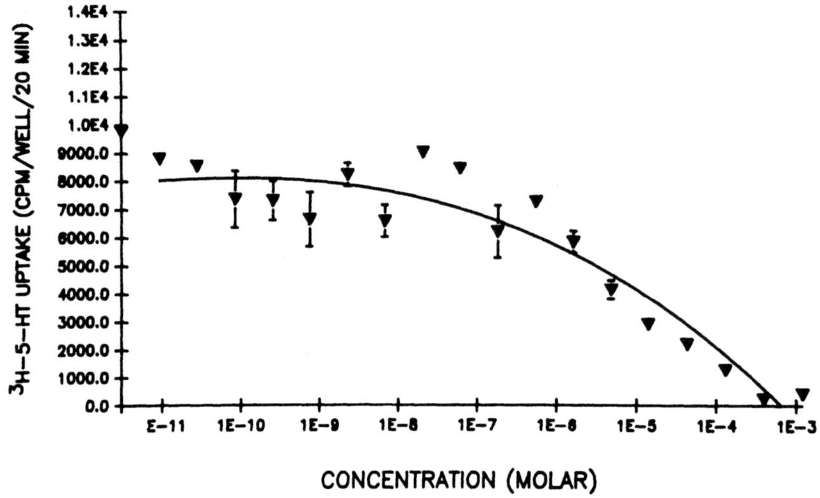

Figure 3. Effects of various concentrations of p-chloramphetamine (PCA) on specific uptake of [^{3}H]-serotonin into primary cultures of serotonin neurons maintained in culture for four days. Each point represents the mean and SEM from four cultures.

Table 2. Effect of the serotonin uptake inhibitor fluoxetine on MDMA toxicity to serotonin neurons.

	Uptake of [^{3}H]-serotonin, cpm		
		Fluoxetine	
	Control	10^{-7}M	10^{-5}M
Control	773 ± 93	723 ± 86	289 ± 57
S-MDMA (10^{-4}M)	184 ± 35	969 ± 32	384 ± 148
R-MDMA (10^{-4}M)	267 ± 68	661 ± 84	–

Table 3. Effects of various pharmacological agents on 10^{-4}M MDMA-induced toxicity.

Specific uptake of [^{3}H]-serotonin cpm/well

Experiment A:	Control	Mianserin 10^{-8}	Chlorpheniramine 10^{-6}	Terfenadine 10^{-6}
Control	5299 ± 595	5743 ± 958	5804 ± 539	5888 ± 477
S(+)MDMA	611 ± 79	731 ± 65	898 ± 54	702 ± 79
R(−)MDMA	1180 ± 54	998 ± 77	762 ± 70	1048 ± 148

Experiment B:	Control	Nomifensine 10^{-6}	BHT-920 10^{-8}
Control	3028 ± 250	3285 ± 50	3659 ± 560
S(+)MDMA	458 ± 59	610 ± 163	1707 ± 259
R(−)MDMA	1403 ± 86	928 ± 202	1788 ± 98

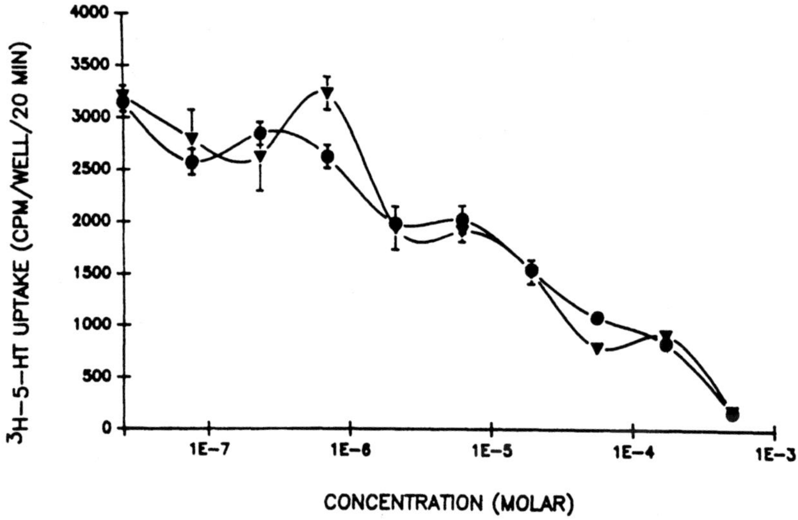

Figure 4. Effects of the $S(+)$-isomer (triangles) and the $R(-)$-isomer (circles) of MDMA on the specific uptake of [^{3}H]-serotonin into serotonergic neurons grown in tissue culture in the presence of hippocampus for four days. Each point represents the mean and SEM of four cultures.

which it has an apparent K_i of 3.6 micromolar) can only be speculated at this time.

Our results with fluoxetine also suggest a role for release in the toxicity. Concentrations of fluoxetine that protect the neurons from MDMA presumably inhibit its transport into the neurons. Thus internalization of the drug appears to be important. It has also been shown in vivo that serotonin uptake inhibitors, such as citalopram [3] and fluoxetine [8], inhibit the toxic actions of MDMA. The higher concentrations of fluoxetine are presemably toxic due to an effect of this drug on systems other than transport.

Our morphological studies indicate the actual loss of cells, although this finding is not clearly shown in vivo. The cell death in our cultures could be due to the fact that these are immature cells that are highly dependent on terminal integrity for survival. However, it could also be indicative of a process that occurs with repeated exposure to the drug.

Our work with neurotransmitter receptor blockers is not yet complete; however, we have eliminated the possible contribution to toxicity made by a number of receptors. Specifically, the lack of protection afforded by mianserin has ruled out an action on the $serotonin_2$ receptor. Similarly, the lack of attenuation by the histamine antagonists terfenadine and chlorpheniramine seems to rule out these receptors.

Our final experiments on the cellular localization of toxicity indicate that the effects of MDMA are directly on the serotonergic nerve terminal and that an intermediate toxin is not produced by other cells. This mechanism of toxicity

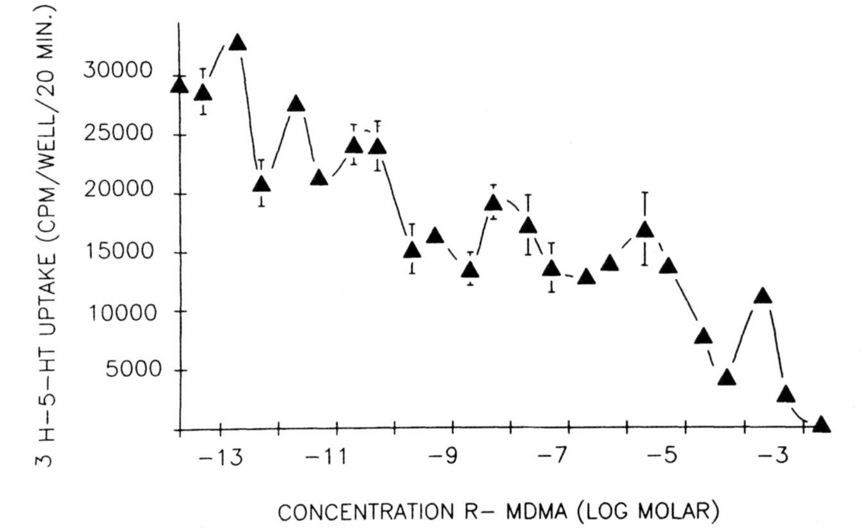

Figure 5. Effects of various concentrations of *R*(−)-MDMA on specific uptake of [^{3}H]-serotonin into primary cultures of serotonin neurons grown in culture for four days after the addition of the antimitotic agent FUDR. Each point represents the mean and SEM of four cultures.

had been shown for MPTP, which is metabolized by astroglial cells into the dopamine toxin MPP^+. Interestingly, the work with FUDR-treated cultures indicates that astroglial cells are, in fact, protective of the serotonin terminal. This may be due to uptake of MDMA into astrocytes, which contain specific high affinity carrier systems for a number of neurotransmitters, including serotonin [9].

The role of target tissue in attenuating the toxicity of the *S*(+)-isomer is not readily explained. In previous studies [1], we have shown that the presence of target tissue is stimulatory to growth of the neurons either by providing trophic factors or by inhibiting toxic factors. The results with MDMA suggest that inhibition of toxic substances is indeed possible. Regardless of the mechanism of toxicity attenuation by target, our findings further point out the possibility that two mechanisms of toxicity are in force and that the *S*(+)-isomer has more long-lasting effects by virtue of eliciting both, whereas the *R*(−)-isomer only elicits one [3].

In conclusion, we have applied the methodology of tissue culture to the problem of MDMA-induced serotonergic neuronal degeneration. We have been able to replicate several in vivo observations in our model, such as the differences between the isomers and the protective effects of fluoxetine. In addition, we have made some new observations on the role of target tissue and astrocytes in attenuating the toxicity, and raised the possibility that chronic exposure may actually lead to cell death. Finally, we have shown a role for release-regulating receptors and that actions on $serotonin_2$ or histamine receptors are not involved.

REFERENCES

1. Azmitia, E.C. and Whitaker–Azmitia, P.M., 1987. Target cell stimulation of dissociated serotonergic neurons in culture. Neuroscience 26:93.
2. Ricaurte, G., Bryan, G., Strauss, L., Seiden, L., and Schuster, C., 1985. Hallucinogenic amphetamine selectively destroys brain serotonin nerve terminals. Science 229:986.
3. Schmidt, C.J., Levin, J.A., and Lovenberg, W., 1987. In vitro and in vivo neurochemical effects of methylenedioxyamphetamine on striatal monoaminergic systems in the rat brain. Biochem. Pharmacol. 36:747.
4. Steele, T.D., Nichols, D.E., and Yim, G.K., 1987. Stereochemical effects of 3,4-methylenedioxymethamphetamine (MDMA) and related amphetamine derivatives on inhibition of uptake of ^{3}H-monoamines into synaptosomes from different regions of rat brain. Biochem. Pharmacol. 36:2297.
5. Battaglia, G., Brooks, B.P., Kulsakdinun, C., and De Souza, E.B., 1988. Pharmacologic profile of MDMA (3,4-methylenedioxymethamphetamine) at various brain recognition sites. Eur. J. Pharmacol. 149:159.
6. Azmitia, E.C. and Segal. M., 1978. An autoradiographic analysis of the differential ascending projections of the dorsal and median raphe nuclei in the rat. J. Comp. Neurol. 179:641.
7. Raiteri, M., Maura, G., Gemignani, A., and Pittaluga, A., 1983. Differential blockade by (−)mianserin of the $alpha_2$-adrenoceptors mediating inhibition of noradrenaline and serotonin release from rat brain synaptosomes. Naunyn–Schmied.'s Arch. Pharmacol. 322:180.
8. Schmidt, C.J., Neurotoxicity of the psychedelic amphetamine, methylenedioxymethamphetamine. J. Pharmacol. Exper. Ther. 240:1.
9. Whitaker, P.M., Vint, C.K., and Morin, R., 1983. ^{3}H-Imipramine labels sites on brain astroglial cells not related to serotonin uptake. J. Neurochem. 41:1319.

12. EFFECT OF MDMA-LIKE DRUGS ON CNS NEUROPEPTIDE SYSTEMS

GLEN R. HANSON, KALPANA M. MERCHANT, MICHEL JOHNSON, ANITA A. LETTER, LLOYD BUSH AND JAMES W. GIBB

1. INTRODUCTION

The ring-substituted amphetamine analogue, 3,4-methylenedioxymethamphetamine (MDMA, "ecstasy") causes in humans psychoactive responses described as a combination of euphoria, enhanced empathy, and central stimulation [1]. This combination of pharmacological effects has caused MDMA to become a popular recreational drug, resulting in its classification as a Schedule I agent. Comparisons with other psychoactive drugs have demonstrated that MDMA and another amphetamine analogue, 3,4-methylenedioxyamphetamine (MDA), possess both stimulant properties, resembling more traditional amphetamine congeners, and hallucinogenic activity, like LSD [2]. This somewhat unique combination of effects has caused some investigators to claim that these so-called "designer" amphetamine analogues represent a new class of pharmacological agents [3, 4].

All findings to date suggest that the MDMA-like drugs exert their pharmacological activity by altering monoaminergic systems. In particular, stimulation of serotonergic and dopaminergic pathways are throught to mediate the psychoactive effects of these compounds. While the relative involvement of serotonin (5-HT) and dopamine (DA) in mediating the pharmacological properties of such drugs is still an issue of controversy, most investigators consider serotonergic pathways as the primary effector transmitter system for these agents. In support of this hypothesis are the findings (1) that these designer amphetamine analogues are more potent at releasing 5-HT than at releasing dopamine [5, 6] and (2) that high doses of both MDMA and MDA selectively damage serotonergic neurons, causing long-term and dramatic decreases in brain levels of 5-HT and the activity of its synthesizing enzyme, tryptophan hydroxylase, while having little detectable impact on dopamine concentration or the activity of its synthesizing enzyme, tyrosine hydroxylase [7, 8, 9].

In spite of the somewhat selective effects on serotonergic pathways by MDMA and MDA, there is evidence that these drugs also significantly increase central nervous system dopaminergic activity. In vitro studies reveal that MDMA is a potent releaser of dopamine through action on the carrier mechanism [5]. In vivo voltametric detection of extracellular dopamine has been used to demonstrate that MDMA administered systemically results in substantial dopamine release from striatum and nucleus accumbens [10]. These findings have been supported by behavioral studies that suggest that MDMA and MDA have amphetamine-like stimulant properties thought to be dopaminergically-mediated [2]. However, in contrast to the serotonergic responses, traditional neurochemical markers used to monitor changes in the activity of the dopaminergic system are little affected by administration of the designer amphetamines. For example, the only dopaminergic changes of significance in the rat include a transient elevation in striatal levels of dopamine and its metabolite, homovanillic acid, following a single dose of MDMA and MDA [9]. Such a change likely reflects a drug-mediated increase in dopamine release. In addition, Stone and coworkers [11] reported that two weeks following multiple doses of MDMA, the striatal contents of dopamine metabolites were slightly lower than corresponding controls, but in animals treated with MDA or another designer amphetamine, N-ethyl-3,4-methylenedioxyamphetamine (MDE), these metabolites remained unaltered. In neither study were changes in tyrosine hydroxylase activity seen following drug treatments.

Interestingly, single and possibly multiple doses of amphetamine or its methylated analogue, methamphetamine, also result in initial transient rises in the levels of dopamine metabolites. These rises are comparable to those observed following MDMA and MDA administration [12, 13]; however, unlike the designer amphetamines, profound long-term decreases (thought to be related to neurotoxic drug effects) occur in tyrosine hydroxylase activity and in the concentrations of dopamine and its metabolites, following multiple doses of both amphetamine or methamphetamine [14, 15].

Based on the above cited changes in dopaminergic neurochemical parameters, there are uncertainties as to how the designer amphetamines compare to amphetamine and methamphetamine in their ability to activate dopaminergic pathways. Apparently, neurochemical evaluation of the dopaminergic systems themselves is not sufficient to answer many of the questions concerning the impact of the designer amphetamines on these pathways. Another approach worth investigating is proposed here and consists of determining the response of transmitter systems that are "downstream" from the dopaminergic pathways. In other words, these downstream systems, due to their neuronal linkage to dopamine projections, are altered by drug-induced fluctuations in postsynaptic dopaminergic activity. Consequently, if such downstream pathways could be identified and their response to changes in dopaminergic activity could be characterized, neurochemical analysis of these pathways following treatments with drugs such as the designer amphetamines would be useful tools for establishing the dopaminergic consequences of these agents.

Within extrapyramidal structures there exist neuropeptide pathways that are dramatically influenced by changes in the postsynaptic activity of the nigral-striatal dopaminergic neurons. However, these same extrapyramidal peptide systems do not appear to respond directly to fluctuations in serotonergic activity [16, 17, unpublished observations]. The transmitter substances of these responsive peptidergic projections include substance P (SP), neurotensin (NT), and dynorphin A_{1-17} (Dyn). Specifically, SP-containing neurons, which originate within the striatum and terminate in the substantia nigra, are thought to serve an excitatory feedback function to the nigral-striatal dopamine pathway. In support of this hypothesis is the observation that intranigral injections of SP cause striatal release of dopamine [18] and stimulate locomotion [19]. As would be expected, nigral administration of SP has no locomotor effects in animals that have received 6-hydroxydopamine lesions to their mesostriatal dopamine pathway [19]. The state of this SP pathway is reciprocally controlled by dopaminergic activity causing striatal and nigral content of SP to increase or decrease in response to repeated increases or decreases, respectively, in nigral-striatal dopaminergic activity [16, 20].

The interactions between neurotensin pathways and the extrapyramidal dopaminergic system are somewhat more complex. As the vast majority of striatal and nigral neurotensin receptors are associated with dopamine neurons [21], neurotensin pathways certainly contribute to the regulation of extrapyramidal dopamine activity. The overall CNS pharmacology of neurotensin has been compared to that of the neuroleptic drugs [22], whereas intraventricular administration of this peptide is reported to antagonize some of the behavioral activity of powerful stimulants, such as amphetamine and cocaine [23]. Short-term or long-term increases in mesostriatal dopaminergic activity cause dramatic but transient elevations in striatal and nigral levels of neurotensin [24]. Finally, dynorphin A is associated with striatal-nigral neurons, which, like the SP pathway, have been postulated to be part of a feedback system to the nigral-striatal dopamine neurons [25]. Increases in activity of the mesostriatal dopamine pathway cause a relatively rapid rise in both striatal and nigral dynorphin concentrations [26]. However, such a feedback role for dynorphin has recently been questioned, as the locomotor activity induced by nigral dynorphin injections is not influenced by elimination of the nigral-striatal dopamine pathway [19].

The effects of the designer amphetamine analogues on SP, neurotensin, and dynorphin pathways associated with extrapyramidal structures have been evaluated. For comparison, the responses of these peptide systems in the nucleus accumbens, a limbic structure, were also determined in some of the experiments. Following the systemic administration of drugs such as MDMA and MDA, changes in the neuropeptide systems were assessed by radioimmunoassay, measuring striatal and nigral content of SP [20], neurotensin [24], and dynorphin [26] immunoreactivity. The antibodies employed in the assays were raised in our laboratory and are highly selective and able to detect 2-8 pg of neuropeptide per sample. The mean neuropeptide contents (pg/mg protein) detected were as follows: striatum — SP = 1,740, neurotensin = 148, dynorphin = 380; substantia nigra — SP = 12,400, neurotension = 501, dynorphin = 760; nucleus accumbens — neurotensin = 377, dynorphin = 823. Differences between treatment groups were assessed employing either the Student's *t*-test or an ANOVA analysis followed by a multiple comparisons test. Significance was judged to occur at the 0.05 level.

2. EFFECT OF MDMA AND RELATED AGENTS ON NEUROPEPTIDES

The responses of SP, neurotensin, and dynorphin extrapyramidal systems to the designer amphetamines were determined and compared to the changes induced by methamphetamine treatment. Sprague–Dawley rats received five doses (six-hour intervals between doses) of MDMA (10 mg/kg/dose), MDA (10 mg/kg/dose), and methamphetamine (15 mg/kg/dose). Approximately 18 hours following treatment, the animals were decapitated and brains were removed. The striata and substantia nigras were dissected out, and for some experiments, the nucleus accumbens was also excised. The tissues were prepared for radioimmunoassay analysis as previously described [20, 24, 26], and the peptide levels were normalized by the total protein content in each sample.

The striatal content of SP was increased to approximately 200% of control by the administration of all three drugs examined. In the substantia nigra, both MDA and methamphetamine administration increased SP content to 150%, whereas MDMA treatment elevated SP levels only to an insignificant 127% of control (Figure 1).

The responses of the neurotensin systems were examined by giving multiple administrations (as described for Figure 1) of two doses of both MDMA (10 or 15 mg/kg/dose) and MDA (5 or 10 mg/kg/dose), as well as one dose of methamphetamine (15 mg/kg/dose). In the striatum and nucleus accumbens, 200–300% increases in neurotensin levels were observed following all drug treatments (Figure 2). However, increases in nigral NT content following both doses of MDMA were about half that found in animals treated with either dose of MDA or methamphetamine.

The responses by the dynorphin projections were studied in animals that were treated with either five administrations of MDMA (10 mg/kg-dose) or of methamphetamine (15 mg/kg/dose). Treatment with either drug increased nigral dynorphin levels to approximately 250% of control (Figure 3). However, MDMA treatment appeared to have substantially less effect on dynorphin

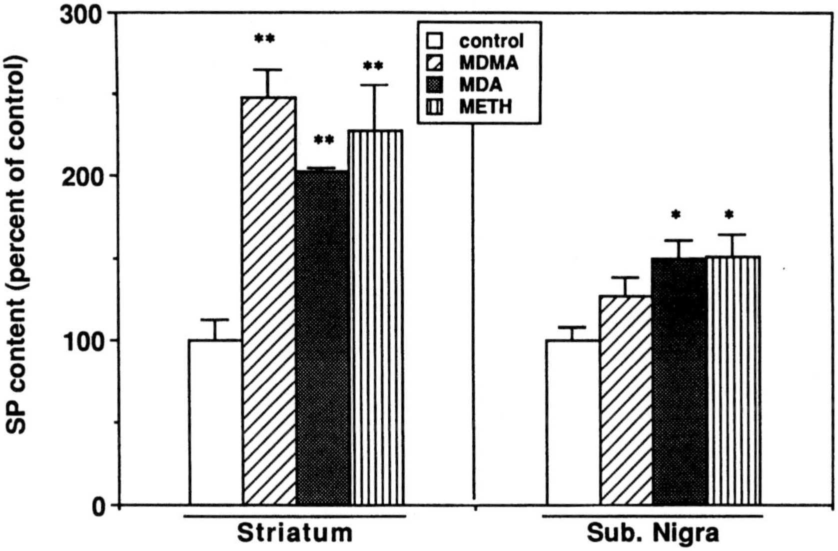

Figure 1. Effects of MDMA, MDA, and methamphetamine on the levels of substance P in extrapyramial structures. Rats were sacrificed 18 hours following 5 doses (6-hour intervals between doses) of MDMA, MDA, or METH. $^{*}P < 0.01$ and $^{**}P < 0.005$ compared to corresponding controls.

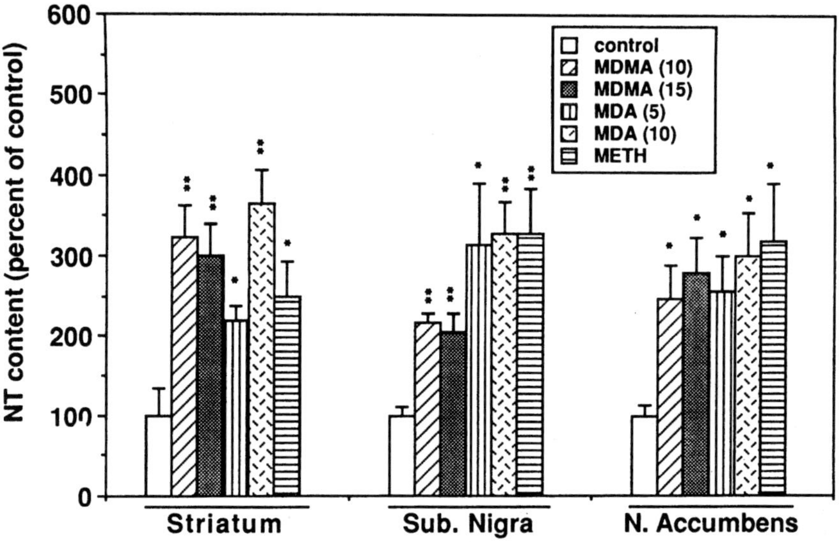

Figure 2. Effects of MDMA, MDA, and methamphetamine on neurotensin levels. Rats received multiple admininstrations of 2 different doses of MDMA (10 or 15 mg/kg/dose) or MDA (5 or 10 mg/kg/dose) or a single dose of METH (15 mg/kg/dose), as described for Figure 1. $^{*}P < 0.01$ and $^{**}P < 0.001$ compared to respective controls.

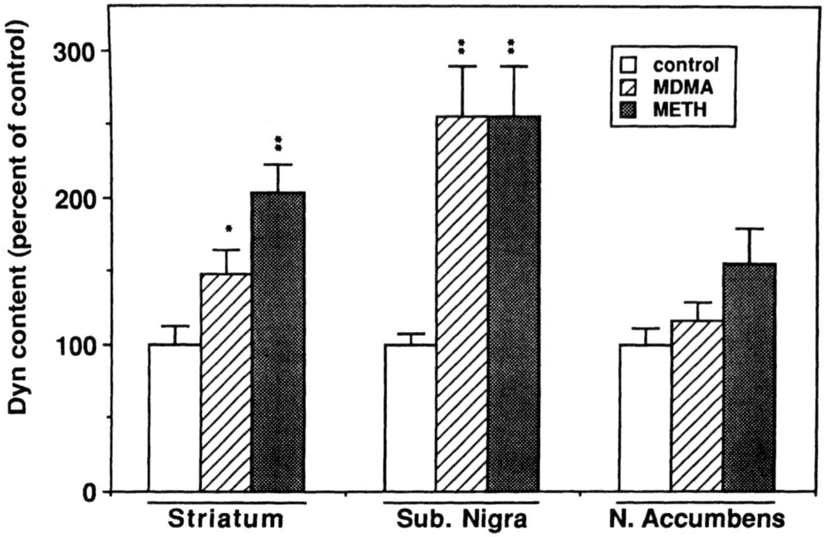

Figure 3. Effect of MDMA and methamphetamine on tissue content of dynorphin A_{1-17}. Rats received multiple administrations of MDMA and METH as described for Figure 1. $^*P < 0.05$ and $^{**}P < 0.01$ compared to respective controls.

content in both the striatum and nucleus accumbens than did methamphetamine.

In order to determine how quickly the peptide changes occur in response to MDMA treatment, a single dose of this drug (10 mg/kg) was administered and animals were sacrificed 12 hours later (Figure 4). No significant changes in either striatal or nigral SP content were detected. In contrast, levels of striatal and nigral neurotensin and dynorphin increased from 150% to 400% of controls following the acute MDMA treatments. Although not shown here, similar patterns of peptide responses have been observed after a single dose of methamphetamine [24, 28, 29].

The role of dopamine in mediating the peptide changes after a single MDMA administration was also examined (Figure 5). Selective antagonists for dopamine D-1 (SCH 23390; 0.5 mg/kg) and D-2 (Sulpiride; 80 mg/kg) receptors were administered 15 minutes prior to MDMA injection. The animals were sacrificed 12 hours following treatment, and striata and substantia nigras were removed and analyzed for neurotensin and dynorphin content. In general, blockade of the D-1 receptor totally prevented the MDMA-induced increases in striatal and nigral neurotensin and dynorphin levels. In contrast, the blockade of D-2 receptors did not reduce the MDMA-related changes. In fact, the presence of sulpiride actually enhanced the increase in striatal neurotensin levels caused by MDMA administration. A similar pattern of responses by extrapyramidal neurotensin systems following methamphetamine treatment

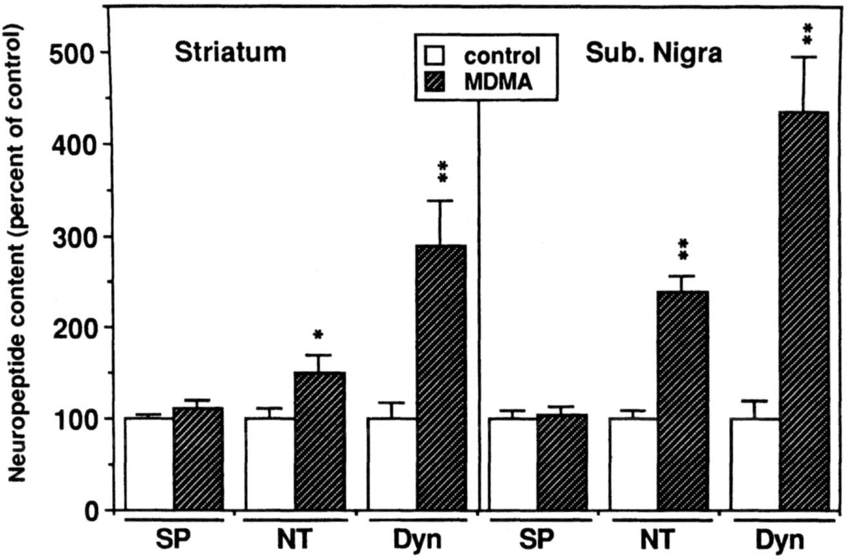

Figure 4. Effect of a single dose of MDMA on neuropeptide levels in extrapyramidal structures. Rats were given a single injection of MDMA (10 mg/kg) and sacrificed 12 hours after treatment. *P < 0.05 and **P < 0.005 compared to respective controls.

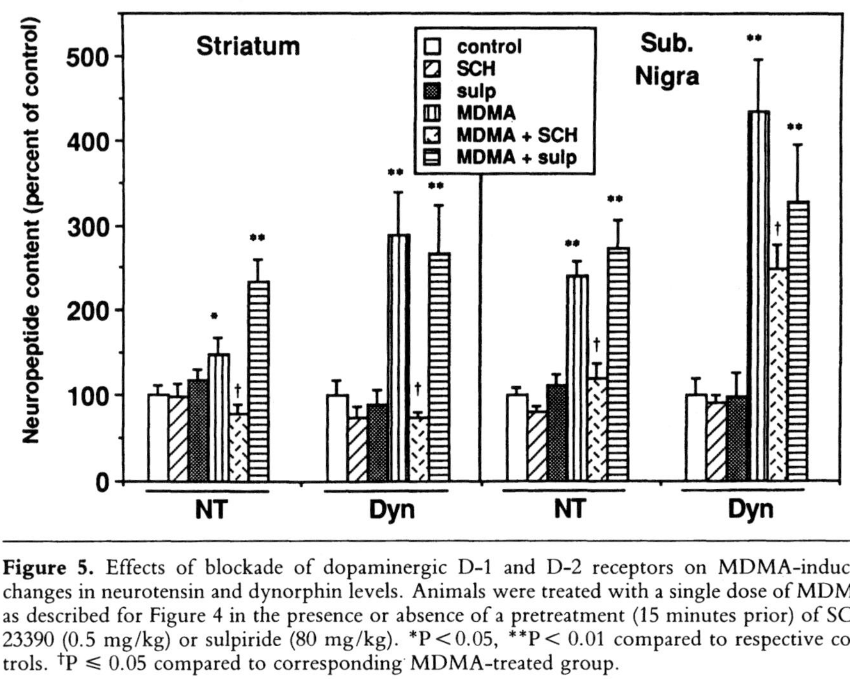

Figure 5. Effects of blockade of dopaminergic D-1 and D-2 receptors on MDMA-induced changes in neurotensin and dynorphin levels. Animals were treated with a single dose of MDMA as described for Figure 4 in the presence or absence of a pretreatment (15 minutes prior) of SCH 23390 (0.5 mg/kg) or sulpiride (80 mg/kg). *P < 0.05, **P < 0.01 compared to respective controls. †P ⩽ 0.05 compared to corresponding MDMA-treated group.

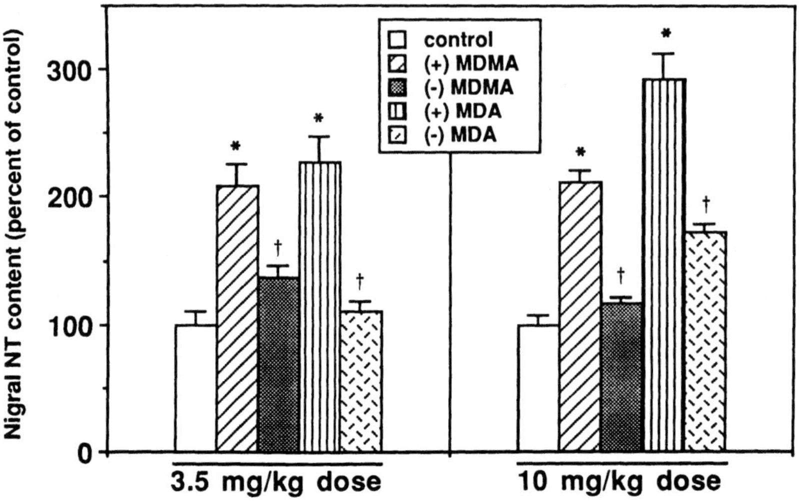

Figure 6. Effects of MDMA isomers on nigral neurotensin levels. Rats received 5 doses of the (+) and (−) isomers of MDMA and MDA according to the treatment schedule described for Figure 1. *$P<0.01$ compared to corresponding control. †$P<0.01$ compared to corresponding isomer.

has been reported [27]. One noteworthy exception was the response by the nigral dynorphin system. Antagonism of the D-1 receptor appeared to reduce the MDMA effect only by 50%; D-2 blockade might also have attenuated the MDMA effect, although the change was not statistically significant.

It has been demonstrated by Schmidt and coworkers [5] that the (+) enantiomers of MDMA and MDA are more active on dopaminergic systems than their (−) enantomeric counterparts. Consequently, as the changes in peptide content are likely a result of drug-mediated dopamine release, the effects of two doses of the (+) and (−) enantiomers of both MDMA and MDA were evaluated. Rats received five injections (six-hour intervals between administrations) of 3.5 and 10 mg/kg/dose of both enantiomers of MDMA and MDA. Changes in nigral neurotensin content were determined (Figure 6). For both doses and with both drugs the (+) enantiomer caused a substantially greater increase in nigral neurotensin levels than the corresponding (−) enantiomer.

The effects on neuropeptide systems by one other designer amphetamine was evaluated. Compared to MDMA and MDA, N-ethyl-3,4-methylenedioxyamphetamine (MDE) exerts smaller effects on both the dopaminergic and serotonergic systems, with a quicker recovery [30]. The responses by neurotensin systems of the striatum, substantia nigra, and nucleus accumbens reflected a similar pattern of response (Figure 7). Rats were sacrificed 3 and 18 hours after receiving 5 administrations of 10 mg/kg/dose of MDE. Significant increases in neurotensin content were observed in both the striatum and

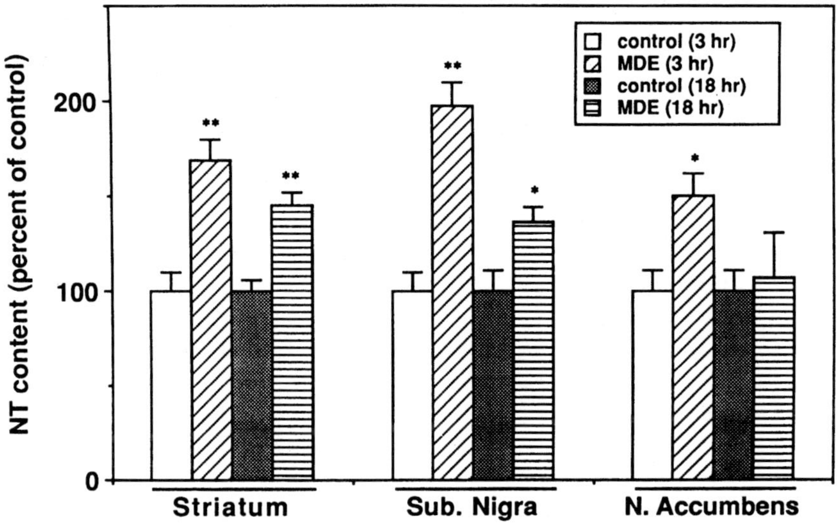

Figure 7. Effects of MDE administration on neurotensin levels. Rats received multiple doses of MDE (10 mg/kg/dose) as described for Figure 1 and animals were sacrificed either 3 or 18 hours following treatment. *P < 0.02 and **P < 0.005 compared to the corresponding controls.

substantia nigra with the greater increase occurring three hours after treatment. The neurotensin levels were significantly elevated only at the three-hour time in the nucleus accumbens. These data suggest that significant recovery by the neurotensin systems occurs within 18 hours following MDE treatment. Compared with the neurotensin response to similar treatments by MDMA and MDA (Figure 2), MDE was considerably less potent.

These findings demonstrate that at least three peptide systems, specifically SP, neurotensin, and dynorphin pathways in extrapyramidal and limbic structures, are particularly responsive to MDMA, MDA, and, to some extent, MDE administrations. Even a single injection of MDMA caused 150-400% increases in striatal and nigral levels of neurotensin and dynorphin (Figure 4). These peptide changes are almost certainly dopamine-mediated, as coadministration of the D-1 antagonist, SCH 23390, totally blocked or at least substantially attenuated these peptide responses (Figure 5). In addition, those amphetamine analogues that were less potent as dopamine-releasing agents (e.g., MDE less than MDMA and MDA, or (−) isomers less than (+) isomers of MDMA and MDA) caused smaller peptide changes (Figures 6, 7).

For the most part, the effects of MDMA and MDA on the peptides were comparable to those caused by methamphetamine administration (Figures 1,2,3). This finding was somewhat surprising as methamphetamine is thought to have more dopamine-releasing action than either MDMA or MDA [5]. The responses by the peptide systems suggest, in general, that drug-induced increased activity at the dopaminergic postsynaptic receptors (particularly the D-1 subtype) is similar following treatments with methamphetamine, MDMA, and MDA. However, subtle and possibly important differences were observed that may have relevance to the distinctive pharmacological features of each of these drugs. For example, (1) MDMA appeared to be less effective at elevating nigral SP levels than was MDA or methamphetamine (Figure 1); (2) MDMA was only half as effective as MDA or methamphetamine in elevating nigral neurotensin levels (Figures 2,6); and (3) MDMA was less effective than methamphetamine at increasing the levels of dynorphin in the striatum and nucleus accumbens.

The significance of these selective differences is unclear. One possible explanation is that multiple dopaminergic systems are involved in the regulation of the peptide pathways, and these dopamine neuronal networks respond differentially to the designer amphetamines and methamphetamine. Perhaps some dopamine neurons are equally affected by three drugs, thus, the peptide systems linked to these dopaminergic projections are equally influenced by MDMA, MDA, and methamphetamine. However, other dopamine neurons might be more sensitive to the actions of one of the drugs (usually methamphetamine), causing a more dramatic response to treatment with that drug by the peptide systems linked to such dopamine projections. Even so, the present data demonstrate that MDA and MDMA do have an important dopaminergic component to their neurochemical activity, which in some ways appears to be comparable to that of the more traditional amphetamines, such as methamphetamine. Although to a lesser extent, MDE also has significant dopamine-enhancing action.

3. CONCLUSIONS

The significance of the peptide changes in response to drug treatments has not been elucidated. It is possible that decreases in peptide release following treatment with these drugs results in accumulation of the neuropeptides and a rise in peptide tissue content. The dramatic increases in nigral and striatal neurotensin and dynorphin levels following a single dose of MDMA would be suggestive of such a mechanism. Another factor to consider is a drug-induced increase in the synthesis of the peptide. In fact, Bannon et al. [31] have reported that a single dose of methamphetamine causes a substantial rise in the levels of mRNA for preprotachykinin, the SP precursor, suggesting a stimulation of SP synthesis. It is possible that similar drug-induced increases in peptides synthesis are at least a factor in the observed peptide responses.

In order to understand fully the contribution of these peptidergic pathways in mediating the effects of the designer amphetamines, it is necessary to understand their synaptic connections. It is likely that these peptide projections serve at least two major roles. First, they are feedback systems that help to regulate dopaminergic activity, second, they serve an efferent function and transport dopamine-initiated messages to pathways that originate in extrapyramidal or limbic structures and project to other brain regions. Consequently, the responses of the peptide systems described herein might be essential in the regulation of feedback phenomena such as drug tolerance and sensitization, as well as serve as important links to conduct drug-induced dopaminergic messages to brain regions crucial for mediating drug effects. As both extrapyramidal and limbic peptidergic pathways appear to be influenced by the designer amphetamines, it is likely that these peptide pathways contribute to the effects of the drugs on locomotion and mental states. These possibilities warrant investigation.

ACKNOWLEDGEMENTS

This work was supported by U.S. Public Health Service Grants DA 00869 and DA 04222. The MDMA, MDA, MDE, and methamphetamine were generously supplied by the National Institute on Drug Abuse, Rockville, MD and SCH 23390 was a gift from Schering Corp., Bloomfield, NJ.

REFERENCES

1. Adler, J., Abramson, P., Katz, S., and Hager, M., 1985. Getting high on "Ecstasy." Newsweek, April 15, p. 96.
2. Glennon, R., Yousif, M., and Patrick, G., 1988. Stimulus properties of 1-(3,4-methylenedioxyphenyl)-2-aminopropane (MDA) analogs. Pharmacol. Biochem. Behav. 29:443–449.
3. Nichols, D., 1986. Differences between the mechanism of action of MDMA, MBDB and the classic hallucinogens. Identification of a new therapeutic class: Entactogens. J. Pscyhoactive Drugs 18:305–313.
4. Oberlender, R. and Nichols, D., 1988. Drug discrimination studies with MDMA and amphetamine, Psychopharmacol. 95:71–76.
5. Schmidt, C.J., Levin, J., and Lovenberg, W., 1987. In vitro and in vivo neurochemical effects of methylenedioxymethamphetamine on striatal monoaminergic systems in the rat brain. Biochem. Pharmacol. 36:747–755.
6. Johnson, M., Hoffman, A., and Nichols, D., 1986. Effects of enantiomers of MDA, MDMA and analogues on ^{3}H-serotonin and ^{3}H-dopamine release from superfused rat brain slices. Eur. J. Pharmacol. 132:269–276.
7. Schmidt, C.J., Wu, L., and Lovenberg, W., 1986. Methylenedioxymethamphetamine: A potentially neurotoxic amphetamine analog. Eur. J. Pharmacol. 124:175–178.
8. Ricaurte, G., Strauss, L., Seiden, L., and Schuster, C., 1985. Hallucinogenic amphetamine selectively destroys brain nerve terminals. Science 229:986–988.
9. Stone, D.M., Stahl, D., Hanson, G.R., and Gibb, J.W., 1986. The effects of 3,4-methylenedioxyamphetamine (MDA) and 3,4-methylenedioxymethamphetamine (MDMA) on monoaminergic systems in the rat brain. Eur. J. Pharmacol. 128:41–48.
10. Yamamoto, B. and Spanos, L., 1988. The acute effects of methylenedioxymethamphetamine on dopamine release in the awake-behaving rat. Eur. J. Pharmacol. 148:195–203.
11. Stone, D.M., Johnson, M., Hanson, G.R., and Gibb, J.W., 1987. A comparison of the neurotoxic potential of methylenedioxyamphetamine (MDA) and its N-methylated and N-ethylated derivatives. Eur. J. Pharmacol. 134:245–248.
12. Johnson, M., Hanson, G.R., and Gibb, J.W., 1988. Effects of dopaminergic and serotonergic receptor blockade on neurochemical changes induced by acute administration of methamphetamine and 3,4-methylenedioxymethamphetamine. Neuropharmacol. 27:1089–1096.
13. Roffler–Tarlov, S., Sharman, D., and Tegerdine, P., 1971. 3,4-dihydroxyphenylacetic acid and 4-hydroxy-3-methoxyphenylacetic acid in the mouse striatum: A reflection of intra-and extra-neuronal metabolism of dopamine? Br. J. Pharmacol. 42:343–351.
14. Schmidt, C.J., Sonsalla, P., Hanson, G.R., Peat, M., and Gibb, J.W., 1985. Methamphetamine-induced depression of monoamine synthesis in the rat: Development of tolerance. J.

Neurochem. 44:852–855.
15. Kogan, F., Nichols, W., and Gibb, J.W., 1976. Influence of methamphetamine on nigral and striatal tyrosine hydroxylase activity and on dopamine levels. Eur. J. Pharmacol. 36:363–371.
16. Ritter, J., Schmidt, C., Gibb, J.W., and Hanson, G.R., 1985. Dopamine-mediated increases in nigral substance P-like immunoreactivity. Biochem. Pharmacol. 34:3161–3166.
17. Sonsalla, P.K., Gibb, J.W., and Hanson, G.R., 1986. Nigrostriatal dopamine actions on the D-2 receptors mediate methamphetamine effects on the striatonigral substance P system. Neuropharmacol. 25:1221–1230.
18. Reid, M., Herrera–Marschitz, M., Hokfelt, T., Terenius, L., and Ungerstedt, U., 1988. Differential modulation of striatal dopamine release by intranigral injection of gammaaminobutyric acid (GABA), dynorphin and substance P. Eur. J. Pharmacol. 147:411–420.
19. Herrera–Marschitz, M., Christensson–Nylander, I., Sharp, T., Stainis, W., Reid, M., Hokfelt, T., Terenius, L., and Ungerstedt, U., 1986. Striatonigral dynorphin and substance P pathways in the rat: II. Functional analysis. Exp. Brain Res. 64:193.
20. Hanson, G.R., Alphs, L., Wolf, W., Levine, R., and Lovenberg, W., 1981. Haloperidol-induced reduction of nigral substance P-like immunoreactivity: A probe for the interactions between dopamine and substance P neurons. J. Pharmacol. Exp. Ther. 218:568–574.
21. Quirion, R., Chiueh, C., Everist, H., and Pert, A., 1985. Comparative localization of neurotensin receptors on nigrostriatal and mesolimbic dopaminergic terminals. Brain Res. 327:385–389.
22. Nemeroff, C., 1986. The interaction of neurotensin with dopaminergic pathways in the central nervous system: Basic neurobiology and implications for the pathogenesis and treatment of schizophrenia. Psychoneuroendocrinology 11:15–37.
23. Skoog, K., Cain, S., and Nemeroff, C., 1986. Centrally administered neurotensin suppresses locomotor hyperactivity induced by d-amphetamine but not by scopolamine or caffeine. Neuropharmacology 25:777–782.
24. Letter, A., Merchant, K., Gibb, J.W., and Hanson, G.R., 1987. Effect of methamphetamine on neurotensin concentration in rat brain regions. J. Pharmacol. Exp. Ther. 241:443–447.
25. Herrera–Marschitz, M., Hokfelt, T., Ungerstedt, U., and Terenius, L., 1983. Functional studies with the opioid peptide dynorphin: Acute effects of injections into the substantia nigra reticulata of naive rats. Life Sci. 33:555–558.
26. Hanson, G.R., Merchant, K.M., Letter, A., Bush, L., and Gibb, J.W., 1987. Methamphetamine-induced changes in the striatal-nigral dynorphin system: Role of D-1 and D-2 receptors. Eur. J. Pharmacol. 144:245–246.
27. Letter, A., Matsuda, L., Merchant, K., and Hanson, G.R., 1987. Characterization of dopaminergic influence on striatal-nigral neurotensin systems. Brain Res. 422:200–203.
28. Ritter, J., Schmidt, C., Gibb, J.W., and Hanson, G.R., 1984. Increases of substance P-like immunoreactivity within striatal-nigral structures after subacute methamphetamine treatment. J. Pharmacol. Exp. Ther. 229:487–492.
29. Hanson, G.R., Merchant, K.M., Letter, A., Bush, L., and Gibb, J.W., 1988. Characterization of methamphetamine effects on the striatal-nigral dynorphin system. Eur. J. Pharmacol., in press.
30. Johnson, M., Hanson, G.R., and Gibb, J.W., 1987. Effects of N-ethyl-3,4-methylenedioxyamphetamine (MDE) on central serotonergic and dopaminergic systems of the rat. Biochem. Pharmacol. 36:4085–4093.
31. Bannon, M.J., Elliot, P.J., and Bunney, E.B., 1987. Striatal tachykinin biosynthesis: Regulation of mRNA and peptide levels by dopamine agonists and antagonists. Mol. Brain Res. 3:31–37.

13. NEUROENDOCRINOLOGICAL EFFECTS OF MDMA IN THE RAT

J. FRANK NASH AND HERBERT Y. MELTZER

1. INTRODUCTION

The amphetamine analogue 3,4-methylenedioxymethamphetamine (MDMA) has attracted public attention, as well as that of the medical community, because of its effects on mood and cognition and the possibility that it is a neurotoxin. Clinical and anecdotal case reports suggest that MDMA elevates mood, relieves dysphoria, and enhances insight that lasts far beyond possible direct effects of the drug [1, 2]. A number of studies in rodents and primates have reported that single and multiple administrations of MDMA selectively deplete brain concentrations of serotonin (5-HT) and its major metabolite, 5-hydroxyindoleacetic acid (5-HIAA) [3–7]. Since it has been postulated that diminished serotonergic function may increase the vulnerability to depression or actually precipitate it [8], the 5–HT-depleting effects of MDMA and its reputed antidepressant properties appear difficult to reconcile, raising the possibility that current theories of the role of 5-HT in depression or the mechanism of action of MDMA, or both, may need revision.

It has been reported that MDMA releases 5-HT and, to a lesser extent, dopamine (DA), in in vitro preparations, such as whole brain synaptosomes [9] and rat hippocampal slices [10]. MDMA has been reported to block the uptake of [^{3}H]-5-HT and [^{3}H]-norepinephrine (NE) into synaptosomes prepared from the hypothalamus, hippocampus, and striatum of rat brain [11]. Although a preliminary report suggested that [^{3}H]-MDMA bound selectively to membrane preparations prepared from rat brains [12], it was subsequently reported that [^{3}H]-MDMA selectively labeled glass fiber filters [13]. These authors concluded that [^{3}H]-MDMA has limited usefulness in radioligand binding studies. Finally, Lyon et al. [14] found that MDMA had little affinity for 5-HT_1, 5-HT_2, or D_2 binding sites. Collectively, these studies suggest that MDMA is not a direct 5-HT or dopamine agonist but rather stimulates the release of endogenous 5-HT and dopamine, as well as blocks the reuptake of 5-HT, dopamine, and norepinephrine.

MDMA has been studied quite extensively in drug discrimination studies. For the most part, these studies clearly support subjective reports in humans that MDMA is not an hallucinogen [15, 16]. For example, in rodents trained to discriminate either LSD or DOM from saline in two-lever drug discrimination paradigms, MDMA did not generalize to either hallucinogen [16, 17]. Moreover, MDMA has been reported to substitute completely for amphetamine in rodents [18, 19], pigeons [20], and primates [21]. These data suggest that MDMA is more amphetamine-like than hallucinogenic, at least with respect to drug discrimination properties.

The most thoroughly investigated effect of MDMA is the ability of single or repeated administration of MDMA to deplete the concentration of 5-HT and 5-HIAA in the brain. Single administration of MDMA produces a dose-dependent depletion of 5-HT and 5-HIAA in 5-HT terminal brain areas, including frontal cortex, striatum, hippocampus, and, to a lesser extent, the hypothalamus [3, 22–24]. In addition, parenteral administration of MDMA produces a dose-dependent reduction in the activity of tryptophan hydroxylase, the rate limiting step in 5-HT synthesis [25, 26], but in vitro, MDMA had no effect on tryptophan hydroxylase activity [25]. The ability of MDMA to deplete biogenic amines has been found to be specific to 5-HT [25, 26]. Complimentary studies have found that single or repeated administration of MDMA reduces the number of 5-HT uptake sites labeled by [^{3}H]-paroxetine, a marker of 5-HT nerve terminals [27–29]. The most elegant and definitive study of the selective toxicity of MDMA to date is the report of O'Hearn et al. [5]. In this study, immunocytochemistry was used to demonstrate that MDMA selectively damages the axon terminals associated with 5-HT, without affecting 5-HT cell bodies. The ability of MDMA to selectively damage 5-HT axon terminals in rodents and nonhuman primates raises the possibility that this compound could produce similar consequences in humans.

It is well documented that 5-HT plays a role in the secretion of several pituitary hormones [30]. For example, pharmacological studies both in vivo [31] and in vitro [32] support a stimulatory role of 5-HT in the secretion of ACTH and, as a consequence, corticosterone. Similarly, the selective 5-HT uptake inhibitor fluoxetine has been reported to increase corticotropin-releasing factor and vasopressin concentrations in hypophysial portal plasma [33]. In addition, lesions of the medial basal hypothalamus significantly attenuate p-chloroamphetamine-induced and quipazine-induced corticosterone secretion in rodents [34, 35]. A recent histological study found that serotonergic fibers originating from the dorsal and medial raphe nuclei synapse directly with CRF synthesizing neurons located in the hypothalamus [36]. Thus, neuroendocrine response patterns have been used to evaluate the mechanism of action of numerous drugs, as well as the functional state of serotonergic mechanisms [37].

The identification of multiple 5-HT binding sites in the brain [38, 39] raises the possibility that different 5-HT receptors control the secretion of pituitary hormones. Koenig et al. [40, 41] reported that both 5-HT_{1A} agonists, such as 8-OH-DPAT, buspirone, gepirone, and ipsapirone, and the 5-HT_2 agonists, MK-212 and DOI (Nash et al., unpublished observation), stimulate the secretion of corticosterone in rodents.

The results presented in this chapter are an extension of previous studies conducted in this laboratory [42]. The present studies were conducted to further understand the mechanism by which MDMA stimulates the secretion of corticosterone in rodents.

2. NEUROENDOCRINE EFFECTS OF MDMA

Male Sprague–Dawley rats weighting 170–200 g upon delivery were purchased from Zivic Miller Laboratory (Hillson, PA) and were used in all experiments. The animals were housed six per cage in a temperature-controlled room (22–24°C) with a light/dark cycle of 12/12 hours (lights on at 0600). Food (Wayne Lab Blox) and tap water were available *ad libitum*.

The racemic mixture of MDMA was generously provided by Dr. David E. Nichols of Purdue University. Para-chlorophenylalanine (PCPA) hydrochloride was purchased from Sigma Chemical Co., St. Louis, MO. The other drugs were generously provided by the manufacturers: fluoxetine hydrochloride and LY 53857 (Eli Lilly and Co., Indianapolis, IN), ketanserin tartrate (Janssen Pharmaceutica, Beerse, Belgium), and xylamidine tosylate (Wellcome Research Laboratories, Beckenham, England). All drugs were dissolved in distilled water and injected i.p.

On the day before an experiment in which plasma ACTH and corticosterone were determined, the animals were housed six per cage and transferred to the experimental room. All experiments were performed between 0800–1000 hours.

In the dose-response study, MDMA (3,10, and 20 mg/kg) was administered i.p. 30 minutes prior to decapitation. In the interaction studies, the antagonist was administered at times and doses indicated in the results section, followed by MDMA (3 mg/kg, i.p.) 30 minutes prior to sacrifice. In all cases, trunk blood was collected in chilled plastic tubes containing EDTA/aprotinin. The blood samples were immediately centrifuged (3000 rpm × 20 minutes), and plasma was obtained and transferred to plastic tubes. All plasma samples were stored at −80°C until the time of assay.

In the 5-HT depletion study, PCPA (150 mg/kg, i.p.) or vehicle was administered for three consecutive days. Twenty-four hours following the last injection, each rat was challenged with saline or MDMA (3 mg/kg, i.p.) and sacrificed 30 minutes later. Trunk blood was obtained for hormone determinations. In addition, the hypothalamus from saline-and PCPA-pretreated animals challenged with vehicle was removed and homogenized in 0.1N perchloric acid for determination of 5-HT and 5-HIAA concentrations.

Plasma concentration of corticosterone was determined by RIA. [^{3}H]-Corticosterone was purchased from Dupont New England Nuclear, Boston, MA, and the antiserum was purchased from Radioassay Systems Laboratories, Inc., Carson, CA. The unlabeled corticosterone used in preparing the RIA standard was obtained from Steraloids, Inc., Whiliter, NH.

ACTH concentration was measured in unextracted plasma samples following the RIA procedure described by Nicholson et al. [43]. Plasma samples were collected in chilled polypropylene tubes containing 0.15 ml of the protease inhibitor aprotinin (Sigma Chemical Co., St. Louis, MO) and 0.1 ml 5% EDTA as an anticoagulant. The reference ACTH was generously provided by Dr. Raiti at the NIDDK and the National Hormone and Pituitary Program (University of Maryland School of Medicine). The primary antiserum used in this assay, IgG-ACTH-1, was purchased from IgG Corporation, Nashville, TN. [^{125}I]ACTH was purchased from Radioassay Systems Laboratories, Inc., Carson, CA. The concentration of plasma ACTH for all samples was measured in the same assay, in duplicate, with an intraassay variation of less than 10% and a sensitivity of 5 pg/ml.

The concentration of 5-HT and 5-HIAA in the hypothalamus was determined by high pressure liquid chromatograph with electrochemical detector using previously described methods [42].

The data were analyzed using one-way and two-way analysis of variance. Differences between treatment groups were assessed with the Student–Newman–Keul's test and were considered significant at $p < .05$.

Table 1 presents the effect of MDMA administration on plasma concentrations of ACTH and corticosterone. MDMA produced a dose-dependent increase in the secretion of both hormones. No difference was observed in plasma concentrations of ACTH and corticosterone following the administration of 10 and 20 mg/kg MDMA. MDMA (3 mg/kg) produced a signi-

Table 1. Effect of MDMA on ACTH and corticosterone secretion in rodents.

Txt	Dose (mg/kg)	N	Corticosterone (ug/dl)	ACTH (pg/ml)
Vehicle	–	10	7.1 ± 1.7	38.1 ± 8.8
MDMA	3	15	25.4 ± 2.0*	314.8 ± 43.0*
MDMA	10	6	35.1 ± 3.5*	497.4 ± 61.8*
MDMA	20	6	37.4 ± 1.8*	585.8 ± 44.0*

MDMA was administered i.p. 30 minutes prior to decapitation. Each value is the mean ± SE of 6–15 rats.
*$p < .05$ compared to vehicle treated group.

Table 2. Effect of vehicle or PCPA pretreatment on 5-HT and 5-HIAA concentrations in the hypothalamus.

	Hypothalamus	
	5-HT	5-HIAA
Vehicle	7.39 ± 0.30	5.91 ± 0.41
PCPA	1.77 ± 0.08 (−76%)	1.07 ± 0.08 (−82%)

Percent changes are compared to corresponding vehicle group.
Vehicle or PCPA (150 mg/kg) were administered for 3 consecutive days and the animals were sacrified 24 hours following the last injection. Each value is the mean ± SE of 6 rats. The values are presented as ng/mg protein.

ficant ($p < .05$) increase in plasma ACTH and corticosterone concentration, which was significantly less than that produced by either the 10 or 20 mg/kg dose of MDMA.

The administration of PCPA (150 mg/kg, i.p.) for three days significantly ($p < .05$) reduced the hypothalamic concentration of 5-HT and 5-HIAA by 75% and 82%, respectively (Table 2). PCPA treated rats challenged with vehicle had significantly ($p < .05$) elevated basal plasma corticosterone concentrations, as compared with the rats treated only with the solvent vehicle (Table 3). More importantly, PCPA-pretreatment completely abolished MDMA-induced corticosterone secretion (Table 3).

Table 4 presents the effect of ketanserin and fluoxetine pretreatment on MDMA-induced corticosterone secretion. Administration of ketanserin produced a dose-dependent inhibition of MDMA-induced corticosterone secretion. Similarly, pretreatment with fluoxetine 16 hours prior to MDMA administration significantly ($p < .05$) reduced MDMA-induced corticosterone secretion. Neither ketanserin nor fluoxetine significantly altered basal concentrations of plasma corticosterone (Table 4).

Figure 1 illustrates the effect of the selective $5\text{-}HT_2$ antagonist LY 53857 on MDMA-induced ACTH and corticosterone secretion. LY 53857 (0.3, 1.0,

Table 3. Effect of vehicle of PCPA pretreatment on MDMA-induced corticosterone secretion.

	Dose	Corticosterone (ug/dl)	
Txt	(mg/kg)	Vehicle	PCPA
Vehicle	–	4.7 ± 0.7	11.1 ± 3.0[a]
MDMA	3.0	28.8 ± 1.9[a]	14.7 ± 2.1[a,b]

Vehicle or PCPA (150 mg/kg) were administered for consecutive days and 24 hours following the last injection, each rat was challenged with either vehicle or MDMA (3 mg/kg). Each value represents the mean ± SE of 6 rats.
[a] significantly ($p < .05$) greater than vehicle plus vehicle group.
[b] significantly ($p < .05$) less than vehicle plus MDMA group.

Table 4. Effect of ketanserin and fluoxetine on MDMA-induced corticosterone secretion.

	Dose	Corticosterone (ug/dl)	
Txt	(mg/kg)	Vehicle	MDMA
Vehicle	–	6.7 ± 1.8	32.7 ± 3.3
Ketanserin	0.3	–	23.1 ± 1.7
Ketanserin	1.0	–	13.9 ± 2.0*
Ketanserin	3.0	7.0 ± 3.0	12.6 ± 2.7*
Fluoxetine	10.0	7.1 ± 2.0	23.2 ± 2.4*

Ketanserin (0.3, 1.0, and 3.0 mg/kg) was administered 90 minutes prior to decapitation and 60 minutes before MDMA (3 mg/kg) admininstration. Fluoxetine (10 mg/kg) was administered 16 hours prior to the administration of MDMA (3 mg/kg); animals were sacrificed 30 minutes later. Each value is the mean ± SE of 6 rats. *$p < .05$ versus the vehicle plus MDMA group.

and 3.0 mg/kg) was administered 90 minutes prior to sacrifice and 60 minutes before MDMA (3 mg/kg) was administered. The 1.0 and 3.0 mg/kg dose of LY 53857 significantly ($p < .05$) reduced MDMA-induced ACTH and corticosterone secretion. The 3.0 mg/kg dose of LY 53857 had no effect on basal concentrations of ACTH and corticosterone.

Pretreatment with the peripheral 5-HT antagonist xylamidine had no effect on MDMA-induced ACTH or corticosterone secretion (Figure 2). Xylamidine (1.0 and 5.0 mg/kg) was administered 90 minutes prior to sacrifice and 60 minutes before MDMA (3 mg/kg) administration. The basal concentrations of ACTH and corticosterone were also unaffected by xylamidine (5 mg/kg) pretreatment.

Finally, Table 5 presents the effect of pretreatment with haloperidol (0.3 mg/kg) on MDMA-induced corticosterone secretion. Haloperidol was administered 90 minutes prior to decapitation and 60 minutes before MDMA administration. Haloperidol had no effect on either basal or MDMA-induced corticosterone secretion.

3. CONCLUSIONS

The administration of MDMA results in a dose-dependent increase in plasma ACTH and corticosterone concentrations in rodents. Depletion of 5-HT in the hypothalamus by 80% following pretreatment with the tryptophan hydroxylase inhibitor PCPA completely abolished MDMA-induced corticosterone response. Pretreatment with the selective 5-HT uptake inhibitor fluoxetine or the 5-HT_2 antagonists, ketanserin and LY 53857, significantly inhibited MDMA-induced ACTH and corticosterone secretion. Conversely, neither the peripheral 5-HT antagonist xylamidine nor the D_2 antagonist haloperidol significantly affected MDMA-induced neuroendocrine responses. These data suggest that MDMA is taken up by a fluoxetine-sensitive carrier, releases endogenous 5-HT, which interacts with 5-HT_2 receptor mechanisms most likely in the hypothalamus resulting in the secretion of ACTH and corticosterone.

Although a great many studies have been conducted over the past three

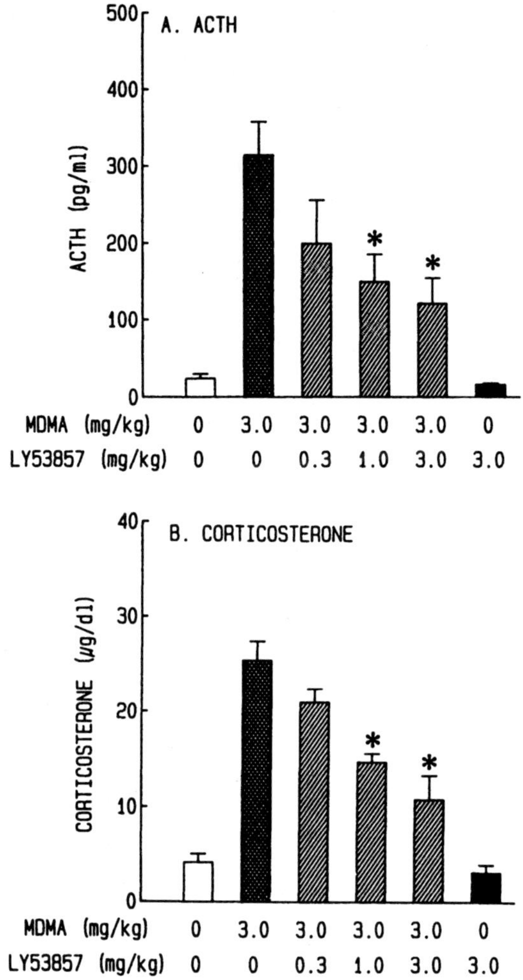

Figure 1. Effect of MDMA on concentrations of plasma ACTH (A) and corticosterone (B) in rats pretreated with vehicle or LY 53857. Rats were injected with either vehicle or LY 53857 (0.3, 1.0, and 3.0 mg/kg) 90 minutes prior to sacrifice and 60 minutes before the admininstration of MDMA (3 mg/kg). All drugs were administered i.p. in a 1 ml/kg volume. Each value represents the mean ± SE of 6 rats. The asterisk (*) represents a significant ($p < .05$) difference as compared with the vehicle + MDMA group.

years designed to further understand the pharmacological/toxicological properties of MDMA, the mechanism of action of this amphetamine analogue remains unsolved. For example, it seems quite clear from our studies that at relatively low doses (~3 mg/kg), MDMA potently stimulates serotonergic systems. This finding agrees with in vitro studies that demonstrate the 5-HT releasing properties of MDMA in synaptosomes [9] and hippocampal slice

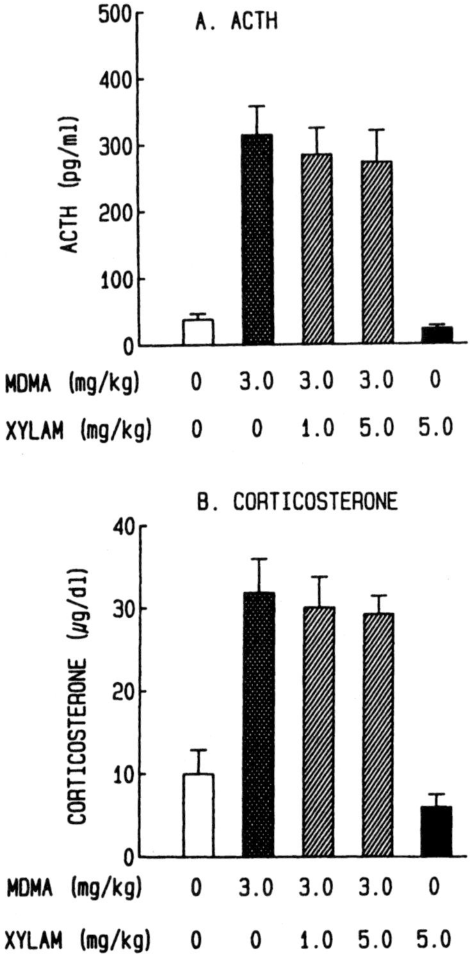

Figure 2. Effect of MDMA on plasma concentrations of ACTH (A) and corticosterone (B) in rats pretreated with xylamidine. Rats were injected with xylamidine (1.0 and 5.0 mg/kg) 90 minutes prior to sacrifice and 60 minutes before the admininstration of MDMA (3 mg/kg). All drugs were administered i.p. in a 1 ml/kg volume. Each value represents the mean ± SE of 6 rats.

preparations [10]. Yet drug discrimination studies suggest that MDMA possesses amphetamine-like stimulus properties [44], which inferentially supports a dopamine-mediated response. Moreover, the lack of stimulus generalization with hallucinogens argues against the involvement of 5-HT_2 receptor mechanisms by which MDMA acts, based on studies suggesting that these hallucinogens are potent 5-HT_2 agonists [45–47]. Finally, with the notable exception of the studies of Steele et al. [11] and Johnson et al. [26], little attention has been directed towards examining the effect of MDMA on other neurotransmitter systems.

Table 5. Effect of haloperidol on MDMA-induced corticosterone secretion.

	Dose	Corticosterone (ug/dl)	
Txt	(mg/kg)	Vehicle	MDMA
Vehicle	–	6.4 ± 1.8	19.9 ± 2.0
Haloperidol	0.3	5.0 ± 1.6	20.4 ± 1.8

Haloperidol (0.3 mg/kg, i.p.) was administered 90 minutes prior to decapitation and 60 minutes before MDMA (3.0 mg/kg, i.p.). Each value is the mean ± SE of 5–6 rats.

The ability of ketanserin and the more selective $5\text{-}HT_2$ antagonist LY 53857 [48, 49] to block MDMA-induced ACTH and corticosterone secretion suggests that a part of the effect of MDMA is mediated by this 5-HT receptor subtype. Although it has been reported that the $5\text{-}HT_2$ antagonist pirenpirone fails to attenuate the disruptive effects of MDMA on operant responding in mice [50], there have been reported differences between rats and mice in response to MDMA [51, 52], which could account for this discrepancy. It is probable that the action of MDMA is dependent upon both serotonergic and catecholaminergic mechanisms. The drug discrimination cue and neuroendocrine responses may arise from catecholaminergic and serotonergic mechanisms, respectively. This explanation is supported by the fact that the 5-HT agonists MK-212 and quipazine elevate serum corticosterone via $5\text{-}HT_2$ receptor mechanisms [40, 53] without producing visual hallucinations or perceptual distortions in humans [54, 55].

Pretreatment with fluoxetine significantly antagonized MDMA-induced corticosterone secretion. This finding is consistent with reports in which fluoxetine pretreatment blocked the 5-HT depleting effects of MDMA [3]. These data suggest that MDMA is taken up into axon terminals via a fluoxetine-sensitive uptake carrier. These studies suggest that MDMA releases endogenous 5-HT, which interacts with postsynaptic $5\text{-}HT_2$ receptors, since pretreatment with PCPA, ketanserin, or LY 58357 abolished MDMA-induced Corticosterone secretion. An indirect effect of MDMA on serotonergic mechanisms is supported by the weak affinity of MDMA for $5\text{-}HT_2$ and $5\text{-}HT_1$ binding sites [14].

The inability of the peripheral 5-HT antagonist xylamidine [56] to block MDMA-induced ACTH and corticosterone secretion suggests that this response is centrally mediated. The doses of xylamidine used in this study have been reported to block peripherally-mediated responses to 5-HT agonists [57] and thus could not account for the negative findings. Similarly, in the present study, haloperidol had no effect on MDMA-induced increase in plasma corticosterone concentration even though the dose used previously had been reported to block the corticosterone response to the D_2 agonist pergolide [58]. Collectively, these findings argue against an involvement of either peripheral

5-HT or central mechanisms in MDMA-induced activation of the hypothalamic-pituitary-adrenocortical axis.

As mentioned in the introduction, 5-HT has been suggested to play a role in the etiology of depression. Specifically, it has been postulated that a decrease in central 5-HT function may increase the susceptibility to depression [8]. Based on this hypothesis, drugs that enhance serotonergic neurotransmission should be effective antidepressants. For example, drugs that block the reuptake of monamines, such as the tricyclics, imipramine, and amitriptyline, as well as newer more selective 5-HT uptake inhibitors, (e.g., fluoxetine), appear to enhance 5-HT transmission and are effective antidepressants [59, 60]. Similarly, electroconvulsive shock (ECS) reduces 5-HT uptake and increases postsynaptic receptor function [60]. Chronic administration of the 5-HT precursors tryptophan or 5-hydroxytryptophan have been reported to be effective antidepressants when administered alone, or with much greater efficacy when administered in combination with tricyclic antidepressants [61, 62]. Collectively, these data support the hypothesis that serotonergic function is diminished in depression and that drugs that enhance serotonergic neurotransmission effectively alleviate the symptoms of depression.

The acute administration of MDMA elevates mood. Based on the proceeding discussion, a probable explanation for the mood-elevating properties of MDMA is the release of 5-HT. In support of this hypothesis is the study of Ward et al. [63] in which the 5-HT releaser fenfluramine was reported to significantly improve the mood of subjects with major depression following acute administration. The authors concluded that the acute effect of fenfluramine mimicked long-term antidepressant effects in these subjects. It is tempting to speculate that the chronic administration of MDMA may effectively enhance 5-HT neurotransmission by releasing endogenous 5-HT and blocking its reuptake.

In this regard, p-chloroamphetamine (PCA) has previously been reported to be an effective antidepressant [64, 65]. In these studies, the daily administration of 75 mg of PCA (~1.0 mg/kg) significantly improved the mood of depressed patients. The antidepressant effect of PCA was suggested to be the result of enhanced serotonergic transmission. It is interesting to note that, like MDMA, both PCA and fenfluramine have been previously reported to produced long-term depletion of 5-HT and 5-HIAA in rodents [66, 67]. Finally, although the administration of fenfluramine, PCA, and MDMA potently release 5-HT from nerve terminals, all of these amphetamine analogues interact with other neurotransmitter systems, such as dopamine and norepinephrine, which may contribute to their complex pharmacological effects.

The study of the 5-HT depleting properties of MDMA in primates [7] clearly illustrates the potential neurotoxic effects of this agent in humans. However, determining the consequences of MDMA-induced 5-HT depletion in laboratory animals is needed. We plan to study whether MDMA administration affects serotonergic function, using neuroendocrine challenge studies.

As reviewed by Meltzer and Nash [37], the secretion of hormones, including prolactin, ACTH, and cortisol, is dependent, in part, upon serotonergic stimulation. These studies are believed to provide inferential data regarding the functional status of 5-HT systems. For example, the i.v. infusion of the 5-HT precursor tryptophan stimulates the secretion of prolactin and growth hormone. Several studies have found that tryptophan-induced prolactin and growth hormone secretion is blunted in depressed patients as compared to normal volunteers [68, 69]. These blunted responses have been attributed to an abnormality in serotonergic function. Several studies have found that the cortisol response following the administration of 5-HTP is enhanced in depressed patients as compared to normal volunteers [70–72]. These results were interpreted as an enhancement of postsynaptic receptor sensitivity as a result of diminished presynaptic 5-HT function. We propose that if MDMA destroys axon terminals, then a compensatory change of postsynaptic 5-HT receptor function may occur, resulting in an enhanced cortisol response to direct-acting 5-HT agonists such as MK-212. Studies such as these will provide valuable information about the potential prolonged effects of MDMA on serotonergic function.

In conclusion, the neuroendocrine responses to MDMA administration appear to be mediated by endogenous 5-HT via 5-HT_2 receptor mechanisms. Since 5-HT_2 antagonists inhibit MDMA-induced hormone secretion, it would be interesting to determine whether a 5-HT_2 antagonist, such as ketanserin or ritanserin, would block the subjective effects of MDMA in humans. Studies such as this would provide extremely important data regarding the role of 5-HT_2 receptors in the regulation of mood. MDMA may possess antidepressant properties, but this should be rigorously tested in preclinical studies in comparison to established antidepressants.

ACKNOWLEDGEMENTS

The research reported was supported in part by USPHS MH 41684, MH 41683, MH 41594, and grants from the Cleveland and Sawyer Foundations. J.F.N. is the recipient of a NARSAD Fellowship extension award. H.Y.M. is the recipient of a USPHS Research Career Scientist Award MH 47808. We are grateful to Ms. Lee Mason for her secretarial assistance in preparing this manuscript.

REFERENCES

1. Greer, G. and Strassman, R.J., 1985. Information on "Ecstasy." Am. J. Psychiat. 142:1391.
2. Grinspoon, L. and Bakalar, J.B., 1986. Can drugs be used to enhance the psychotherapeutic process? Am. J. Psychother. 40:393–404.
3. Schmidt, C.J., 1987. Neurotoxicity of the psychedelic amphetamine, methylenedioxymethamphetamine. J. Pharmacol. Exp. Ther. 240:1–7.
4. Commins, D.L., Vosmer, G., Virus, R.M., Woolverton, W.L., Schuster, C.R., and Seiden, L.S., 1987. Biochemical and histological evidence that methylenedioxymethamphetamine (MDMA) is toxic to neurons in the rat brain. J. Pharmacol. Exp. Ther. 241:338–345.
5. O'Hearn, E., Battaglia, G., De Souza, E.B., Kuhar, M.J., and Molliver, M.E., 1988.

Methylenedioxyamphetamine (MDA) and methylenedioxymethamphetamine (MDMA) cause selective ablation of serotonergic axon terminals in forebrain: Immunocytochemical evidence for neurotoxicity. J. Neurosci. 8:2788–2803.
6. Ricaurte, G.A., DeLanney, L.E., Irwin, I., and Langston, J.W., 1988. Toxic effects of MDMA on central serotonergic neurons in the primate: Importance of route and frequency of drug administration. Brain Res. 446:165–168.
7. Ricaurte, G.A., Forno, L.S., Wilson, L.E., DeLanney, L.E., Irwin, I., Molliver, M.E., and Langston, J.W., 1988. (±)3,4-methylenedioxymethamphetamine (MDMA) selectively damages central serotonergic neurons in non-human primates. JAMA 260:51–55.
8. Meltzer, H.Y. and Lowy, M.T., 1987. The serotonin hypothesis of depression. In *Psychopharmacology: The Third Generation of Progress.* New York: Raven Press, pp. 513–526.
9. Nichols, D.E., Lloyd, D.H., Hoffman, A.J., Nichols, M.B., and Yim, G.K.W., 1982. Effects of certain hallucinogenic amphetamine analogues on the release of [^{3}H]serotonin from rat brain synaptosomes. J. Med. Chem. 25:530–535.
10. Johnson, M.P., Hoffman, A.J., and Nichols, D.E., 1986. Effects of the enantiomers of MDA, MDMA and related analogues on [^{3}H]serotonin and [^{3}H]dopamine release from superfused rat brain slices. Eur. J. Pharmacol. 132:269–276.
11. Steele, T.D., Nichols, D.E., and Yim, G.K.W., 1987. Stereochemical effects of 3,4-methylenedioxymethamphetamine (MDMA) and related amphetamine derivatives on inhibition of uptake of [^{3}H]monoamines into synaptosomes from different regions of rat brain. Biochem. Pharmacol. 36:2297–2303.
12. Gehlert, D.R., Schmidt, C.J., Wu, L., and Lovenberg, W., 1985. Evidence for specific methylenedioxymethamphetamine (Ecstasy) binding sites in the rat brain. Eur. J. Pharmacol. 119:135–136.
13. Wang, S.S., Ricaurte, G.A., and Peroutka, S.J., 1987. [^{3}H]3,4-Methylenedioxymethamphetamine (MDMA) interactions with brain membranes and glass fiber filter paper. Eur. J. Pharmacol. 138:439–443.
14. Lyon, R.A., Glennon, R.A., and Titeler, M., 1986. 3,4-Methylenedioxymethamphetamine (MDMA): Stereoselective interactions at brain 5-HT_1 and 5-HT_2 receptors. Psychopharmacol. 88:525–526.
15. Anderson, G.M., Braun, G., Braun, V., Nichols, D.E., and Shulgin, A.T., 1978. Absolute configuration and psychotomimetic activity. In *Quantitative Structure Activity Relationships of Analgesics, Narcotic Antagonists and Hallucinogens.* Washington, DC: U.S. Government Printing Office, pp. 8–15.
16. Nichols, D.E., Hoffman, A.J., Oberlender, R.A., Jacob III, P., and Sulgin, A.T., 1986. Derivatives of 1-(1,3-benzodioxol-5-yl)-2-butanamine. Representatives of a novel therapeutic class. J. Med. Chem. 29:2009–2015.
17. Glennon, R.A., Young, R., Rosecrans, J.A., and Anderson, G.M., 1982. Discriminative stimulus properties of MDA analogs. Biol. Psychiat. 17:807–814.
18. Glennon, R.A., and Young, R. 1984. Further investigation of the discriminative stimulus properties of MDA. Pharmacol. Biochem. Behav. 20:501–505.
19. Oberlender, R. and Nichols, D.E., 1988. Drug discrimination studies with MDMA and amphetamine. Psychopharmacol. 95:71–76.
20. Evans, S.M. and Johanson, C.E., 1986. Discriminative stimulus properties of (±)3,4-methylenedioxymethamphetamine and (±)3,4-methylenedioxyamphetamine in pigeons. Drug Alc. Depend. 18:159–164.
21. Kamien, J.B., Johanson, C.E., Schuster, C.R., and Woolverton, W.L., 1986. The effects of (±)methylenedioxymethamphetamine and (±)methylenedioxyamphetamine in monkeys trained to discriminate (+)amphetamine from saline. Drug Alc. Depend. 18:139–147.
22. Schmidt, C.J., Wu, L., and Lovenberg, W., 1986. Methylenedioxymethamphetamine: A potentially neurotoxic amphetamine analogue. Eur. J. Pharmacol. 124:175–178.
23. Stone, D.M., Merchant, K.M., Hanson, G.R., and Gibb, J.W., 1987. Immediate and long-term effects of 3,4-methylenedioxymethamphetamine on serotonin pathways in brain of rat Neuropharmacol. 26:1677–1683.
24. Mokler, D.J., Robinson, S.E., and Rosecrans, J.A., 1987. (±)3,4-Methylenedioxymethamphetamine (MDMA) produces long-term reductions in brain 5-hydroxytryptamine in rats. Eur. J. Pharmacol. 138:265–268.
25. Schmidt, C.J., Levin, J.A., and Lovenberg, W., 1987. *In vitro* and *in vivo* neurochemical

effects of methylenedioxymethamphetamine on striatal monoaminergic systems in the rat brain. Biochem. Pharmacol. 36:747–755.

26. Johnson, M., Letter, A.A., Merchant, K., Hanson, G.R., and Gibb, J.W., 1988. Effects of 3,4-methylenedioxyamphetamine and 3,4-methylenedioxymethamphetamine isomers on central serotonergic, dopaminergic and nigral neurotensin systems of the rat. J. Pharmacol. Exp. Ther. 244:977–982.
27. Battaglia, G., Yeh, S.Y., O'Hearn, E., Molliver, M.E., Kuhar, M.J., and De Souza, E.B., 1987. 3,4-Methylenedioxymethamphetamine and (±)3,4-methylenedioxyamphetamine destroy serotonin terminals in rat brain: Quantification of neurodegeneration by measurement of [^{3}H]paroxetine-labeled serotonin uptake sites. J. Pharmacol. Exp. Ther. 242:911–916.
28. De Souza, E.B. and Kuyatt, B.L., 1987. Autoradiographic localization of [^{3}]H-paroxetine-labeled serotonin uptake sites in rat brain. Synapse 1:488–496.
29. Battaglia, G., Yeh, S.Y., and De Souza, E.B., 1988. MDMA-induced neurotoxicity: Parameters of degeneration and recovery of brain serotonin neurons. Pharmacol. Biochem. Behav. 29:269–274.
30. Tuomisto, J. and Mannisto, P., 1985. Neurotransmitter regulation of anterior pituitary hormones. Pharmacol. Revs. 37:251–332.
31. Fuller, R.W., 1981. Serotonergic stimulation of pituitary-adrenocortical function in rats. Neuroendocrinology 32:118–127.
32. Holmes, M.C., Di Renzo, G.D., Beckford, B., Gillham, B., Jones, M.T., 1982. Role of serotonin in the control of secretion of corticotropin releasing factor. J. Endocrinol. 93:151–160.
33. Gibbs, D.M. and Vale, W. 1983. Effect of serotonin reuptake inhibitor fluoxetine on corticotropin releasing factor and vasopressin secretion into hypophysial portal blood. Brain Res. 280:176–179.
34. Meyer, J.S., McElroy, J.F., Yehuda, R., and Miller, J., 1984. Serotonergic stimulation of pituitary-adrencortical activity in rats: Evidence for multiple sites of action. Life Sci. 34:1891–1898.
35. Van de Kar, L.D., Karteszi, M., Bethea, C.L., and Ganong, W.F., 1985. Serotonergic stimulation of prolactin and corticosterone secretion is mediated by different pathways from the mediobasal hypothalamus. Neuroendocrinology 41:380–384.
36. Liposits, Z.S., Phelix, C., and Paull, W.K., 1987. Synaptic interaction of serotonergic axons and corticotropin releasing factor (CRF) synthesizing neurons in the hypothalamic paraventricular nucleus of the rat. Histochemistry 86:541–549.
37. Meltzer, H.Y. and Nash, J.F., 1988. Serotonin and mood: Neuroendocrine aspects. In *Current Topics in Neuroendocrinology*, Vol. 8. Berlin-Heidelberg, Germany: Springer–Verlag, pp. 182–210.
38. Peroutka, S.J. and Snyder, S.H., 1979. Multiple serotonin receptors: Differential finding of [^{3}H]5-hydroxytryptamine, [^{3}H]lysegic acid diethylamide and [^{3}H]spiroperidol. Mol. Pharmacol. 16:687–699.
39. Peroutka, S.J., 1987. Serotonin receptors. In *Psychopharmacology: The Third Generation of Progress*. New York: Raven Press, pp. 303–311.
40. Koenig, J.I., Gudelsky, G.A., and Meltzer, H.Y., 1987. Stimulation of corticosterone and β-endorphin secretion in the rat by selective 5-HT receptor subtype activation. Eur. J. Pharmacol. 137:1–8.
41. Koenig, J.I., Meltzer, H.Y., and Gudelsky, G.A., 1988. 5-hydroxytryptamine$_{1A}$ receptor-mediated effects of buspirone, gepirone, and ipsapirone. Pharmacol. Biochem. Behav. 29:711–715.
42. Nash, J.F., Meltzer, H.Y., and Gudelsky, G.A., 1988. Elevation of serum prolactin and corticosterone concentrations in the rat after the administration of 3,4-methylenedioxymethamphetamine. J. Pharmacol. Exp. Ther. 245:873–879.
43. Nicholson, W.E., Davis, D.R., Sherrel, B.J., and Orth, D.N., 1984. Rapid radioimmunoassay for corticotropin in unextracted human plasma. Clin. Chem. Acta 30:259–265.
44. Glennon, R.A., 1986. Discriminative stimulus properties of phenylisopropylamine derivates. Drug Alc. Depend. 17:119–134.
45. Glennon, R.A., Titeler, M., and McKenney, J.D., 1984. Evidence for 5-HT$_2$ involvement in the mechanism of action of hallucinogenic agents. Life Sci. 35:2505–2511.
46. Rasmussen, K., Glennon, R.A., and Aghajanian, G.K., 1986. Phenethylamine hallucinogens

in the locus coerulus: Potency of action correlates with rank order of 5-HT_2 binding affinity. Eur. J. Pharmacol. 132:79–82.
47. Titeler, M., Lyon, R.A., and Glennon, R.A., 1988. Radioligand binding evidence implicates the brain 5-HT_2 receptor as a site of action for LSD and phenylisopropylamine hallucinogens. Psychopharmacol. 94:213–216.
48. Cohen, M.L., Fuller, R.W., and Kurz, K.D., 1983. LY 53857, a selective and potent serotonergic (5-HT_2) receptor antagonist, does not lower blood pressure in the spontaneously hypertensive rat. J. Pharmacol. Exp Ther. 227:337–332.
49. Cohen, M.L., Kurz, K.D., Mason, N.R., Fuller, R.W., Marzoni, G.P., and Garbrecht, W.L., 1985. Pharmacological activity of the isomers of LY 53837, potent and selective 5-HT_2 receptor antagonists. J. Pharmacol. Exp. Ther. 235:319–323.
50. Rosecrans, J.A. and Glennon, R.A., 1987. The effect of MDA and MDMA ("Ecstasy") isomers in combination with pirenpirone on operant responding in mice. Pharmacol. Biochem. Behav. 28:39–42.
51. Stone, D.M., Hanson, G.R., and Gibb, J.W., 1987. Differences in the central serotonergic effects of methylenedioxymethamphetamine (MDMA) in mice and rats. Neuropharmacology 26:1657–1661.
52. Logan, B.J., Laverty, R., Sanderson, W.D., and Yee, Y.B., 1988. Differences between rats and mice in MDMA (methylenedioxymethamphetamine) neurotoxicity. Eur. J. Pharmacol. 152:227–234.
53. Fuller, R.W. and Snoddy, H.D., 1984. Central serotonin antagonist activity of ketanserin. Res. Comm. Chem. Path. Pharmacol. 46:151–154.
54. Parati, E.A., Zanardi, P., Cocchi, D., Caraceni, T., and Müller, E.E., 1980. Neuroendocrine effects of quipazine in man in health state or with neurological disorders. J. Neural Transm. 47:273–297.
55. Lowy, M.T. and Meltzer, H.Y. 1988. Stimulation of serum cortisol and prolactin secretion in man by MK-212, a centrally active serotonin agonist. Biol. Psychiat. 23:818–828.
56. Fuller, R.W., Kurz, K.D., Mason, N.R., and Cohen, M.L., 1986. Antagonism of a peripheral vascular but not an apparently central serotonergic response by xylamidine and BW 501C67. Eur. J. Pharmacol. 125:71–77.
57. Fletcher, P.J. and Burton, M.J., 1986. Dissociation of the anorectic actions of 5-HTP and fenfluramine. Psychopharmacol. 89:216–220.
58. Fuller, R.W. and Snoddy, H.D., 1984. Central dopamine receptors mediating pergolide-induced elevation of serum corticosterone in rats. Characterization by the use of antagonist. Neuropharmacol. 23:1389–1394.
59. Asberg, M., Eriksson, B., Martensson, B., Traskman-Bendz, L., and Wagner, A., 1986. Therapeutic effects of serotonin uptake inhibitors in depression. J. Clin. Psychiat. 47:4 (Suppl), pp. 23–34.
60. Willner, P., 1985. Antidepressants and serotonergic neurotransmission: An integrative review. Psychopharmacol. 85:387–404.
61. van Praag, H.M., 1981. Management of depression with serotonin precursors. Biol. Psychiat. 16:291–310.
62. van Praag, H.M., 1983. In search of the mode of action of antidepressants. 5-HTP/Tryosine mixtures in depressions. Neuropharmacol. 22:433–440.
63. Ward, N.G., Ang, J., and Pauinich, G., 1985. A comparison of the acute effects of dextro-amphetamine and fenfluramine in depression. Biol. Psychiat. 20:1090–1097.
64. van Praag, H.M., Korf, H., and van Woudenberg, F., 1970. Investigation into the possible influence of chlorinated amphetamine derivates on 5-hydroxytryptamine synthesis in man. Psychopharmacol. 18:412–420.
65. van Praag, H.Y. and Korf, J., 1973. 4-Chloroamphetamines. Chance and trend in the development of new antidepressants. J. Clin. Pharmacol. 13:3–14.
66. Sanders-Bush E., Bushing, J.A., and Sulser, F., 1975. Long-term effects of p-chloroamphetamine and related drugs on central serotonergic mechanisms. J. Pharmacol. Exp. Ther. 192:33–39.
67. Clineschmidt, B.V., Totaro, J.A., McGuffin, J.C., and Pflueger, A.B., 1976. Fenfluramine: Long-term reduction in brain serotonin (5-hydroxytryptamine). Eur. J. Pharmacol. 35:211–214.
68. Heninger, G.R., Charney, D.J., and Sternberg, D.E., 1984. Serotonergic function in de-

pression. Prolactin response to intravenous tryptophan in depressed patients and healthy subjects. Arch. Gen. Psychiat. 41:398–402.
69. Koyama, T. and Meltzer, H.Y., 1986. A biochemical and neuroendocrine study of the serotonergic system in depression. In *New Results in Depression Research*. Berlin-Heidelberg, Germany: Springer–Verlag, pp. 169–188.
70. Meltzer, H.Y., Umberkoman–Wiita, B., Robertson, A., Tricou, B.J., Lowy, M., and Perline, R., 1984a. Effect of 5-hydroxytryptophan on serum cortisol levels in major affective disorders. I. Enhanced response in depression and mania. Arch. Gen. Psychiat. 41:366–374.
71. Meltzer, H.Y., Perline, R., Tricou, B.J., Lowy, M., and Robertson, A., 1984b. Effect of 5-hydroxytryptophan on serum cortisol levels in major affective disorders. II. Relation to suicide, psychosis and depressive symptoms. Arch. Gen. Psychiat. 41:379–387.
72. Maes, M., DeRuyter, M., Hobin, P., Claes, R., Bosma, G., and Suy, E., 1987. The cortisol responses to 5-hydroxytryptophan, orally, in depressive inpatients. J. Affective Disord. 13:23–30.

INDEX

9 780792 303053

Printed by Books on Demand, Germany